Synthesis of Inorganic Materials

Ulrich Schubert and Nicola Hüsing

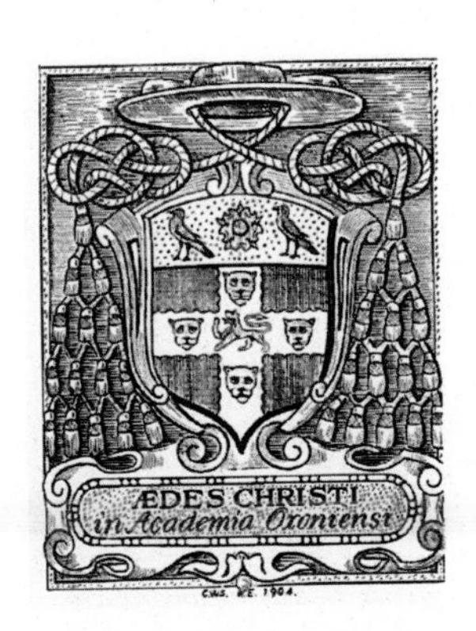

Related Titles

Schmid, G., Krug, H., Waser, R., Vogel, V., Fuchs, H., Grätzel, M., Kalyanasundaram, K., Chi, L. (eds.)

Nanotechnology
9 Volumes

2012
ISBN: 978-3-527-31723-3

Comba, P., Kerscher, M.

Coordination Chemistry
Concepts and Applications

2012
ISBN: 978-3-527-32300-5

Lalena, J. N., Cleary, D. A.

Principles of Inorganic Materials Design

2010
ISBN: 978-0-470-40403-4

Cademartiri, L., Ozin, G. A.

Concepts of Nanochemistry

2009
Softcover
ISBN: 978-3-527-32597-9

Cademartiri, L., Ozin, G. A.

Concepts of Nanochemistry

2009
Hardcover
ISBN: 978-3-527-32626-6

Ulrich Schubert and Nicola Hüsing

Synthesis of Inorganic Materials

3rd completely revised and enlarged edition

WILEY-VCH Verlag GmbH & Co. KGaA

The Authors

Prof. Dr. Ulrich Schubert
Technische Universität Wien
Institut für Materialchemie
Getreidemarkt 9/165
1060 Wien
Österreich

Prof. Dr. Nicola Hüsing
Universität Salzburg
FB Materialforschung & Physik
Hellbrunner Straße 34
5020 Salzburg
Österreich

1st Reprint 2012

Library of Congress Card No.: applied for

British Library Cataloguing-in-Publication Data
A catalogue record for this book is available from the British Library.

Bibliographic information published by the Deutsche Nationalbibliothek
The Deutsche Nationalbibliothek lists this publication in the Deutsche Nationalbibliografie; detailed bibliographic data are available on the Internet at <http://dnb.d-nb.de>.

Typesetting MPS Limited, Chennai
Printing and Binding Markono Print Media Pte Ltd, Singapore
Cover Design Adam-Design, Weinheim

Print ISBN: 978-3-527-32714-0

Contents

Synthesis of Inorganic Materials, Third Edition. Ulrich Schubert and Nicola Hüsing.

Published 2012 by WILEY-VCH Verlag GmbH & Co. KGaA

Preface

When the first edition of "Synthesis of Inorganic Materials" was published in 2000, we had hoped that the book would make the role of chemistry in materials science more visible, especially to students, and would provide an overview of chemical methods for the synthesis of inorganic materials. In the meantime, the textbook has become an established source in many chemistry and materials science curricula. The success of the first and second edition and the very positive feedback encouraged us to revise and update the book.

In the third edition, an own chapter (Chapter 6) is dedicated to template-assisted syntheses to porous materials to demonstrate the versatility and generality of this approach. Due to the rapid development of methodologies in the chemistry of nanomaterials, Chapter 7 was completely rewritten and refocused. Other chapters were restructured as well. We have also included new, emerging topics, such as mechanochemical syntheses, chemical vapor infiltration, non-hydrolytic sol–gel processes, and syntheses of graphene or nanorods. Other sections were significantly updated, such as the sections on biological crystallization and metal-organic frameworks. New figures and new examples were included, and the recommended literature was updated.

Within a textbook, there are several possible ways in which materials may be treated, according to:

- their chemical composition (organic polymers, metals, oxides, nitrides, carbides, etc.);
- their physical state (ceramics, glass, composites, polymers, etc.);
- their properties and applications (electronic materials, magnetic materials, optical materials, etc.);
- technological aspects (powder preparation, sintering methods, preparation of films or coatings, etc.); and
- the methods for their preparation (solid-state reactions, polycondensations, gas-phase reactions, etc.).

In most materials science and solid-state chemistry textbooks – even those that are highly recommended – there is a regrettable lack of chemical information. Materials science is often reduced to physical and technological aspects, and chemistry is only introduced to discuss bonding and to describe structures.

Synthesis of Inorganic Materials, Third Edition. Ulrich Schubert and Nicola Hüsing.

Published 2012 by WILEY-VCH Verlag GmbH & Co. KGaA

Processes for the preparation and modification of materials – the most important contribution of chemistry to material science – are mostly treated just in passing – or perhaps not at all!

With this textbook we are attempting to fill this gap. The book is not intended as a substitute for existing, physically or technologically oriented textbooks on materials science, but rather to complement chemical aspects. The nucleus of this book was a lecture course on "Inorganic materials from molecular precursors" given by the authors at the Vienna University of Technology. The selection of suitable precursors and the development of correct conditions to obtain a product with the desired composition and properties and a suitable (micro-)structure is what chemists can contribute to materials science.

Nevertheless, this textbook is intended for use not only by chemistry students (for whom, we have tried to keep the number of physical formulae to a minimum), but also by physics and materials science students (for whom, we have tried to keep the required chemical prerequisites to a minimum). The glossary at the end of the book may help to bridge the gap between chemical/physical/materials science fundamentals. Terms in the text preceeded by an arrow refer to the glossary.

In an up-to-date textbook, SI units should be used exclusively. However, different scientific communities still have their own habits concerning physical units (e.g., ceramists prefer Pascal, while CVD people prefer bar or atmosphere as the pressure unit). We therefore decided to leave some transformations of physical units to the reader – with the help of a table on the inner cover of the book.

Since the main focus of this book is on syntheses, we do not treat all major inorganic materials comprehensively, neither do we discuss naturally occurring materials such as lime, asbestos, gypsum, and so on. Instead, some materials were selected as examples to discuss the ways in which (natural or artificial) chemical compounds are transformed into materials. For the same reason, materials properties and technological aspects are discussed only exemplarily.

The book is organized according to preparation processes. Since many materials can be prepared using several methods, this organization inevitably has the consequence that some materials are treated in more than one chapter. For example, perovskites can be prepared by solid-state reactions, by sol–gel processing, by hydrothermal processes, or by CVD, and are therefore discussed in different chapters.

One difficulty we were facing was to avoid writing a textbook on preparative inorganic chemistry, that is, to discuss the preparation of inorganic *materials* instead of inorganic *compounds*. In fact, almost any chemical compound is a potential "material," and therefore the distinction is not always obvious. Not being too conservative, we considered materials as compounds that are used technically, or that have the potential for being used. In order to stay as close as possible to the real world of material science, we have tried to introduce a relevant technically applied material in most sections. The properties and uses of this material are discussed exemplarily.

At the end of each chapter, a selection of more recent books and review articles is provided, which may help the reader to study those processes under discussion in greater detail.

We hope that the third edition will continue to have an impact on the teaching of materials chemistry and to bridge the gap between the various disciplines in materials science.

Vienna and Salzburg, May 2011

Nicola Hüsing
Ulrich Schubert

Acknowledgement

We thank Harald Schauer who made invaluable contributions to this book with his careful and patient drawing of many figures.

We thank the following companies and institutions who have provided us with information on, and photographs of, their products that allowed us to illustrate the technical relevance of the methods discussed:

- *Austria Microsysteme Intl.*, Unterpremstätten (Dr. Holger Wille), on CVD processes for microelectronics (Figure 3.15).
- *Degussa*, Hanau, on the Aerosil process (Figures 3.26, 3.27, and 3.28).
- *Department of Materials Science, University of Erlangen-Nürnberg* (Prof. Dr. Peter Greil) on biomorphic SiC ceramics (Figure 2.11).
- *Fraunhofer-Institut für Silicatforschung*, Würzburg (Dieter Sporn and Dr. Klaus Rose) on sol–gel materials (Figures 4.44, 4.45, 4.63, and 4.65).
- *Hoechst AG*, Frankfurt, on aerogels (Figure 4.69).
- *Institut für Anorganische Technologie, Vienna University of Technology* (Prof. Dr. Benno Lux and Dr. Roland Haubner) on diamond films (Figure 3.20).
- *Philips Forschungslaboratorien*, Aachen (Dr. Ulrich Nieman), on halogen lamps (Figure 3.3).
- *Wacker-Chemie*, Burghausen (Dr. Johann Weis), on silicones (Figures 5.4 and 5.5).

The following publishers transferred copyrights for the reproduction of figures from their publications:

Academic Press: C. J. Brinker, G. W. Scherer, *Sol–Gel Science: The Physics and Chemistry of Sol-Gel-Processing*: Figures 4.42, 4.47, 4.52, and 4.55.

American Chemical Society, Advances in Chemistry, Series: Figure 4.29 (*245*, **1995**, 535).

Chemical Reviews: Figure 2.23 (*105*, **2005**, 1103).

Chemistry of Materials: Figure 2.23 (*10*, **1998**, 2898), Figure 2.27 (*8*, **1996**, 1752), Figure 4.17 (*4*, **1992**, 549), Figure 6.21 (*8*, **1996**, 1682), Figure 6.33 (*18*, **2006**, 6069), Figure 6.34 (*17*, **2005**, 4262 and *18*, **2006**, 6069).

Journal of the American Chemical Society: Figure 6.32 (*128*, **2006**, 15606), Figure 7.5 (*115*, **1993**, 8706).

Langmuir: Figure 6.31 (*23*, **2007**, 3996).

Synthesis of Inorganic Materials, Third Edition. Ulrich Schubert and Nicola Hüsing.

Published 2012 by WILEY-VCH Verlag GmbH & Co. KGaA

Chapman & Hall: a. W. Weimer (Ed.), *Carbide, Nitride and Boride Materials – Synthesis and Processing*: Figures 2.10, 2.20, and 3.5.

CRC Press: Handbook of Optical Properties Vol. II: Figure 7.8.

Elsevier: Aerosol Science and Technology: Figures 3.31 and 3.32 (*19*, **1993**, 434), Figure 7.3 (*19*, **1993**, 416).

Chemical Physics Letters: Figure 7.10 (*321*, **2000**, 163).

Endeavour: Figures 4.37 and 4.39 (**1969**, 114).

Microelectronic Engineering: Figure 4.43 (*29*, **1995**, 161).

Microporous Mesoporous Materials: Figures 6.29 and 6.30 (*44–45*, **2001**, 227).

Thermochimica Acta: Figure 3.4 (*299*, **1997**, 50).

Gordon and Breach Publishers: B. Marciniec, J. Chojnowski (Eds.), *Progress in Organosilicon Chemistry*: Figure 5.6.

Harry Deutsch Verlag: K. T. Wilke, *Kristallzüchtung*: Figure 4.36.

International Union of Pure and Applied Chemistry: Pure Appl. Chem.: Figure 6.18 (*73*, **2001**, 381).

Institute of Electrical and Electronics Engineers: IEEE Journal on Selected Topics in Quantum Electronics: Figure 3.24 (*8*, **2002**, 264).

Kluwer: Journal of Materials Science: Figure 6.7 (*18*, **1983**, 1899).

Materials Research Society: Journal of Materials Research: Figure 7.2 (*9*, **1994**, 1307).

MRS Bulletin: Figure 3.6 (*18*(6), **1993**, 25), Figure 5.19 (*32*, **2007**, 549), Figure 5.23 (*34*, **2008**, 682), Figure 6.2 (*14*(4), **1994**, 24), Figure 6.5 (*14*(4), **1994**, 15), Figure 7.21 (*30*, **2005**, 85).

Nature Publishing Group (Nature): Figure 5.22 (*402*, **1999**, 278).

Pergamon Press: C. Suryanarayana, *Non-Equilibrium Processing of Materials*: Figures 4.12, 4.13, and 4.14.

Current Opinion in Solid State & Materials Science: Figure 3.12 (*3*, **1998**, 147), Figure 6.23 (*1*, **1996**, 798).

Progress in Materials Science: Figure 2.12 (*37*, **1993**, 123).

The Royal Society of Chemistry: J. E. Shelby, *Introduction to Glass Science and Technology*: Figure 4.1.

Chemical Society Reviews: Figure 5.23 (*37*, **2009**, 191).

Journal of Materials Chemistry: Figure 2.31 (*19*, **2009**, 2494), Figure 7.13 (*19*, **2009**, 1063).

Springer: Marine Biology: Figure 4.21 (*19*, **1973**, 323).

van Nostrand Reinhold: R. Szostak, *Molecular Sieves – Principles of Synthesis and Identification*: Catalysis Series, **1989**, Figures 6.15, 6.17, 6.19, and 6.20.

VEB-Verlag Leipzig: A. Petzold, *Anorganisch-nichtmetallische Werkstoffe*: Figure 4.7.

Vulkan-Verlag: Silicone: Chemie und Technologie: Figure 5.8.

W. B. Saunders Company: Orthopedic Clinics of North America: Figure 4.28 (*20*, **1989**, 1).

Wiley/Wiley-VCH: Advanced Materials: Figures 4.31 and 4.32 (*7*, **1995**, 609), Figure 6.12 (*5*, **1993**, 127), Figure 6.28 (*18*, **2006**, 1793).

Angewandte Chemie Int. Ed.: Figure 2.29 (*34*, **1995**, 2311), Figures 3.29 and 3.30 (*28*, **1989**, 794), Figure 4.18 (*45*, **2006**, 4597), Figure 4.33 (*42*,

2003, 2350), Figure 5.12 (*28*, **1989**, 1733), Figures 5.20 and 5.21 (*43*, **2004**, 2334), Figures 6.1 and 6.27 (*42*, **2003**, 3604), Figures 7.16 and 7.17 (*41*, **2002**, 1853).

D. M. Adams, *Inorganic Solids:* Figure 2.22.

D. W. Bruce, D. O'Hare (Eds.), *Inorganic Materials*: Figures 4.24 and 4.25.

Chemie in unserer Zeit: Figures 4.20 and 4.22 (*33*, **1999**, 6), Figure 6.16 (*42*, **1995**, 1).

Encyclopedia of Chemical Technology, Kirk-Othmer: Figure 4.40.

European Journal of Inorganic Chemistry: Figure 4.34 (**1999**, 1643).

R. K. Iler, *The Chemistry of Silica*: Figure 4.51.

L. V. Interrante, M. Hampden-Smith (Eds.), *Chemistry of Advanced Materials – an Overview*: Figures 3.14 and 3.21, Table 4.2.

K. J. Klabunde (Ed.), *Nanoscale Materials in Chemistry*: Figure 7.12.

C. Kittel, *Introduction to Solid State Physics*: Figure 7.11.

T. Kodas, M. Hampden-Smith (Eds.), *The Chemistry of Metal CVD*: Figures 3.16 and 3.18.

S. Mann (Ed), *Biomimetic Materials Chemistry*: Figures 4.25, 4.26, 4.27, and 4.30.

Small: Figures 7.23 and 7.25 (*1*, **2005**, 180).

Ullmann's Encyclopedia of Technical Chemistry: Figure 4.41.

M. T. Weller, *Anorganische Materialien*: Figure 2.1.

J. Zarzycki (Ed.), *Materials Science and Technology*: Figures 4.8 and 4.11.

Zeitschrift für Anorganische und Allgemeine Chemie: Figure 2.21 (*277*, **1954**, 156).

Zeitschrift für Naturforschung: Figure 7.2 (*64b*, **2009**, 1246).

The following figures were taken from Internet pages:

Figure 2.2: www.spectraweb.ch/~hwiehl/supra.html
Figure 2.9: www.ifm.liu.se/Matephys/new_page/research/sic/Chapter3.html
Figure 2.13: www.benchmarkceramics.com/Technologies/CCS/techinfo.html
Figure 2.14: www.railway.technology.com/contractors/track/elektro/elektro1.html
Figure 4.19: www.physics.ucsb.edu/~bettye/research/biomin/biomin.html
Figure 4.23: www.soton.ac.uk/~serg/biotech/mtb-main.htm
Figure 4.35b: webmineral.com/data/quartz.shtml
Figure 4.53: www.metu.edu.tr/home/www70/who/ctas/lab/cordi.htm
Figure 4.70: http://stardust.jpl.nasa.gov/tech/aerogel.html
Figure 6.6: indigo4.gi.rwth-aachen.de/flyer/feinguss/omg_e.htm
Figure 7.6: http://www.sigmaaldrich.com/materials-science/nanomaterials/silver-nanoparticles.html
Figure 7.10: http://en.wikipedia.org/wiki/Ferrofluid
Figure 7.15: http://people.ccmr.cornell.edu/~uli/pages/diblock.htm
Figure 7.16: http://people.ccmr.cornell.edu/~uli/pages/aluminosilicates.htm
Figure 7.22: www.phys.ttu.edu/~tlmde/thesis/CARBON_NANOTUBES.html
Figure 7.26: http://www.ucm.es/info/fullerene/CarbonNanotubes.html
Figure 7.29: http://www.npl.co.uk/science-+-technology/nanoscience/surface-+-nanoanalysis/dip-pen-nanolithography

List of Abbreviations

AACVD	aerosol-assisted chemical vapor deposition
a.c.	alternating current
Ac	acetyl
acac	acetylacetonate = 2,4-pentanedionate
AFM	atomic force microscopy
AIBN	azobis(isobutyronitrile)
ALD	atomic layer deposition
ALE	atomic layer epitaxy
AO	atomic orbital
aq	aqueous
Ar	aryl
a.u.	arbitrary units
BBU	basic building unit
b.p.	boiling point
Bu	butyl
C	critical point
cat	catalyst
CBE	chemical beam epitaxy
CBU	composite building unit
CD	compact disc
CDJP	controlled double jet precipitation
cmc	critical micelle concentration
CMC	ceramic matrix composite
COD	cyclooctadiene
Cp	cyclopentadienyl, or heat capacity
CTAB	cetyltrimethylammonium bromide
CVD	chemical vapor deposition
CVI	chemical vapor infiltration
D	dimensional or diffusion coefficient
d.c.	direct current
DLICVD	direct liquid injection CVD
DMF	dimethylformamide
DMSO	dimethylsulfoxide

Synthesis of Inorganic Materials, Third Edition. Ulrich Schubert and Nicola Hüsing.

Published 2012 by WILEY-VCH Verlag GmbH & Co. KGaA

DRAM	dynamic random access memory
diglyme	ethyleneglycol dimethylether
diphos	1,2-bis(diphenylphosphino)ethane
dpm	dipivaloylmethanate (= thd or tmhd)
dppp	1,3-bis(diphenylphosphino)propane
EDTA	ethylenediaminetetraacetic acid
E_F	Fermi energy (Fermi level)
Eq	equation
Et	ethyl
fcc	face-centered cubic
G	free energy
g	gaseous
GMR	giant magnetoresistance
H_c	coercivity
HA	hydroxylapatite
Hex	hexyl
hdp	hexagonal dense packing
hfac	1,1,1,5,5,5-hexafluoroacetylacetonate (= 1,1,1,5,5,5-hexafluoro-2,4-pentanedionate)
HIP	hot isostatic pressing
HOMO	highest occupied molecular orbital
HTV	high-temperature vulcanizing
IEP	isoelectric point
IR	infrared
k_B	Boltzmann constant
L	ligand; or Lewis base
l	liquid
LC	liquid crystal; or liquid crystalline
LCVD	laser-assisted or laser-induced CVD
LDH	layered double hydroxide
LED	light-emitting diode
Ln	lanthanoid
LPCVD	low-pressure CVD
LPS	liquid-phase sintering
LR	liquid rubber
LUMO	lowest unoccupied molecular orbital
μCP	microcontact printing
M	molar; or metal
MBE	molecular beam epitaxy
MCM	Mobil composition of matter
Me	methyl
MLE	molecular layer epitaxy
MMC	metal matrix composite
MPCVD	microwave plasma-assisted CVD
MWNT	multiwalled nanotube

MO	molecular orbital
MO...	metal-organic...
MOF	metal-organic framework
n_D	refractive index
NLO	non-linear optic
nm	nanometer
OAc	acetate
OM...	organometallic...
p	para; or pressure
p_c	critical pressure
PACVD	plasma-assisted CVD
PE	polyethylene
PECVD	plasma-enhanced CVD
Ph	phenyl
phen	phenanthroline
PMMA	poly(methylmethacrylate)
PMC	polymer matrix composite
POSS	polyhedral oligomeric silsesquioxane
Pr	propyl
PTFE	poly(ethyleneterephthalate)
PZC	point of zero charge
PZT	lead zirconate titanate
PVD	physical vapor deposition
PVP	poly(vinylpyrollidone)
py	pyridine
R	organic group
RAM	random access memory
r.f.	radio frequency
ROP	ring-opening polymerization
RPCVD	remote-plasma chemical vapor deposition
RTV	room-temperature vulcanizing
s	solid
S	solubility
SAM	self-assembled monolayer
SBU	secondary building unit
SCF	supercritical fluid
sec	secondary
SEM	scanning electron microscopy
SET	single-electron transfer
SHS	self-propagating high-temperature synthesis
SIMIT	size-induced metal-insulator transition
SSM	solid-state metathesis
STM	scanning tunneling microscope
SWNT	single-walled nanotube
t_{gel}	gel time

T_{ad}	adiabatic temperature
T_b	boiling temperature
T_c	critical temperature, or Curie temperature
T_g	glass-transition (glass-transformation) temperature
T_{ig}	ignition temperature
T_m	melting temperature
TBA	*tert.* butylarsine
TEM	transmission electron microscopy
TEOS	tetraethoxysilane (tetraethylorthosilicate)
TOPO	trioctylphosphine oxide
tert	tertiary
thd	tetramethylheptanedionate (= tmhd or dpm)
THF	tetrahydrofuran
tmhd	tetramethylheptanedionate (= thd or dpm)
TMOS	tetramethoxysilane (tetramethylorthosilicate)
Tr	triple point
TTT curve	time–temperature-transformation curve
UHV	ultrahigh vacuum
UHVCVD	ultrahigh vacuum CVD
UV	ultraviolet
V_m	molar volume
Vi	vinyl
VLS	vapor–liquid–solid
VPE	vapor phase epitaxy
VTMS	vinyltrimethylsilane
wt	weight
XRD	X-ray diffraction
YAG	yttrium aluminum garnet
YBCO	yttrium barium copper oxide

1
Introduction

Although efforts to make new materials, or to improve existing materials, are as old as mankind, only in our age has "materials science" matured into a unique area. In the earliest ages of civilization, materials were made using a "learning by doing" approach. For example, the first piece of iron – which does not occur in elemental form in nature – was most probably made by chance, when an iron ore and a carbon source were accidentally heated together. From today's point of view, the properties of this "new material" would be regarded as minimal, but at the time of its discovery, this was a quantum leap for mankind. Over the millennia, the iron- and steel-making process has been advanced to a highly sophisticated state, as has the quality of the materials produced. The manufacture of glass and ceramics has developed in a similar manner.

The improved properties and better performance of the traditional materials were achieved by advances in materials technology. However, this was only made possible through a better understanding of the underlying basic chemical and physical principles. The route from the first primitive man-made piece of glass to the highest-quality glass currently used in the lenses of large telescopes, for example, was paved by a detailed understanding of the glass chemistry, the structure of glass, crystallization phenomena, and the physics of melts.

During the twentieth century, materials science has led to major changes in the way that we live. For example, only 20 years ago the manuscript of this book would have been written on a typewriter after making time-consuming searches in several scientific libraries, the figures would have been drawn with ink, and the book would have been eventually produced after time-consuming cut-and-paste corrections, typesetting by the printer and photographic reproduction of the figures. Instead, writing of the book and drawing of the figures was done on personal computers and laptops, with much of the literature research having been done using electronic databases and by the Internet, and eventually the manuscript was delivered electronically to the publishing company. This major change in how we worked was only made possible by the dramatic advances in information technology and data-storage capacities that have been developed in a rather short period. As in many other areas, the rate of progress was, and is, determined by the speed at which materials are newly developed or improved, and by advances in their processing. Further development in information technology

Synthesis of Inorganic Materials, Third Edition. Ulrich Schubert and Nicola Hüsing.

Published 2012 by WILEY-VCH Verlag GmbH & Co. KGaA

will to a large degree depend on how materials science allows further miniaturization, and on the development of materials capable of optical or optoelectronic data transfer.

In our generation, many new materials have been discovered with properties that were beyond imagination, such as the high-temperature superconductors, "intelligent" and "adaptive" materials, or nanostructured materials. Furthermore, the demand for longer product lifetimes, higher quality and efficiency, and so on has also placed new challenges on the synthesis and processing methods of known materials. Major changes in materials production technologies were induced by the impact of environmental issues (→ "sustainable development"). An increasingly important goal for materials developments is the minimization of energy and raw materials consumption. For example, materials for moving components in motors or gas turbines with better thermomechanical properties or a lower mass improve the efficiency of the motor or turbine and, as a consequence, save fuel and reduce the emission of waste gases.

Many of the scientific and technological findings that have been discovered in a variety of laboratories, together with the worldwide increase in knowledge, form the basis for new technologies in the twenty-first century. Several trends are apparent:

- Materials become increasingly specialized and multifunctional. In contrast to many materials that have a broad range of applications, many modern materials are tailor-made for only one special application.
- The borders between traditional materials types, such as ceramics, organic polymers, or "natural" materials (wood, bones, wool, etc.) begin to disappear, as well as the border between "natural" and "artificial." New classes of materials are being developed that bridge the gaps between the traditional materials, such as inorganic–organic hybrid materials or biomimetic materials.
- It is realized that the transition between molecules and solids results in materials with unique physical or chemical properties. The so-called nanomaterials (see Chapter 7) with a characteristic dimension in the lower nanometer range have the potential of revolutionizing materials design for many applications.
- Theoretical methods have matured in a way that they are able to predict materials properties. A visionary example is carbon nitride (C_3N_4), which is reputed to be harder than diamond and to have interesting applications as a high-temperature semiconductor. However, this prediction has still to be verified by the synthesis of this compound.

Although modern materials science is interdisciplinary, and more than a blend of some established communities, such as chemistry, physics, or engineering, the experience from these disciplines is essential to carry the development of some material from the raw materials or precursors to the final application.

The triangle *Synthesis and Processing – Composition and Structure – Properties and Performance* represents the essential relations in material science. The properties of a material of a given composition depend to a very high degree on the way that it was made or processed, this being a consequence of different structures (on any length scale). Vice versa, certain applications require certain structural features and certain chemical compositions of the employed materials, which again require the deliberate design or modification of the synthesis and processing procedures.

For example, SiO_2 can be crystallized as quartz (for oscillator crystals, for example) by hydrothermal treatment (see Section 4.4.1). SiO_2 as an insulating layer in a microelectronic device would be made by chemical vapor deposition (see Section 3.2.4.1). Biogenic processes produce amorphous silica, for example as the aesthetically pleasing exoskeletons of diatoms (see Section 4.3.1.1). Silica with a high surface area, used as adsorbent or for thermally insulating materials, for example, is produced either by the aerosol process, where agglomerated spherical, amorphous particles are obtained (see Section 3.3) or as aerogels with a highly porous network structure via sol–gel processing (see Section 4.5.6). Sol–gel processing also allows the preparation of amorphous SiO_2 powders or dense films. Although the composition of the obtained material is SiO_2 in each case, completely different routes of preparation are required.

Traditional classification of materials (natural materials, metals, nonmetallic inorganic materials, organic polymers, composite materials, etc.) is based on the fundamental properties and somehow also reflects the different structural and bonding features. In this book, we wish to emphasize the chemical aspects of materials science. Chemistry is the science of transforming compounds into others on a molecular level, and of investigating the concomitant changes in composition, structure, and properties. We therefore organize the book according to methods of how compounds can be synthesized.

When does an inorganic compound qualify as a "material," in other words, how does this book differ from one on preparative inorganic chemistry? We do not attempt to define the term "material," because many chemical compounds are potential materials, and thus the distinction is not always obvious. Not being too conservative, we considered materials to be compounds that are utilized for some technical application, or that have at least the potential for being used in such a manner. We will, as examples, select several types of materials in order to discuss the ways in which (natural or artificial) chemical compounds are transformed into materials, and show both the options and problems that originate from the various preparation methods, independently of a particular chemical composition.

2
Solid-State Reactions

In this chapter, reactions will be discussed in which at least one of the reactants is in the solid state. A large variety of inorganic solids has been prepared by reacting a solid with another solid, a liquid (melt), or a gas, usually at high temperatures. We will first deal with reactions between two (or more) solid compounds (Section 2.1) followed by a section covering reactions between a solid and a gas (Section 2.2). No separate section is devoted to solid/liquid reactions, because in many reactions that start with solid compounds, a liquid phase (melt) is formed at the reaction temperature, that is, many "solid/solid" reactions are actually "solid/liquid" reactions. It is sometimes difficult to determine what physical phases are involved in a given reaction. Section 2.3 describes intercalation reactions, in which guest species (atoms, molecules, or ions) are inserted into a crystalline host lattice. Intercalation can occur with compounds of any physical state.

2.1
Reactions Between Solid Compounds

When solid compounds are employed to react with each other at high temperatures, this does not necessarily imply that all components are still in the solid state at the temperatures required for the reaction to occur. A liquid phase (melt) or even gaseous intermediates may be involved to provide mass transport. We will first discuss some general principles of solid-state reactions, which are often called the "ceramic method" (Section 2.1.1). Reaction between solid compounds can be stimulated by applying mechanical forces; "mechanochemical syntheses" are treated in Section 2.1.2. We will then turn to two important industrial processes, carbothermal reduction (Section 2.1.3) and combustion synthesis (Section 2.1.4). Fundamental processes during sintering, by which powders are converted to dense solid bodies, are dealt with in Section 2.1.5.

2.1.1
Ceramic Method

The oldest and still most common method of preparing multicomponent solid materials is by direct reaction of solid components at high temperatures.

Synthesis of Inorganic Materials, Third Edition. Ulrich Schubert and Nicola Hüsing.

Published 2012 by WILEY-VCH Verlag GmbH & Co. KGaA

Since solids do not react with each other at room temperature – even if thermodynamics favors product formation – high temperatures are necessary to achieve appreciable reaction rates. The advantage of solid-state reactions is the ready availability of the precursors and the low cost for powder production on the industrial scale. We will not cover all the classes of compounds that can be prepared by solid–solid reactions. Rather, we will discuss some fundamental issues of this method after an introductory example.

2.1.1.1 An Example

$YBa_2Cu_3O_{7-x}$ (YBCO) is → superconducting at about −181 °C ("high-temperature superconductor"), which is well above the temperature of liquid nitrogen (b.p. −196 °C) for cooling of the magnet. The unit cell of $YBa_2Cu_3O_7$ is shown in Figure 2.1. It is derived from the → perovskite structure.

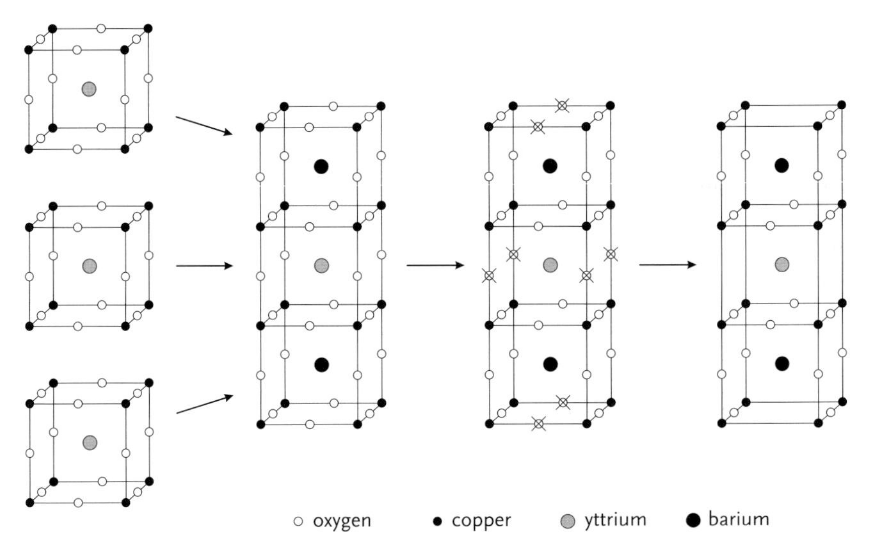

Figure 2.1 Deduction of the unit cell of $YBa_2Cu_3O_7$ (right) from the perovskite structure. Three perovskite unit cells (left) are stacked above each other. Then, the oxygen atoms marked by an x are removed.

Perovskites are a class of solid compounds of the general formula $M^{2+}M^{4+}O_3$ with interesting electrical properties (→ piezoelectricity, ferroelectricity, or high-temperature superconductivity, Figure 2.2). The M^{2+} and oxide ions occupy the anion sites of the rock salt structure, while the M^{4+} ions are in one-quarter of the octahedral sites (those only surrounded by oxide ions). $YBa_2Cu_3O_7$ is an oxygen-deficient variation of this structure type. Its unit cell (Figure 2.1) consists of three stacked perovskite unit cells with Ba^{2+} and Y^{3+} sharing the M^{2+} positions and $^2/_9$ of the oxide ions removed for charge balancing (part of the copper is in the +II and part in the +III oxidation state). The critical temperature for superconductivity depends on the oxygen content x of $YBa_2Cu_3O_{7-x}$. It is highest for $x = 0$; superconductivity breaks down for $x < 0.6$.

Figure 2.2 In 1933, Walter Meissner and Robert Ochsenfeld discovered that a superconducting material will repel a magnetic field (the "Meissner–Ochsenfeld effect"). A body made of a superconducting material thus floats above the surface of a magnet.

$YBa_2Cu_3O_{7-x}$ (YBCO) can be prepared by heating an intimate mixture of yttrium oxide (Y_2O_3), barium peroxide (BaO_2), and cupric oxide (CuO). The starting compounds should be fine grained in order to maximize surface areas and hence reaction rates (see below). The powder is pressed into a pellet to ensure an intimate contact between the → grains. The pellets are placed in an alumina boat and heated in a furnace to 930 °C over a period of 8–12 h, held at this temperature for 12–16 h, allowed to cool to 500 °C and held there for 12–16 h. After heating to 930 °C, the composition is about $YBa_2Cu_3O_{6.5}$. By annealing the material at 500 °C, it reacts further with oxygen from the air, and the final composition is about $YBa_2Cu_3O_{6.9}$. The heating–cooling sequence is shown graphically in Figure 2.3. The oxygen content x in $YBa_2Cu_3O_{7-x}$ depends very much on the oxygen partial pressure during annealing.

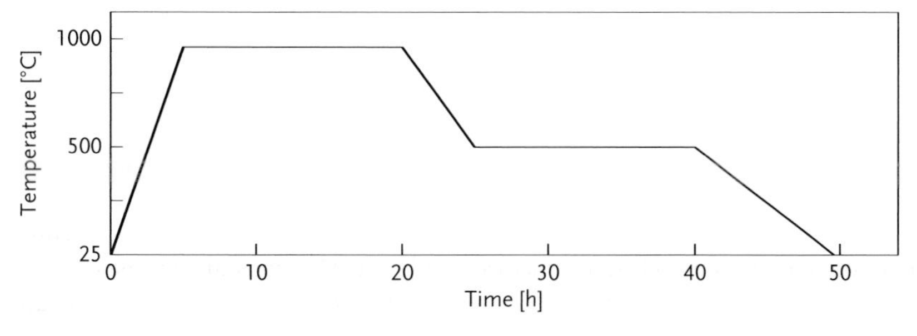

Figure 2.3 Heating protocol for the preparation of $YBa_2Cu_3O_{6.9}$ (see text).

The annealing of $YBa_2Cu_3O_{6.5}$ in an oxygen atmosphere to give superconducting $YBa_2Cu_3O_{6.9}$ is an example of a → topochemically controlled reaction, where reactivity is controlled by the crystal structure rather than by the chemical nature of the reactants.

Other precursor compounds may be employed for the synthesis of YBCO. For example, the metal trifluoroacetates are used for the preparation of YBCO films. Solutions of these salts are prepared by dissolving the metal oxides or metal acetates in trifluoracetic acid (CF_3COOH). An intimate mixture of the metal oxides is obtained by slow heating of the metal trifluoroacetate mixture in moist air from 200 to 780 °C. Carbon dioxide (CO_2) and hydrofluoric acid (HF) are formed as gaseous byproducts. The oxide mixture eventually reacts to give YBCO upon holding at 780 °C for 1 h.

Carbonates, nitrates, hydroxides, and other oxy salts that decompose at high temperatures are often employed in solid-state reactions. The heating program depends very much on the form and reactivity of the reactants. In general, thermal decomposition of solids initiates at some structural defects such as surfaces, → grain boundaries, or dislocations. The gaseous byproducts have to diffuse out of the reaction mixture. If one or more of the reactants is an oxy salt, the mixture is heated first at an appropriate temperature for a few hours to allow decomposition to occur in a controlled manner. If this stage is omitted and the mixture is heated directly at a higher temperature, decomposition may occur very vigorously.

For solid-state reactions in general, some caution is necessary in choosing a suitable container material that is chemically inert to the reactants at the high temperatures. Platinum, silica, stabilized zirconia, and alumina containers are generally used for the synthesis of metal oxides, while graphite containers are employed for sulfides and other → chalcogenides or pnictides. If one of the constituents is volatile or sensitive to the atmosphere, the reaction is carried out in sealed evacuated capsules.

2.1.1.2 **General Aspects of Solid-State Reactions**

In order to understand the difference between reactions in solution and in the solid state, and the problems associated with solid-state reactions, let us consider the thermal reaction of two crystals of the compounds A and B that are in intimate contact across one face (Figure 2.4). When no melt is formed during the reaction, the reaction has to occur initially at the points of contact between A and B, and later by diffusion of the constituents through the product phase.

The first stage of the reaction is the formation of nuclei of the product phase C at the interface between A and B. The process of nucleation will be discussed in detail in Section 4.1.2. This may be difficult if a high degree of structural reorganization is necessary to form the product. After → nucleation of product C has occurred, a product layer is formed. At this stage, there are two reaction interfaces: one between A and C, and another between C and B. In order for further reaction to occur, counterdiffusion of ions from A and B must occur through the existing product layer C to the new reaction interfaces.

As the reaction progresses, the product layer becomes thicker. This results in increasingly longer diffusion paths and slower reaction rates, because the product layer between the reacting particles acts as a barrier. In the simple case where the

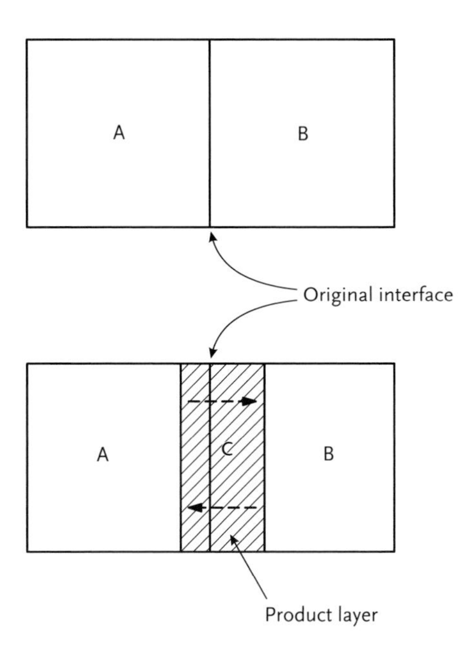

Figure 2.4 Reaction of two crystals (A and B) sharing one face. After initial formation of a product layer C, ions from A and B have to counterdiffuse through the product layer to form new product at the interfaces A/C and B/C.

rate of the reaction is controlled by lattice diffusion through a planar layer, the rate law has a parabolic form:

$$\frac{dx}{dt} = k \cdot x^{-1} \tag{2.1}$$

where x is the amount of reaction (here equal to the thickness of the growing product layer), t is time, and k is the rate constant.

Ions are normally regarded as being trapped on their appropriate lattice sites, and it is difficult for them to move to adjacent sites. Only at very high temperatures do the ions have sufficient energy to diffuse through the crystal lattice. As a rule of thumb, two-thirds of the melting temperature of one component is sufficient to activate diffusion and hence to enable the solid-state reaction.

The formation of the → perovskite barium titanate ($BaTiO_3$) by solid-state reaction of $BaCO_3$ and TiO_2 may serve as an example to illustrate this point, but also to show that such considerations sometimes oversimplify the facts. $BaTiO_3$ is an important material for the fabrication of thermistors, capacitors, optoelectronic devices, and DRAMs. BaO (formed by decomposition of $BaCO_3$) has the rock salt structure (cubic close packing of the oxide ions; Ba^{2+} ions in octahedral sites), while TiO_2 (rutile structure) has a hexagonal close packing of the oxide ions and Ti^{4+} ions in half of the octahedral sites.

The formation of $BaTiO_3$ takes place in at least three stages:

1. First BaO reacts with the outer surface regions of TiO_2 → grains to form nuclei and a surface layer. This requires reorganization of the oxide lattice at the TiO_2/$BaTiO_3$ interface.
2. Further reaction of BaO and the previously formed $BaTiO_3$ leads to the formation of the intermediate Ba-rich phase Ba_2TiO_4. The formation of this phase is necessary for the migration of the Ba^{2+} ions.
3. Ba^{2+} ions from the Ba-rich phase Ba_2TiO_4 migrate into the remaining TiO_2 to form $BaTiO_3$.

From the above discussion it is clear that reaction between two solids may not occur even if thermodynamic considerations favor product formation. There are three important factors that influence the *rate* of reaction between solids:

1. The area of contact between the reacting solids and hence their surface areas.
2. The rate of → nucleation of the product phase.
3. Rates of diffusion of ions through the various phases, and especially through the product phase.

Apart from the problems arising from nucleation and diffusion, the ceramic method suffers from several additional disadvantages:

- Undesirable phases may be formed, such as $BaTi_2O_5$ during the synthesis of $BaTiO_3$.
- The homogeneous distribution of dopants, important for many ceramic materials, is sometimes difficult to achieve.
- There are only limited possibilities for an *in-situ* monitoring of the progress of the reaction. Instead, physical measurements (such as X-ray diffraction) are periodically carried out. Because of this difficulty, mixtures of reactants and products are frequently obtained. Separation of the desired product from these mixtures is generally difficult, if not impossible.
- In many systems the reaction temperature cannot be raised as high as necessary for reasonable reaction rates, because one or more components of the reacting mixture may volatilize.

In order to overcome some of the problems associated with the ceramic method, particularly to reduce the reaction times, it is necessary to optimize the critical parameters.

Surface area of solids The surface area of a given amount of solid depends on the particle size. This is shown by a simple calculation. A cubic crystal with a volume of $1\,cm^3$ has six faces each with an area of $1\,cm^2$ and, therefore, a total surface area of $6\,cm^2$. When this crystal is now cut 10 times parallel to each face (Figure 2.5), 10^3 cubic crystallites are obtained with a dimension of $0.1 \times 0.1 \times 0.1$ cm each. The

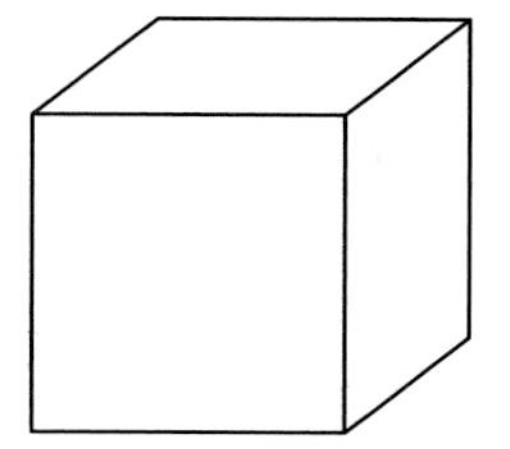

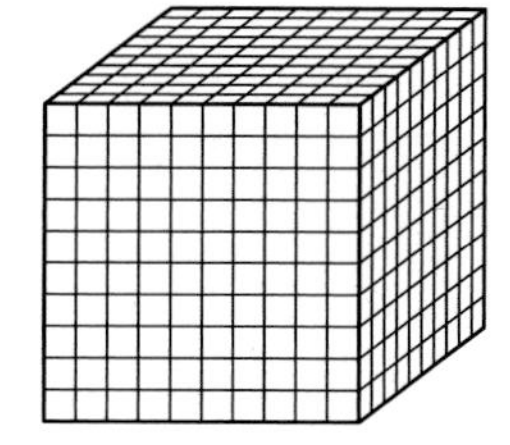

Figure 2.5 When a cubic crystal of $1 \times 1 \times 1$ cm (total surface area $6\,cm^2$) is cut 10 times parallel to each face, 10^3 cubic crystallites of $0.1 \times 0.1 \times 0.1$ cm are obtained. The total volume is the same but the total surface area increases to $60\ cm^2$.

10^3 smaller cubes have the same mass and volume as the large cube, but their total surface area now is 10 times larger ($10^3 \times 6 \times 0.01\,cm^2$).

Grinding of the $1\,cm^3$ crystal or ball-milling for some time typically results in particles with an average size of about $10\,\mu m$ (10^{-3} cm). Hence, the total surface area of the powder would be $6 \times 10^3\,cm^2$ ($0.6\,m^2$) if all grains are cubic crystallites. Note that a $10\,\mu m$ particle size still represents diffusion distances of about 10^4 unit cells! The advantage of further decreasing the crystallite size in the nanometer range is obvious from these considerations (see Chapter 7).

Although the surface area of solids largely controls the area of contact between reacting → grains in a mixture, it does not appear directly in the equation for the rate of reaction, such as Eq. (2.1). However, it is included indirectly since there is an inverse correlation between the thickness of the product layer, x, and the area of contact. For example, when two cubic particles of $10\,\mu m$ size react with each other, the product layer at 50% conversion is $10\,\mu m$ thick. When the dimension of the crystals is decreased to $1\,\mu m$, the total surface area for a given mass of reactants is increased by a factor of 10 (see above), but the thickness of the product layer at 50% conversion is only $1\,\mu m$. According to Eq. (2.1), this will result in a faster reaction.

In practice, it is rather unlikely that all the surfaces of the reacting solids will be in intimate contact, and usually the contact area is considerably less than the total surface area. The area of contact may be increased somewhat by pressing the reacting powder into a pellet. However, even at relatively high pressures, the crystal contacts are not maximized. A further increase in contact area and reduction in pellet porosity may be achieved by compressing the pellet at high temperatures, that is, by hot pressing. However, the densification process is usually slow and may require several hours.

Various modifications have been employed to increase diffusion rates (decrease of diffusion path lengths) in solid–solid reactions:

- Small particle sizes, achieved through grinding, ball milling (Section 2.1.2), spray drying (Section 3.3), freeze drying, and so on. It is possible to reduce the particle size to several tens of nanometers. Solid-state reactions are often greatly facilitated by cooling and grinding the sample periodically. This is because sintering and grain growth of both reactant and product phases may occur during heating, causing a reduction in the surface area of the mixture. Grinding maintains a high surface area and brings fresh surfaces into contact.

- Intimate mixture of the reactants by coprecipitation, sol–gel processing (Section 4.5), and so on.
- Reduction of the diffusion distances by incorporating the cations in the same solid precursor.
- Performing solid-state reactions in molten fluxes or high-temperature solvents (e.g., Bi, Sn, halide salts, or alkaline metals).

Nucleation The overall process in solid-state reactions may not only be controlled by diffusion of reactants, as discussed above, or by the rate of the reaction at the phase boundary, but also by nuclei growth. Nucleation-limited reactions are represented by the Avrami–Erofeyev equation [Eq. (2.2)] and Figure 2.6,

$$x(t) = 1 - e^{kt^{n}} \tag{2.2}$$

where n is a real number, usually between 1 and 3. For $n > 1$, the function has a sigmoid shape. Initially a large number of nuclei are formed, and then the reaction front expands with the growth of the nuclei. When the resulting product regions touch each other, the reaction rate starts to decrease. Note that for most solid-state reactions, it is usually incorrect and misleading to think in terms of reaction order, since the reactions do not involve molecules. Nevertheless, the data may still be represented empirically in this way.

The nucleation step is easier if there is a structural similarity between the product and at least one of the reactants, because this reduces the degree of structural reorganization necessary for nucleation to occur. For instance, in the reaction of MgO and Al_2O_3 to form → spinel, the spinel has a similar oxide ion arrangement to that in MgO (cubic close packing). Spinel nuclei may therefore

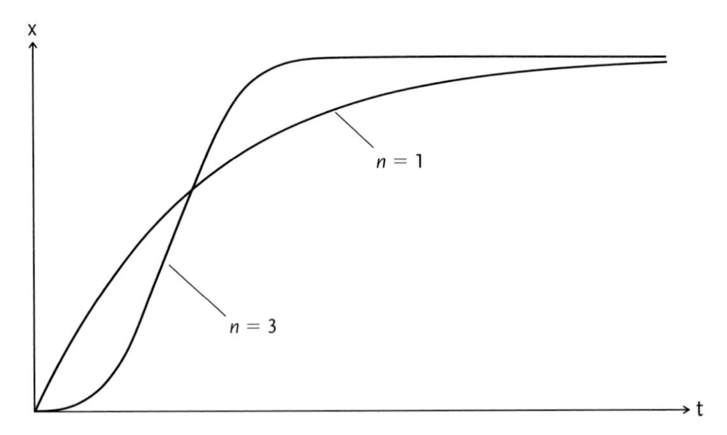

Figure 2.6 Avrami equation for two different values of n. The curve for $n = 3$ graphically shows the induction period at the beginning of the reaction.

form at the surface of the MgO crystals such that the oxide arrangement is essentially continuous across the MgO/spinel interface.

When nucleation is facilitated by a structural similarity, there is usually a clear orientational relationship between the structures of the reactant and product. Furthermore, the interatomic distances, should be similar. If the two structures have quite different interatomic distances then they cannot be matched over a large area of contact. For example, the oxygen–oxygen separations in MgO and BaO are quite different, although both compounds have the rock salt structure. A difference in interfacial lattice parameters of about 15% between nucleus and substrate is the most that can be tolerated for oriented nucleation.

There are two types of oriented reactions: → epitactic reactions and topotactic reactions. In epitactic reactions, the structural relationship is restricted to the actual interface between the two crystals. For example, the two structures may have a common arrangement of oxide ions at the interface, but the structures at both sides of the interface may be different. Epitactic reactions therefore require only a two-dimensional structural similarity at the crystal interface. Topotactic reactions are more specific than epitactic ones because they require not only the structural similarity at the interface but also that this similarity continues into the bulk of both crystalline phases.

Topotactic and epitactic reactions are quite common since their → nucleation step is usually easier than that for reactions in which there is no structural relationship between reactant and product.

The ease of nucleation of product phases also depends on the actual surface structure of the reacting phases. From a consideration of crystal structures it can be shown that in most crystals the structure cannot be the same over the entire crystal surface. For example, the (100) planes of BaO contain Ba^{2+} and O^{2-} ions while a (111) surface will be either a complete layer of Ba^{2+} ions or a complete layer of O^{2-} ions (Figure 2.7). Since different surfaces have different structures, their reactivity is likely to differ considerably. It is difficult, however, to give general rules about which surfaces are the most reactive.

Diffusion of ions Equation (2.1) relates the rate of solid-state reactions to the diffusion of ions through the bulk of the crystals and, especially, through the

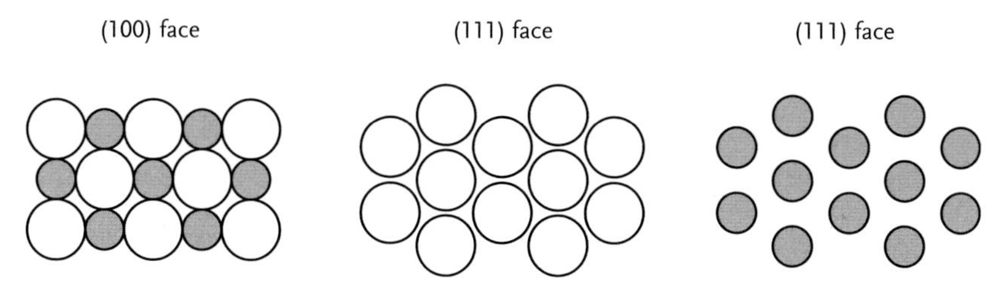

Figure 2.7 Different planes of the BaO (rock salt) structure. The open circles represent the oxygen atoms, the smaller dark circles represent the metal atoms.

product phase. Diffusion path lengths are influenced, *inter alia*, by the particle size of the reactants, the degree of homogenization achieved during mixing, and the intimacy of contact between the → grains. Diffusion of ions is also enhanced greatly by the presence of crystal defects, especially vacancies and interstitials, but also occurs via structural defects such as dislocations and grain boundaries. In general, the rate of diffusion and, therefore, the reactivity of solids depends greatly on the types of crystal defects.

Diffusion path lengths are greatly reduced when the cations are brought in close contact. Instead of heating powder mixtures, solid compounds are used that already contain the different cations in an ideally atomic dispersion. These methods are nearly exclusively employed for the preparation of oxide materials. There are two modifications of this approach:

1. In the *coprecipitation* method, salts of different metals are precipitated together from a common solution. The precipitate then consists of an intimate mixture of two salts or a → solid solution.
2. In the *precursor* method, the cations are incorporated in the same solid precursor.

In both methods, the precipitate is thermally treated to give the desired oxide material. The decomposition temperatures are generally lower than the temperatures employed in the ceramic method due to the shorter diffusion distances of the ions. The precursor method can be subdivided into reactions in which the mixed-metal precursor thermally decomposes without change of the oxidation state of the involved elements, and those where the products are formed by redox reactions.

Coprecipitation Salts of the required metals are dissolved in the same medium (usually water). They are coprecipitated either by concentrating the solution or by adding a precipitating reagent. Hydroxides, carbonates, oxalates, formates, or citrates are often formed by the latter procedure. The obtained precipitate is heated to the required temperature in a desired atmosphere to produce the final product. The coprecipitation method is only successful if the metal salts have similar solubility and the precipitation rate is similar or if solid solutions are formed.

This may be illustrated by the synthesis of $ZnFe_2O_4$ → spinel. Oxalates of zinc and iron are dissolved in water in the ratio of 1 : 1. The solutions are then mixed and heated to evaporate the water. The precipitated fine powder is a → solid solution that contains the cations mixed together, essentially on an atomic scale. The powder is filtered off and heated. Because of the high degree of homogenization, much lower reaction temperatures are sufficient for reaction to occur. The overall reaction may be written as in Eq. (2.3). The method has also been employed for the preparation of → superconducting $YBa_2Cu_3O_{7-x}$.

$$Fe_2(C_2O_4)_3 + Zn(C_2O_4) \longrightarrow ZnFe_2O_4 + 4CO + 4CO_2 \quad (2.3)$$

Another example is the formation of $M_{1-x}M'_xO$ (M, M'=Ca, Mg, Mn, Fe, Co, Zn, Cd). Because their carbonates are isostructural (calcite structure), a large number of carbonate solid solutions $M_{1-x}M'_x(CO_3)$ containing two or more cations can be prepared. These → solid solutions are ideal precursors for the synthesis of solid solutions of the corresponding oxides $M_{1=x}M'_xO$ (rock salt structure). The carbonates are thermally decomposed in vacuum or in flowing dry nitrogen by cleavage of CO_2. The facile formation of the oxides from the carbonates is due to the close relationship between the structures of calcite and rock salt. The $M_{1-x}M'_xO$ solid solutions can be used as precursors for preparing → spinels and other complex oxides.

A number of ternary and quaternary metal oxides can be prepared by employing hydroxide, nitrate and cyanide solid solution precursors as well. For example, hydroxide solid solutions of the general formula $Ln_{1-x}M_x(OH)_3$ (Ln=La or Nd; M=Al, Cr, Fe, Co or Ni) and $La_{1-x-y}Ni_xM_y(OH)_3$ (M=Co or Cu) crystallize in the $Ln(OH)_3$ structure. They are decomposed at relatively low temperatures (around 600 °C) to yield $LnNiO_3$, $LaNi_{1-x}Co_xO_3$, $LaNi_{1-x}Cu_xO_3$, and so on.

Precursor method A variation of the coprecipitation method is that stoichiometric mixed-metal salts are precipitated. A selection is shown in Table 2.1. For example, ferrite spinels such as $NiFe_2O_4$ can be prepared by slowly heating the mixed-metal acetate $Ni_3Fe_6O_3(OH)(OAc)_{17} \cdot 12py$ to 200–300 °C to burn off the organic material, followed by heating in air at ~1000 °C for 2–3 days. The starting mixed-metal salt with the Ni : Fe ratio of exactly 1 : 2 is formed from a basic double acetate hydrate compound.

Another type of mixed-metal precursors are redox compounds. Compounds like $(NH_4)_2Cr_2O_7$ that contain both oxidizing $Cr_2O_7^{2-}$ and reducing NH_4^+ groups when ignited decompose autocatalytically to yield Cr_2O_3 (artificial volcano). The exothermicity of the reaction is due to the oxidation of NH_4^+ to N_2 and H_2O by the dichromate ion that itself is reduced to Cr(III). Fine, voluminous particles are obtained due to the gas-producing reaction. Similar precursors have been used to synthesize chromite spinels, MCr_2O_4 (M=Mg, Zn, Cu, Mn, Fe, Co, Ni) (see Table 2.1). For example, magnesium chromite, $MgCr_2O_4$, is prepared by heating precipitated $(NH_4)_2Mg(CrO_4)_2 \cdot 6H_2O$ gradually to 1100–1200 °C [Eq. (2.4)].

Table 2.1 Examples of ceramic powders obtained from mixed-metal precursors.

Mixed-metal precursor	Ceramic product
$La[Co(CN)_6] \cdot 5H_2O$	$LaCoO_3$
$Ba[TiO(C_2O_4)_2]$	$BaTiO_3$
$M_3Fe_6O_3(OH)(OAc)_{17} \cdot 12py$	MFe_2O_4 (M=Mg, Mn, Co, Ni) (ferrite spinels)
$(NH_4)_2M(CrO_4)_2 \cdot 6H_2O$	MCr_2O_4 (M=Mg, Ni) (chromites)
$(NH_4)_2M(CrO_4)_2 \cdot 2NH_3$	MCr_2O_4 (M=Cu, Zn) (chromites)
$MCr_2O_7 \cdot 4py$	MCr_2O_4 (M=Mn, Co) (chromites)
$MFe_2(C_2O_4)_3 \cdot xN_2H_4$	MFe_2O_4 (M=Mg, Mn, Co, Ni, Zn) (ferrites)

$$(NH_4)_2Mg(CrO_4)_2 \cdot 6H_2O \longrightarrow MnCr_2O_4 + N_2 + 10H_2O \qquad (2.4)$$

By careful control of the experimental conditions, these precursor methods are capable of yielding phases of accurate stoichiometry. This is important since several chromites and ferrites are valuable magnetic materials whose properties may be sensitive to purity and stoichiometry.

2.1.2 Mechanochemical Synthesis

Mechanochemistry studies chemical and physicochemical processes that are stimulated or accelerated by mechanical activation of solids. To this end, fine-grain powders are placed together with a number of hardened steel or tungsten carbide (WC)-coated balls in a sealed container that is vigorously agitated at room temperature, mainly in planetary mills. The → grain size of powder particles can thus be reduced to the lower nanometer scale. When powder mixtures are employed, this process can be used for the synthesis of solid compounds. For example, synthesis of zinc ferrite, $ZnFe_2O_4$, from Fe_2O_3 and ZnO requires heating to 800 °C for 2 h for quantitative reaction. The same result can be achieved within the same period by ball milling at room temperature.

As discussed above, reactions between two solids involve the formation of a product phase at the interfaces of the reactants followed by diffusion of atoms of the reactant phases through the product phase (Figure 2.4). High temperatures are therefore required for the reaction to occur at reasonable speeds. In contrast, during mechanical attrition the powder particles are subjected to severe mechanical deformation from collisions with the hard balls. The basic process of mechanical attrition is illustrated in Figure 2.8.

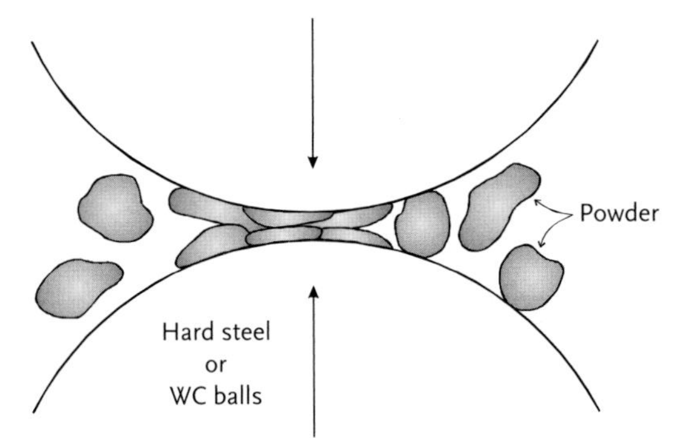

Figure 2.8 Basic process of mechanical attrition of powder particles.

An excess of free energy is produced mainly by plastic deformation arising under the joint action of high pressure and shear, and active surfaces of solids are formed by friction and fracture. Mechanical activation generates clean and fresh surfaces, increases defect density, and reduces particle sizes (increase of the contact area between particles). Diffusion of atoms is thus substantially accelerated and therefore the reaction rates increase even at ambient temperatures. Although part of the mechanical energy is transformed to heat and, consequently, temperature can *locally* increase, overall temperature change plays no significant role in the majority of mechanochemical reactions.

The deformation during mechanical attrition is localized at the early stage in shear bands, with a thickness of about 1 μm, extending throughout the entire particle and consisting of series of high-density dislocations. Nanometer-sized grains are nucleated within these shear bands. This results in an extremely fine-grained microstructure with randomly oriented grains separated by high-angle grain boundaries for longer duration of ball milling.

Mechanochemistry offers many possibilities for low-temperature reactions in the solid phase without dissolution or melting of reagents. Refractory compounds, intermetallic compounds, → composite materials, and nanocrystalline materials were synthesized, as well as organic compounds, metal complexes, polymers, or pharmaceutical preparations. Mechanochemical synthesis also allows the preparation of metastable phases, supersaturated solid solutions, or amorphous phases.

One of the advantages is that solids can be obtained from components with large differences in melting points or melting and boiling points. For example, the boiling point of silicon (2355 °C) is below the melting point of tungsten (3410 °C), and the densities differ by a factor of more than eight (2.31 and 19.35 g cm^{-3}, respectively). This renders preparation of monophase tungsten silicides by standard methods very difficult, while mechanochemical synthesis is possible.

2.1.3 Carbothermal Reduction

Carbothermal reduction is practiced commercially for the synthesis of many nonoxide ceramic powders such as carbides, nitrides or borides.

We will discuss this method for one of the most prominent carbothermal reduction processes, the so-called "Acheson process." Almost all silicon carbide, SiC ("carborundum"), produced worldwide is made by this method. SiC is mainly used in four areas of application:

1. **As an abrasive.** SiC powders are used for cutting and grinding precious and semiprecious stones and for fine grinding and lapping of metals and optical glasses. Bound with synthetic resins and ceramic binders, SiC grits are used in grinding wheels, whetstones, hones, abrasive cutting-off wheels, and monofiles for machining of all types of materials.
2. **As a deoxidizer.** SiC is used in cast iron and steel production as a deoxidant, and for carburization and siliconization.
3. **As a refractory material.** SiC is applied as a structural refractory material with excellent thermal shock, oxidation and corrosion resistance in linings and skid rails for furnaces, and in kiln furniture.

4. **As electric heating elements and electrical resistors**. For example, SiC is used for heating elements operating in oxidizing atmospheres up to 1500 °C owing to its good electric conductivity combined with its excellent oxidation resistance.

The first commercial plant to produce SiC was built in 1896 in Niagara by Acheson to meet the demand for abrasives at this time. The basic design of the original electric furnace (Figure 2.9) has remained unchanged, despite larger sizes and better efficiency. Most Acheson furnaces are shaped like a trough.

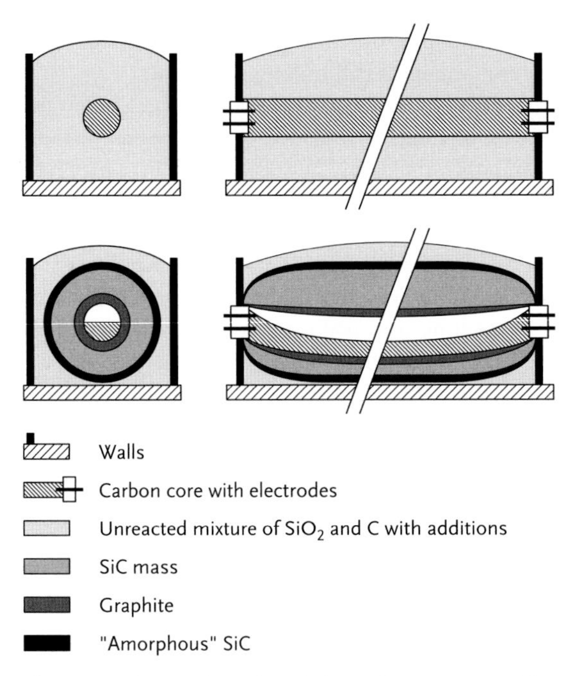

Figure 2.9 Section through an Acheson furnace before (above) and after the reaction (below).

Graphite electrodes connected to a graphite core (that constitutes the initial electrical path through the reaction mixture) are laid in a mixture of carbon and sand. When an electric current is passed through the graphite core, the charge is heated from within by resistive heating. The reaction takes place at temperatures above 1577 °C. This results in the formation of a hollow cylinder of SiC, and the expulsion of carbon monoxide gas. The charge acts as a → refractory container as well as a thermal insulator for the ingot being formed.

Current conventional furnaces are about 12–18 m long, and the power intensities approach 260 kW m^{-1}. The technological limit to furnace size and power intensity is the available electrical power supply technology. The manufacture of 1 kg SiC requires about 12 kW h, and so most plants are located in areas where power is relatively inexpensive.

The overall reaction of the SiC formation is

$$SiO_2 + 3C \longrightarrow SiC + 2CO \quad (2.5)$$

The SiO_2 source can be sand, quartzite, or crystalline rock quartz. The most common carbon source is petroleum coke, the final residue from crude oil refining. Carbon black, graphite, charcoal, pyrolyzed organic polymers, or other carbon sources can also be used. The particle size of both components is in the range of 5–10 mm; the size distribution should be narrow so that the final furnace mix is permeable to the CO gas that must leave the reaction zone. Porosity agents, such as sawdust, bagasse, or rice hulls, are often incorporated into the furnace mix to make it additionally permeable to the escaping CO. High pressures of gas may locally be built up to form voids and channels to more porous parts of the mixture. The product from an Acheson furnace at the end of a run is a large hollow cylinder (Figure 2.9). The unreacted furnace mix is recycled.

Pure SiC is colorless and transparent. However, such pure SiC cannot be made by the Acheson process because nitrogen from the air is soluble in SiC, causing the crystals to take a green color. Black SiC is obtained in the presence of impurities, such as Al (maximum solubility 2%) and B (maximum solubility 0.5%). To make green SiC, the furnace mixture must be made from quartz and low metal content coke. The highest tonnage application for SiC (about 50% of the produced SiC) is "metallurgical SiC," a silicon and carbon source for the iron casting industry. Purity is not important for its function. Purer SiC is needed for abrasive (about 40%), refractory, and other applications.

The SiC-forming reaction is much more complicated than given in Eq. (2.5). Originally, it was assumed that SiC solely results from solid-state reaction between SiO_2 and carbon. However, this is not reasonable for the relatively large particle sizes used in commercial SiC production. One of the current models is that SiC forms as a result of four subreactions [Eqs. (2.6)–(2.9)], which provide mass transport via the vapor phase.

$$C(s) + SiO_2(s) \longrightarrow SiO(g) + CO(g) \quad (2.6)$$

$$SiO_2(s) + CO(g) \longrightarrow SiO(g) + CO_2(g) \quad (2.7)$$

$$C(s) + CO_2(g) \longrightarrow 2CO(g) \quad (2.8)$$

$$2C(s) + SiO(g) \longrightarrow SiC(s) + CO(g) \quad (2.9)$$

SiO_2 initially reacts at the contact points with coke particles in a solid-state reaction to liberate CO and gaseous SiO, [Eq. (2.6)]. Reaction of the CO with further SiO_2 results in the formation of additional SiO and CO_2 [Eq. (2.7)]. Carbon dioxide is reconverted to CO by the Boudouard equilibrium [Eq. (2.8)]. SiC is formed by reaction of gaseous SiO directly at the surface of the carbon particles once the C/SiO_2 contact points are consumed [Eq. (2.9)]. Silicon is transported to the carbon particles in the form of SiO (Figure 2.10). As the product layer grows, the reaction surface decreases and the reaction gets slower.

Porous (cellular) SiC can also be produced by this method. Native wood was converted into carbon preforms by high-temperature pyrolysis in an argon

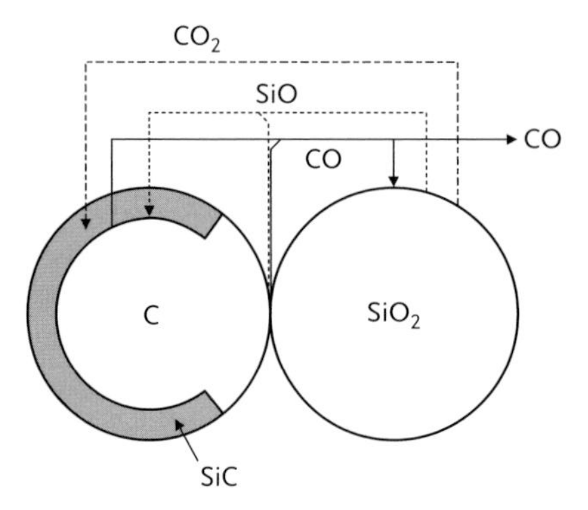

Figure 2.10 Material transport paths during the preparation of SiC by the Acheson process.

atmosphere. Subsequent infiltration and reaction with gaseous SiO at 1600 °C results in a biomorphic SiC ceramic (see also Section 3.2.7 on chemical vapor infiltration). The pore structure of the SiC ceramic reproduces the original wood morphology, as shown in Figure 2.11.

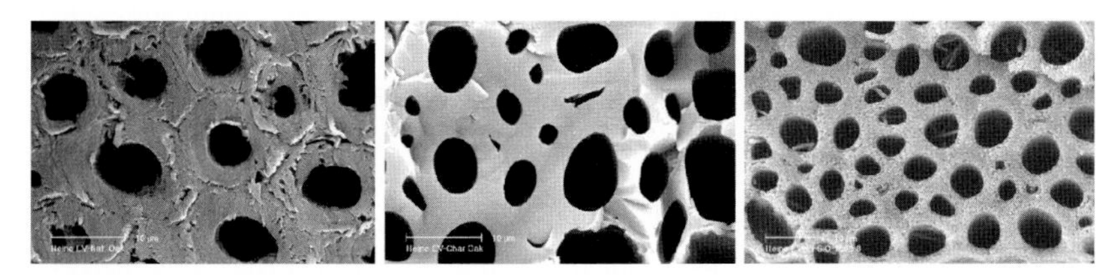

Figure 2.11 Cellular microstructure of native tissue oak (left), the carbon preform obtained by pyrolysis (center), and biomorphic SiC ceramic after reaction with SiO at 1600 °C.

In general, the manufacture of non-oxide ceramic materials by carbothermal reduction can be carried out in a variety of ways, and various types of reactors have been developed (electric-arc furnaces, moving-bed furnaces, rotary-tube reactors, fluidized-bed reactors, etc.). A discussion of the advantages and disadvantages of the various reactor types would be beyond the scope of this book.

Although the final equilibrium products are determined solely from the temperature, pressure, and chemical species present, the mechanism and rate of a given reaction depend on a number of additional variables, such as particle size, degree of mixing of the reactants, diffusion rates, gas concentration, porosity, and the presence of impurities, as already discussed in Section 2.1.1.

Carbides are made by the high-temperature reaction between carbon and metal oxides alone, as exemplified by the Acheson process discussed above. The synthesis of borides requires the presence of elemental boron that is usually formed *in situ* by reduction of B_2O_3. The energy required to manufacture the borides (with B_2O_3 as the boron source) is greater than for the corresponding carbides because carbon has to reduce both the metal oxide and B_2O_3. Nitrides are obtained in the presence of nitrogen sources, usually elemental nitrogen ("carbothermal nitridation"). In all cases, the reactions are highly exothermic and thermodynamically favorable at very high temperatures. Since the reactions are reversible, it is advantageous to remove the byproduct CO. Selected examples are shown in Table 2.2.

The fact that many of the carbothermal reductions are intrinsically fast is evidence that solid–solid reactions do not dominate the process. As already discussed for the Acheson process, gaseous species play a very important role. Some of the solid reactant metal oxides sublime, have a substantial vapor pressure, dissociate, or are reacted into volatile species within the temperature range for reaction.

Reduction processes involving CO are believed to be the primary route for the reduction of certain oxides to produce gaseous suboxides. Carbothermal reductions involving SiO_2 almost always occur through gaseous SiO, as already discussed, Eqs. (2.6) and (2.7). Under reducing conditions, SiO is the dominant gaseous species at temperatures between about 1000 and 1700 °C. Other examples of the formation of suboxides by CO are given in [Eqs. (2.10) and (2.11)].

$$B_2O_3 + CO \longrightarrow B_2O_2(g) + CO_2 \quad (2.10)$$

$$Al_2O_3 + 2CO \longrightarrow Al_2O(g) + 2CO_2 \quad (2.11)$$

For carbothermal reductions involving B_2O_3, there is an increasing vapor pressure of B_2O_3 at temperatures above the melting point of 450 °C. Above tem-

Table 2.2 Examples of the carbothermal reduction, and minimum temperatures.

Reactions	Minimum temperatures (°C) at atmospheric pressure
Carbides	
$2Al_2O_3 + 9C \rightarrow Al_4C_3 + 6CO$	1950
$2B_2O_3 + 7C \rightarrow B_4C + 6CO$	1550
$SiO_2 + 3C \rightarrow SiC + 2CO$	1500
$TiO_2 + 3C \rightarrow TiC + 2CO$	1300
$WO_3 + 4C \rightarrow WC + 3CO$	700
$2MoO_3 + 7C \rightarrow Mo_2C + 6CO$	500
Borides	
$Al_2O_3 + 12B_2O_3 + 39C \rightarrow 2AlB_{12} + 39CO$	1550
$V_2O_5 + B_2O_3 + 8C \rightarrow 2VB + 8CO$	950
$V_2O_3 + 2B_2O_3 + 9C \rightarrow 2VB_2 + 9CO$	1300
$TiO_2 + B_2O_3 + 5C \rightarrow TiB_2 + 5CO$	1300
$2TiO_2 + B_4C + 3C \rightarrow 2TiB_2 + 4CO$	1000
Nitrides	
$Al_2O_3 + 3C + N_2 \rightarrow 2AlN + 3CO$	1700
$B_2O_3 + 3C + N_2 \rightarrow 2BN + 3CO$	1000
$3SiO_2 + 6C + 2N_2 \rightarrow Si_3N_4 + 6CO$	1550
$2TiO_2 + 4C + N_2 \rightarrow 2TiN + 4CO$	1200
$V_2O_5 + 5C + N_2 \rightarrow 2VN + 5CO$	600

peratures of approximately 1450 °C, B_2O_2 (and other B/O species) become more important. For Al_2O_3 systems under reducing conditions, the major gaseous aluminum species above 1200 °C are gaseous Al and Al_2O. Reaction of the sub-oxides (B_2O_2 or Al_2O) with solid carbon provides a mechanism for the formation of the carbides similar to Eq. (2.9).

Liquid phases (melts) may also be involved, particularly for reactions with B_2O_3. In the formation of B_4C, there is a change in mechanism at about 1700 °C. The liquid-phase reduction of B_2O_3 [Eq. (2.12)] dominates at lower temperatures, and the gas-phase reaction at higher temperatures [Eq. (2.13)].

$$7C(s) + 2B_2O_3(l) \longrightarrow B_4C(s) + 6CO(g) \tag{2.12}$$

$$5C(s) + 2B_2O_2(g) \longrightarrow B_4C(s) + 4CO(g) \tag{2.13}$$

In the carbothermal nitridation of silicon, Si_3N_4 can form from SiO either by reaction with solid carbon [Eq. (2.14)] or by reaction with CO in a gas-phase reaction, [Eq. (2.15)].

$$3C + 3SiO + 2N_2 \longrightarrow Si_3N_4 + 3CO \tag{2.14}$$

$$3CO + 3SiO + 2N_2 \longrightarrow Si_3N_4 + 3CO_2 \tag{2.15}$$

2.1.4 Combustion Synthesis

The combustion synthesis approach uses highly exothermic reactions. Such reactions typically have high activation energies and generate substantial amounts of heat. Once the reactions are initiated by the rapid input of energy from an external source, sufficient heat is released to render the reactions self-sustaining. Ignition can be achieved using an electric arc, a laser pulse, a spark, a chemical reaction, and so on. The reactants are thus heated rapidly (10^3–10^6 K s^{-1}) to very high temperatures. The reactions are so fast that they are pseudoadiabatic, that is, all the energy produced by the exothermic reaction is used to heat the sample.

Combustion syntheses can be conducted in the self-propagating mode and the simultaneous combustion mode, although in reality many combustion synthesis reactions lie in between:

- **Self-propagating mode, also referred to as "self-propagating high-temperature synthesis" (SHS):** The combustion reaction is initiated at one point by heating to the temperature at which reaction initiates (ignition temperature, T_{ig}) and then propagates rapidly through the reaction mixture in the form of a combustion wave (Figure 2.12, route A and Figure 2.13). In this mode, the heat of the reaction furnishes more than 90% of the energy required for the synthesis. For example, $T_{ig} = 1300$ °C for $Si + C \longrightarrow SiC$ or $T_{ig} = 630$ °C for $Ni + Al \longrightarrow NiAl$.
- **Simultaneous combustion mode, also referred to as thermal explosion**: Once the entire sample has been heated to T_{ig}, reaction takes place simultaneously through the reactant mixture, rather than as a propagating combustion wave (Figure 2.12, route B).

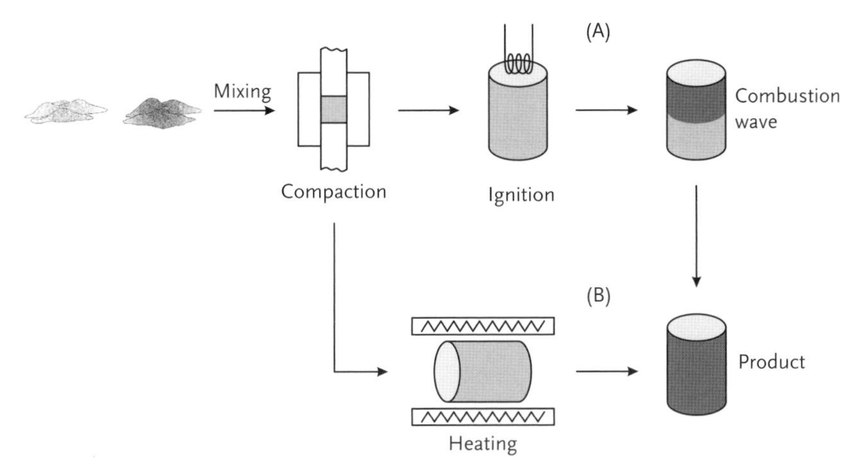

Figure 2.12 Schematic illustration of the self-propagating combustion mode (route A) and the thermal explosion mode (route B).

Figure 2.13 SHS before ignition (front), during reaction (middle), and after reaction (back) of the compacted sample.

An early application of combustion synthesis was the "thermite" reduction of metal oxide powders with aluminum powder yielding either metal or metal/alumina → composites. An example is given in Eq. (2.16). The heat generated by the exothermic reaction ($849\,kJ\,mol^{-1}$) is sufficient for welding railroad tracks, for example (Figure 2.14).

$$3Fe_3O_4 + 8Al \longrightarrow 9Fe + 4Al_2O_3 \tag{2.16}$$

Figure 2.14 A device for welding railroad tracks by the thermite process.

In practice, the aligned ends of the rail are surrounded by a → refractory mold and preheated to 600–1000 °C. A crucible containing the thermite mixture is placed over the butt joint, and the mixture is ignited. The liquid iron and alumina formed in the reaction separate due to their different densities. The heavier iron flows in the butt joint through an opening in the bottom of the crucible.

Combustion synthesis is a versatile method for the synthesis of a variety of technologically useful solid materials, such as binary and ternary metal borides, carbides, silicides, → chalcogenides, nitrides, or hydrides, as well as → alloys, composites, or cemented carbides.

Both mixtures of solids and solid–gas systems are employed as starting compounds in combustion syntheses.

Solid–solid systems The reactant mixture is usually pelletized to increase the intimate contact between individual particles and placed in a → refractory container, degassed and ignited in a vacuum or in an inert atmosphere.

Solid–gas systems Such reactions are often called "filtration combustion" processes, because the gas is supplied to the reaction front from the surrounding atmosphere by "filtration" through the porous mass. They will be discussed in Section 2.2.

Compared with conventional ceramic processing, the main advantages of the combustion syntheses are:

- it requires less energy than conventional methods;
- the short reaction times result in low operating and processing costs and time savings;
- expensive processing equipment is not necessary;
- inorganic materials can be synthesized and consolidated into a final product in one step by utilizing the chemical energy of the reactants;
- the high reaction temperatures may expel volatile impurities and thus result in higher purity products; and
- the high thermal gradients and rapid cooling rates can give rise to nonequilibrium phases.

SHS reactions can be characterized by the adiabatic combustion temperature T_{ad}. This can be calculated by assuming that the enthalpy of the reaction heats up the products and that no energy is lost by heating the surrounding environment. Thus, T_{ad} is a measure of the exothermicity of the reaction and defines the upper limit for any combustion system. Based on T_{ad}, SHS reactions for binary systems can be classified as follows (reactions involving a gaseous reactant are treated in Section 2.2):

- When T_{ad} is less than the boiling point (T_b) of both elements participating in the reaction ($T_{ad} < T_b$), no vapor phase is present (at T_{ad}!) in the reaction. Ideal gas-free reactions are rare ($Mo + B \rightarrow MoB$ is an example). However, reactions can made gasless by applying high pressures.
- When T_{ad} is between the boiling and the melting point of both reactants ($T_m < T_{ad} < T_b$) reaction occurs in liquid phase. An example is $Ni + Al \longrightarrow NiAl$ ($T_{ad} = 1640\,°C$; Ni: $T_m = 1453\,°C$, $T_b = 3100\,°C$; Al: $T_m = 660\,°C$, $T_b = 2500\,°C$).
- Propagation velocities of the combustion wave are highest when T_{ad} is between the melting point of both reactants ($T_{m(1)} < T_{ad} < T_{m(2)}$), because the molten component spreads at a high rate in the compact. An example is $Ti + C \rightarrow TiC$, where Ti is molten at T_{ad} (2940 °C).
- For the reasons discussed above, propagation velocities are lowest, and combustion is difficult, when both reactants are in the solid state at T_{ad} ($T_{ad} < T_m$). An example is $Ta + C \rightarrow TaC$ with $T_{ad} = 2430\,°C$).

As a rule-of-thumb, combustion does not occur if $T_{ad} < 1200\,°C$, and self-propagating combustion occurs if $T_{ad} > 2200\,°C$. In the range $1200\,°C < T_{ad} < 2200\,°C$, a combustion wave cannot propagate but can be made to do so by special techniques such as preheating of the reactants. For example, the reactions $Si + C \longrightarrow SiC$ ($T_{ad} = 1530\,°C$) or $Ti + Al \longrightarrow TiAl$ ($T_{ad} = 1245\,°C$) are not self-sustaining when $T_{start} = 25\,°C$, while the reaction $Ti + C \longrightarrow TiC$ ($T_{ad} = 2940\,°C$) is self-sustaining. The reaction $Ti + Al \longrightarrow TiAl$ can be made self-sustaining by preheating above 100 °C.

Combustion syntheses are difficult to control because of the high reaction rates. However, some control is possible by:

- addition of diluents, which do not take part in the reaction, but increase the thermal mass of the system and lower T_{ad}. The diluent may be the product (e.g., the rate of the highly exothermic reaction of titanium and boron is reduced by addition of TiB_2), or an inert compound. The latter option results in → composite materials.
- an increase in particle size of the reactants. This leads to a decrease of the combustion wave velocity, and the reaction rate decreases, as expected for solid–solid reactions.

The materials obtained by combustion synthesis vary from friable, slightly sintered powders to monolithic porous products that are formed when a fraction of the products is molten at the peak temperature. The inherently high porosity of the products originates both from the density difference between the reactants and the products, and from outgassing or evaporation due to the high reaction temperatures. Important parameters that control the final product composition and morphology are:

- reactant particle-size distribution, morphology, and purity;
- → green density, which influences the thermal conductivity of the reactant mixture;
- ignition techniques, initial temperature and pressure of the reacting mixture;
- the difference between the temperature of the initial reaction mixture and the completely reacted product; and
- size of the sample and reactor configuration.

The usual techniques can be applied to densify the materials after synthesis (see Section 2.1.5). However, an *in-situ* consolidation would provide an important economical method for manufacturing ceramic materials. In order to obtain a high degree of consolidation, it is advantageous to maintain one of the constituents in a partially molten condition. Simultaneous combustion synthesis and densification is currently used in the direct manufacture of → hard-alloy components, such as rollers, drawing blocks, pressing equipment, and cutting inserts. Densification can be achieved by applying a suitable consolidation pressure either during the reaction or immediately after the reaction is completed and while the products are still in a plastic state at the high combustion temperatures.

2.1.4.1 Classification of Combustion Synthesis Reactions

Combustion synthesis reactions can be classified according to the kind of reactants.

The product is synthesized from its elements Many carbides, silicides, borides, nitrides, oxides, or hydrides were prepared by this method, including solid solutions and → composite materials. For example, the reaction between titanium and carbon [Eq. (2.17)], one of the most widely studied combustion synthesis reactions, results in stoichiometric TiC, which is a valuable → refractory and abrasive material. The reaction is highly exothermic, liberating 183 kJ mol^{-1}. The synthesis of 20-kg batches was reported to last about 60–90 s, while cooling of the reactor afterwards took 1.5–2 h.

$$Ti + C \longrightarrow TiC \tag{2.17}$$

Numerous intermetallic compounds were also produced using the SHS method, for example a variety of shape-memory → alloys (such as TiNi). Aluminides have also received some attention, particularly aluminides with nickel, zirconium, and copper.

Thermite-type reactions Thermite reactions are extensions of the → Goldschmidt process for the reduction of an ore to its constituent metals. In these systems, the combustion synthesis reaction involves the reduction of a metallic compound. Most common is the reduction of an oxide by a strongly reducing metal, mainly Al and Mg. Magnesium is more desirable, because the resulting MgO byproduct is easily leachable with hydrochloric acid.

There are two types of thermite reactions. The first involves the reduction of an oxide to the element, for example the well-known "thermite reaction" [Eq. (2.16)], and the second the reduction of an oxide to the element followed by its reaction with another element to form a refractory compound such as borides, carbides, silicides, and nitrides (e.g., Eqs. (2.18) and (2.19)).

$$SiO_2 + C + 2Mg \longrightarrow SiC + 2MgO \tag{2.18}$$

$$TiO_2 + B_2O_3 + 5Mg \longrightarrow TiB_2 + 5MgO \tag{2.19}$$

In each case, a metal oxide is formed as a byproduct. The obtained ceramic–ceramic → composites are often interesting in themselves. For the synthesis of the pure non-oxide powders, the oxide has to be chemically leached. This technique may be used when the cost of the elemental powder is too high (e.g., Ta, Hf, B) or when the direct reaction of the element does not generate a sufficient amount of heat to be self-sustaining.

For example, the heat of formation of B_4C from the elements is only -39 kJ mol^{-1}. Therefore, the T_{ad} (about 730 °C) is not high enough for a self-propagating reaction when the reactant mixture is ignited at room temperature. This problem can be solved by preheating or by the thermal explosion technique. However, since elemental boron is rather expensive, the use of the thermite reaction [Eq. (2.20)] is more economic. The reaction enthalpy of this particular reaction is -1135 kJ mol^{-1}, resulting in T_{ad} of 2470 °C. Ignition occurs when the reactant mixture is heated to 930 °C in an inert gas (argon) atmosphere.

$$2B_2O_3 + C + 6Mg \longrightarrow B_4C + 6MgO \tag{2.20}$$

Solid-state metathesis (SSM) reactions This method involves rapid, low-temperature-initiated solid-state exchange reactions [Eq. (2.21)]. Some examples are given in Table 2.3. Other technologically important materials have also been prepared by this method, such as metal phosphides, arsenides, or antimonides.

$$ME_x + M'E'_y \longrightarrow ME'_u + M'E_v \qquad (2.21)$$

Particularly useful are reactions between (anhydrous!) metal halides (E = halide) and alkaline-metal (or alkaline-earth metal) main-group compounds (M′ = alkaline or alkaline earth metal). The desired product is easily separated from the alkaline metal halide byproduct by washing with alcohol and/or water. Additional byproducts may be formed to balance the reactions (e.g., elemental carbon is additionally formed in the reaction of tantala and calcium carbide (Table 2.3)).

The mechanism of SSM reactions is hardly understood, because the speed of the reactions prevents direct observation. Both recombination of ions and (uncharged) atoms are discussed.

SSM reactions produce very fine scale crystallites, typically in the range 10–100 nm. SSM therefore offers an interesting route to nanoparticles, provided the problems with particle agglomeration can be overcome.

2.1.5 Sintering

When a compacted powder is heated at an elevated temperature that is below its melting point, powder particles fuse together, voids between the particles decrease, and eventually a dense solid body is obtained. This phenomenon is called sintering. Sintering processes have been used extensively for the manufacture of

Table 2.3 Examples of SSM reactions.

Reactants	Solid products
Carbides	
$TiCl_3 + CaC_2$	$TiC + CaCl_2$
$ZrCl_4 + Al_4C_3$	$ZrC + AlCl_3$
$Ta_2O_5 + CaC_2$	$TaC + CaO$
Nitrides and phosphides	
$GaI_3 + Li_3N$	$GaN + LiI$
$NaBF_4 + NaN_3$	$BN + NaF$
$ZrCl_4 + Na_3P$	$ZrP + NaCl$
Borides and silicides	
$VCl_3 + MgB_2$	$VB_2 + MgCl_2$
$V_2O_5 + Mg_2Si/CaSi_2$	$VSi_2 + MgO/CaO$
Chalcogenides	
$ZrCl_4 + Na_2O$	$ZrO_2 + NaCl$
$TiCl_3 + Na_2O_2$	$TiO_2 + NaCl$
$MnCl_2 + Na_2S_2$	$MnS + NaCl$
$AgF + Na_2Se$	$AgSe + NaF$

ceramics and ironware for hundreds of years, and today sintering is still a very important process for the manufacture of a wide variety of industrial materials.

In this section, we will only briefly touch on the technological issues of sintering, which are more thoroughly treated in materials science textbooks. Rather, we will point out some chemical issues related to sintering processes.

The driving force behind sintering is the excess surface free energy of a powder compact. When heated, the system tries to decrease its surface free energy by decreasing its total surface area. This is reached by mass transport that joins the powder particles together.

There is a chemical-potential difference between surfaces of dissimilar curvature within the system. A concave surface has a negative free energy and convex surfaces a positive free energy. As a consequence, mass transport occurs from the particle surface (convex) to the interparticle necks or pores (concave). The greater the curvature, that is, the finer the particle size, the greater the driving force for sintering.

During sintering, mass transport can occur by solid-state, liquid-phase, and vapor-phase mechanisms individually, or in combination (Figure 2.15):

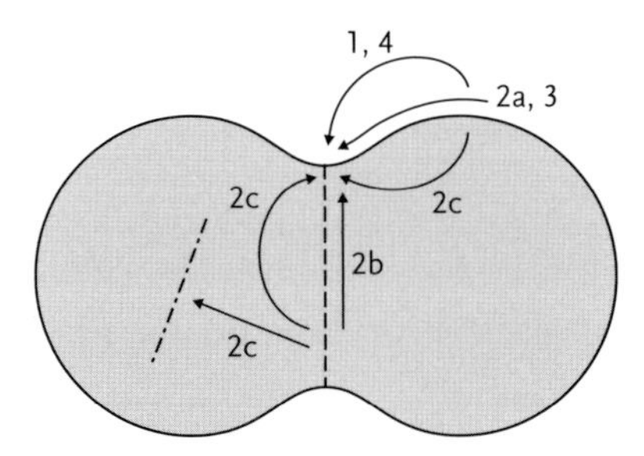

Figure 2.15 Diffusion paths during sintering. The numbers correspond to the numbers in the following paragraph.

- **Path 1: Evaporation–condensation**. Compared to a flat surface, the vapor pressure of a convex surface is higher and that of a concave surface is lower. As a consequence, a substance vaporizes at the particle surface and condenses at the necks and in the pores.
- **Path 2: Diffusion**. The driving force for diffusion are differences in vacancy concentration. The diffusion mechanism can be subclassified into surface diffusion (2a), grain-boundary diffusion (2b), and volume diffusion (2c). The driving force for surface diffusion is the difference in vacancy concentration between convex and concave areas. The vacancy concentration is smaller on a convex surface than that on a flat surface, and higher on a concave surface. As a consequence, vacancies flow from a neck (concave area) to a convex area and thus a mass flow occurs in the reverse direction. Mass flow through the volume of a grain (volume diffusion) may originate at convex surfaces, grain boundaries, or dislocation in the grain matrix. Except in substances with high vapor pressure such as NaCl, the sintering of powders proceeds by the volume diffusion mechanism.

- **Path 3: Flow.** A perpendicular pressure pointing to the center of a powder particle acts on a convex surface, and a pressure pointing away from the center acts on a concave surface. When a substance is fluid, mass transport proceeds under the pressure difference.
- **Path 4: Dissolution–precipitation.** In the initial phase of sintering, the presence of a liquid phase that wets the solid phase allows the rearrangement of particle packing by gliding. Subsequently, the substance dissolved from convex surfaces, where solubility is higher, is transported to concave surfaces, where solubility is lower, and precipitates out. Dissolution, precipitation, or diffusion of a substance in the liquid phase can be rate-controlling.

2.1.5.1 Stages of Sintering

The development of microstructures during sintering is rather complicated, but may be distinguished in three stages (Figure 2.16).

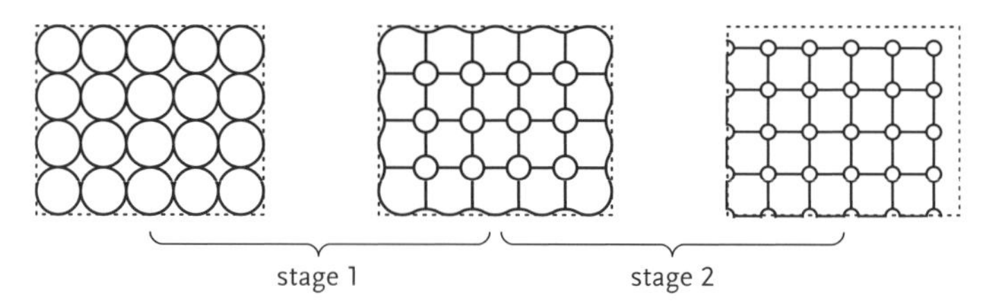

Figure 2.16 A two-dimensional sphere model illustrating the first two stages during sintering.

Initial stage Initially, material is transported from higher-energy convex particle surfaces to the lower-energy, concave intersections between adjacent particles to form necks ("neck growth"). The powder particles fuse together and the area of contact increases gradually. Since mass is only transported from convex to concave areas, the total pore volume and the distance between the particle centers remain about constant, and shrinkage of the → green body is only about 4–5%. In this stage the relative density, which is the density of the powder compact divided by its theoretical density, is about 0.5–0.6.

Intermediate stage In this stage, interparticle necks grow, the area of grain boundaries (the interface plane shared by two grains) increases, interparticle contacts flatten, and the pore diameters decrease. The distance between the particle centers and the volume of the compact decreases (shrinkage of 5–20%), that is, densification occurs. The relative density increases to about 0.95.

Final stage When the relative density increases above 0.95, isolated spherical pores remain only at triple points (intersection lines where three → grains meet) or inside the grain matrix. In the final stage, these pores are gradually eliminated and the relative density increases further.

2.1.5.2 Factors Affecting Sintering

The most important powder physical characteristics that can affect sintering are particle size, particle packing, and particle shape.

Particle size Material transport occurs faster over shorter distances, and less material needs to be transported to fill small pores. Furthermore, very fine particles have high surface energies. Therefore, smaller powder particles speed up the sintering process and lower the sintering temperatures and pressure (see also Chapter 7). Due to the thermodynamic considerations discussed above, larger grains grow at the expense of smaller ones. Consequently, as sintering progresses, the average size of the grains increase ("coarsening"), and the size distribution becomes narrower. Since coarsening is much slower than sintering, grain growth can occur especially in the final sintering stage. Uncontrolled grain growth is usually detrimental to the ceramic's properties.

Particle packing Improved particle packing increases the number of contact points between adjacent particles and the relative density of the compact. Consequently, densification occurs faster (better material transport) and with less volume shrinkage. One of the most important causes of non-uniform particle packing is the formation of aggregates (see Section 3.3).

Particle shape Irregular-shaped particles, which have a high surface area to volume ratio, have a higher driving force for densification and sinter faster than equiaxed particles. Particles that pack poorly sinter poorly.

2.1.5.3 Ceramics Processing

Ceramic processing, that is, the fabrication of ceramics from powders, involves a number of steps (Figure 2.17). Each of these steps must be carefully controlled because ceramics are basically flaw intolerant, and chemical and physical defects severely degrade properties. Mistakes in ceramic processing are cumulative, and generally cannot be corrected during sintering or by postsintering processes.

We will only briefly describe the different steps of ceramic processing. Readers interested in ceramic processing are referred to textbooks on ceramic engineering for details.

Benefication These processes modify the chemical and/or physical properties of raw materials to render them more processable. They may include:

- particle-size reduction by crushing, grinding, or milling;
- purification by washing, extraction, chemical leaching (to extract contaminations), separation, and so on;

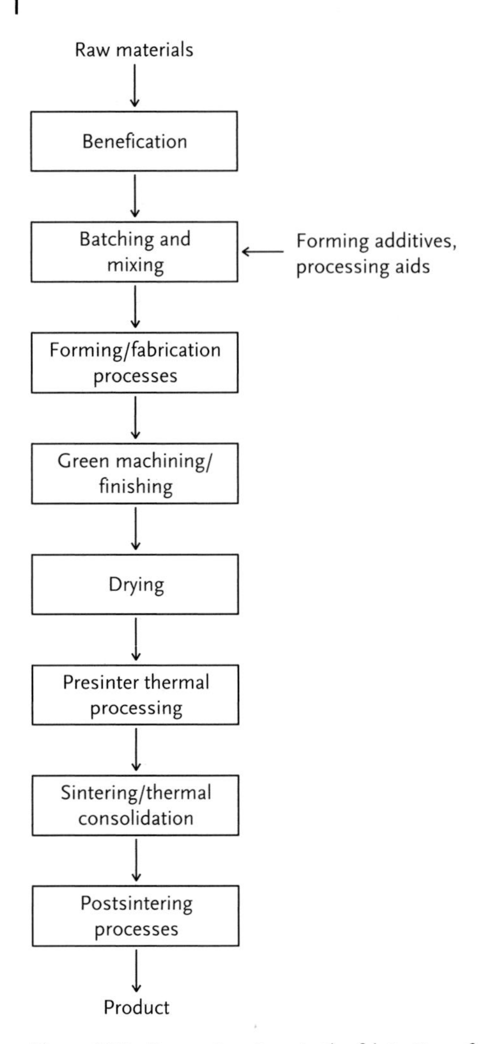

Figure 2.17 Processing steps in the fabrication of ceramics.

- sizing (sieving) and classification;
- → calcination to effect decomposition of precursor compounds, to liberate gases (water), or to achieve structural transformations;
- preparation of a slurry by dispersion in liquids to wet the solid, break down agglomerates or stabilize the system to prevent agglomeration (flocculation); and
- granulation of powders to improve flow, handling, packaging, and compaction.

Batching and mixing The constituents of a ceramic body are combined to produce a more chemically and physically homogeneous system for forming. Forming additives or processing aids are commonly used to render ceramic powders more processable.

- **Binders** are used to impart strength to a → green (unfired) ceramic body for handling and machining. They are mostly organic polymers (poly(vinyl alcohol), poly(ethylene glycol), etc.) that adsorb to the particle surfaces and promote interparticle bridging or flocculation.
- **Plasticizers** are often added to binders to reduce → crosslinking density and increase flexibility. Typical examples are water and ethylene glycol for vinyl binders, glycerin or ethylene glycol for clay bodies, or molten waxes or oils for thermoplastic polymers used in injection molding.
- **Lubricants** are added to lower friction between particles, and between particles and die surfaces. Stearic and oleic acid are good lubricants.
- **Deflocculants**, dispersants or anticoagulants are added to slurries to improve dispersion and dispersion stability. Deflocculants adsorb on particle surfaces and prevent the approach of particles either by electrostatic or steric stabilization (see Section 4.5.1).
- **Surfactants** or wetting agents can also be added to a slurry to improve dispersion. *Antifoams* (defoamers) are added to slurries to remove trapped gas bubbles from the liquid. Typical antifoams are fluorocarbons, silicones (see Section 5.2), stearates, and high molecular weight alcohols.

Forming/fabrication processes A cohesive body of the desired shape and size is obtained by consolidation and molding of ceramic powders (Figure 2.18).

- **Dry forming.** Dry powders can be simultaneously compacted and shaped by pressing in a rigid die or flexible mold. Pressing is the most widely used forming process in ceramic manufacturing. Powders should be free flowing, should have a high bulk density, should consist of deformable particles, and should not stick to the die. *Dry pressing* (also called die pressing) involves compacting a powder between two plungers in a die cavity (Figure 2.17a). *Isostatic pressing* allows the manufacture of complex shapes. Typically, a flexible (rubber) mold is filled with the powder, sealed, placed in a gas- or liquid-filled pressure chamber, and pressurized to compact the powder (Figure 2.17b). The gas or liquid serves to apply the pressure uniformly over the surface of the → green body.
- **Plastic forming.** Plastic ceramic bodies such as plastic clay bodies or ceramic bodies plasticized with organic binders deform inelastically without rupture under a compressive load. The cohesive strength required for plastic deformation is provided by capillary pore pressure and particle agglomeration. In the *extrusion* method (Figure 2.18c), the plastic body is ejected through an opening of the die to shape green bodies of simple geometries (e.g., tubes, bricks, tiles, catalyst supports, electrodes). *Jiggering* involves shaping of a plastic clay body on a spinning porous plaster mold by a moistened shaping tool (Figure 2.18d). This is the most widely used forming technique used in manufacturing small, axial-symmetrical whiteware ceramics such as china or electrical porcelain. In *powder injection molding*, a hot ceramic/binder mixture is injected into a mold with a desired shape (Figure 2.18e). The binder can be removed easily from the green forms by thermal decomposition prior to sintering. It is possible to form fairly complicated shapes such as turbochargers by this method.

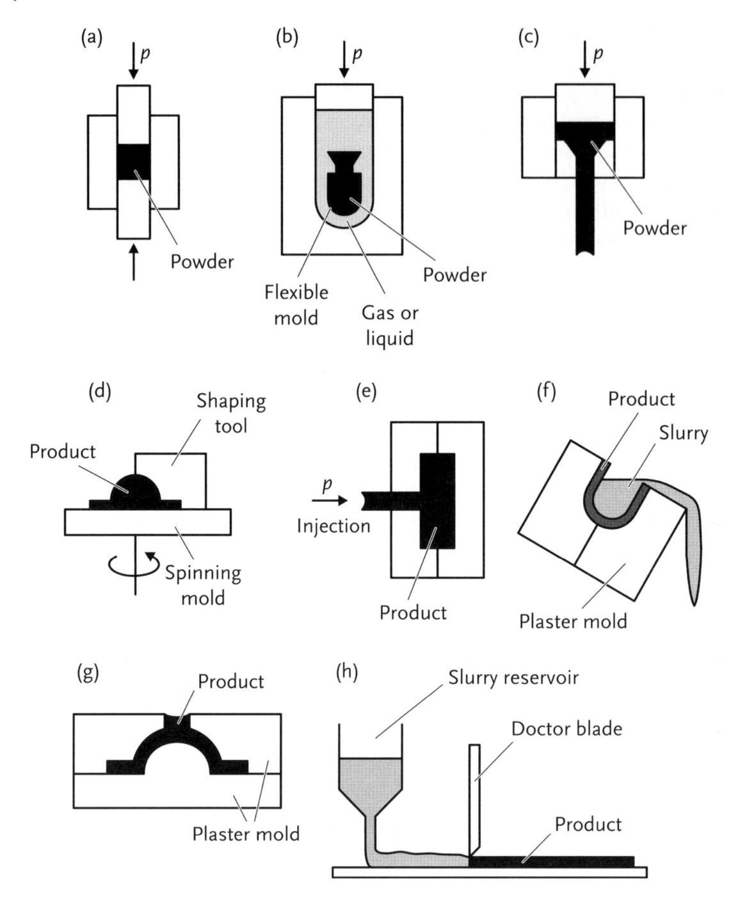

Figure 2.18 Forming processes of ceramic powders (p = pressure). (a) Dry pressing; (b) isostatic pressing; (c) extrusion; (d) jiggering; (e) injection molding; (f) drain casting; (g) solid casting; and (h) tape casting.

- **Paste forming**. Ceramic thick-film pastes, consisting of a ceramic powder, a sintering aid such as borosilicate glass, and organic additives, are used for decorating ceramic tableware or to form capacitors and insulation layers for microelectronic packaging.
- **Slurry forming**. Slurries with proper viscosity are prepared by dispersing powder in a solvent, usually water. A good slurry is well dispersed and free of air bubbles, and the particle setting rate should be low. A high solid concentration is desirable to minimize drying shrinkage and the time and energy required for forming. In the *slip-casting* method the slurries are poured in a porous mold having a cavity of the desired shape. The capillary suction of the porous mold draws the liquid from the slurry. A higher solids content cast is formed on the inner wall of the mold, its thickness increasing with the time. If the excess slurry is drained after a fixed time, hollow shapes are obtained (*"drain casting,"*

Figure 2.18f). Alternatively, the entire amount of slurry can be used to form a dense shape ("*solid casting*"; Figure 2.18g).

- **Gel casting** employs *in-situ* → polymerization of organic monomers to produce a polymer that binds individual particles together to form a cohesive body. *Coagulation casting* is a related process in which the → green body is formed by coagulation of particles. This is usually induced by a pH change. For example, a rise in pH can be achieved by adding urea to the slurry. Decomposition of urea by the enzyme urease liberates ammonia that increases the pH. *Tape casting* (doctor blading, knife coating) is used for producing ceramic sheets or tapes. A slurry is poured on a smooth supporting surface; its thickness is controlled by the doctor blade (Figure 2.18h).

Green finishing/machining This is often required after forming in order to eliminate rough surfaces, smooth forming seams, and modify the size and shape of the green body in preparing for sintering.

Drying Drying of the green body must be carefully controlled in order to avoid cracking, warping, and shape distortions. The different stages during drying are the same as discussed for the drying of gels later in this book (Section 4.5.2.4).

Presinter thermal processing In this step, organic or inorganic additives and impurities are burnt out or decomposed, and residual moisture is removed. This is achieved by heat treatment below the sintering temperature, or in a controlled series of ramps and isothermal holds.

Sintering A ceramic body densifies during sintering. Thermally activated material transport (see above) transforms loosely bound particles into a dense, cohesive body. Porosity or void space between the particles is reduced, and interparticle contact or → grain-boundary area increases. The methods for sintering of powder compacts can be classified into two major categories. One is called *pressureless sintering*, which involves heating of the formed compact to about two-thirds of its melting temperature at ambient pressure and holding for a given time. *Pressure sintering* employs the simultaneous use of pressure and temperature, that is, densification during sintering.

- **Pressureless sintering.** In *solid-state sintering*, densification occurs by solid-state, diffusion-controlled mass transport. Since mass transport occurs faster through a liquid than through a solid, densification occurs faster in the presence of a liquid phase. *Liquid-phase sintering* (LPS) requires a liquid phase (melt) that is chemically compatible, that wets the particles, and that dissolves some of the solid. In a system containing multiple phases, such as cermets or porcelain, the sintering temperature must be between the melting point of the phase that forms the melt and that of the other phase(s). Many non-oxide powders with covalent bonds between the constituting elements are difficult to sinter by solid-state sintering because mass transport by diffusion is low, even at high temperatures. In such cases, sintering aids are added that form a liquid phase. For example, the addition of a few per cent of certain oxides to Si_3N_4 leads to liquid silicate phases during sintering.

- **Pressure sintering.** Compared to pressureless sintering, pressure sintering is more expensive, does not allow production of complicated shapes, and is not suited for mass production. On the other hand, pressure sintering lowers the sintering temperature or allows faster sintering at the same temperature than pressureless sintering. Pressure sintering is usually used in manufacturing ceramics that are difficult to sinter without pressure such as non-oxide ceramics. The most common technique is called *hot pressing*. A powder is compacted in a die and pressed uniaxially between two forming rams while heated to high temperatures. The commonly used graphite dies allow temperatures up to 2500 °C, but limit the maximum pressure to about 100 MPa. *Hot isostatic pressing* (HIP) is similar to cold isostatic pressing used in forming the → green body, except that the pressing is done at the sintering temperature. The "flexible" mold is then made of metals or glasses that soften at high temperatures and transmit the applied pressure uniformly to the ceramic body. The pressuring medium is typically a gas (argon). Complex shapes can be formed by HIP. Si_3N_4 can be sintered by HIP under a nitrogen gas pressure of 1–5 MPa, because high-pressure nitrogen can suppress the decomposition reaction in elemental silicon and nitrogen at temperatures above about 1800 °C.

Postsintering processes These may involve machining to meet dimensional tolerances, surface coating or modification, and so on.

2.2 Solid–Gas Reactions

The use of a combination of solid and gaseous reactants allows the problem of poor contact between two solids to be overcome. Furthermore, unfavorable equilibria involving gaseous species (e.g., the thermal decomposition of nitrides into the elements) will be shifted towards the products because there is a large excess of the gaseous reactant.

We will discuss some important process options for an industrially important solid–gas reaction, the *direct nitridation* of elemental silicon to give silicon nitride (Si_3N_4). Silicon nitride ceramics have high reliability, thermal shock and wear resistance, strength and → hardness as well as good corrosion and oxidation resistance even up to 1200–1300 °C. The excellent thermomechanical properties combined with the low density make Si_3N_4 suitable for applications such as moving automotive parts (valves, turbocharger rotors, piston heads, etc.) or gas-turbine rotors. The excellent wear resistance is utilized in metal machining inserts or ultrahigh speed bearings.

There are two hexagonal modifications of crystalline Si_3N_4, the α- and the β-phase (Figure 2.19). In both modifications, three-dimensional networks are formed from corner-sharing SiN_4 tetrahedra. The covalent Si–N bonds are the

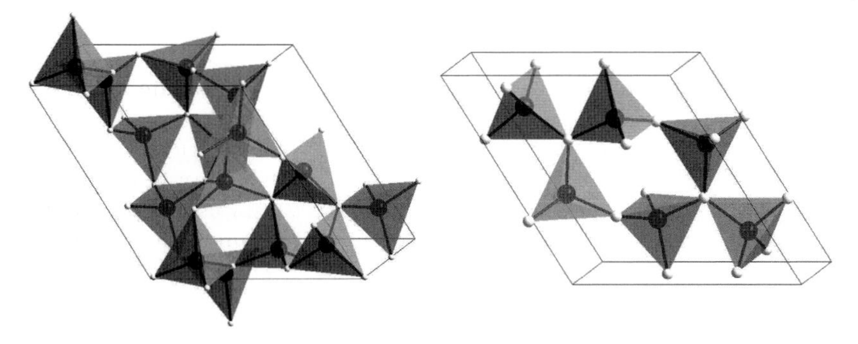

Figure 2.19 The crystal structures of α- (left) and β-Si_3N_4 (right).

reason for the high hardness and strength (the same is true for SiO_2 or SiC, because of the covalent Si–O and Si–C bonds).

The α-phase is the low-temperature modification that transforms to the β-phase at temperatures >1650 °C. The α/β ratio is an important parameter for the characterization of Si_3N_4 powders. The α-phase has a better sintering behavior.

There are three technically important methods for the preparation of Si_3N_4 powders:

1. carbothermal nitridation of SiO_2 (see Section 2.1.3);
2. ammonolysis of $SiCl_4$ either in the liquid state (see Section 5.6) or in the gas phase (see Section 3.3);
3. nitridation of silicon powder.

The latter method, although nearly a century old, is still the dominant technical process. The overall process is represented by Eq. (2.22).

$$3Si + 2N_2 \longrightarrow Si_3N_4 \tag{2.22}$$

The process can be run continuously or discontinuously. The purity of the silicon employed determines the purity of the obtained Si_3N_4 powder. Therefore, → semiconductor-grade silicon is required to obtain highly pure Si_3N_4 powders. Reaction with elemental nitrogen proceeds with sufficient reaction rates only above 1100 °C. The rate not only depends on the particle size of the silicon powder employed, but also on the presence of impurities. For example, traces of iron catalyze the nitridation reaction. Owing to the exothermic nature of the reaction ($\Delta H = -746\,\text{kJ mol}^{-1}$), the process must be carefully controlled. The temperature must be high enough to obtain complete conversion, but too high temperatures favor formation of the undesired β-phase and of aggregates. Optimum results are obtained at about 1250–1350 °C with the nitridation gas diluted by 20–40% H_2 and the silicon powder diluted by α-Si_3N_4. Both measures help to moderate the exothermic nature of the direct nitridation reaction. However, the tight temperature control requires extended periods of reaction times (several hours). Intermediate grinding is needed to expose unreacted silicon to the nitridation atmosphere.

When Si_3N_4 is prepared in a combustion mode, high gas pressures are necessary to reach self-sustaining conditions. Otherwise gas is depleted at the reaction front. Ignition does not occur if the nitrogen pressure is below 50 MPa. At higher pressures, however, the reaction becomes self-propagating. The high pressure is also useful to suppress the thermal decomposition of Si_3N_4. In contrast to the direct nitridation, the reaction is extremely rapid, and since temperatures $>1700\,^{\circ}C$ are reached, a product with a high portion ($>95\%$) of the β-phase is obtained. As was the case with the controlled nitridation, the addition of Si_3N_4 as a diluent is an important process variable.

Many oxides, hydrides, nitrides, and oxynitrides are formed by heating a solid reagent in an oxygen (air), hydrogen, or nitrogen atmosphere, respectively. Ammonia gas is also used as a nitrogen source for the preparation of nitrides and oxynitrides (see also Chapter 3).

As discussed above for Si_3N_4, conventional methods of solid–gas syntheses involve heating a solid in a gas atmosphere below its ignition point. The reaction rates are low under these conditions because metal ions must diffuse through the product layer to the product/gas interface, or gas molecules through the product layer to the metal/product interface.

Only when the ignition conditions (temperature and gas pressure) are met, can the process proceed in a self-sustaining regime (see Section 2.1.4). High gas pressures must often be employed in order to avoid depletion of the gas at the reaction front. The reactant gas flow can be either cocurrent (Figure 2.20, left) or countercurrent (Figure 2.19, right) to the direction of the propagating combustion front.

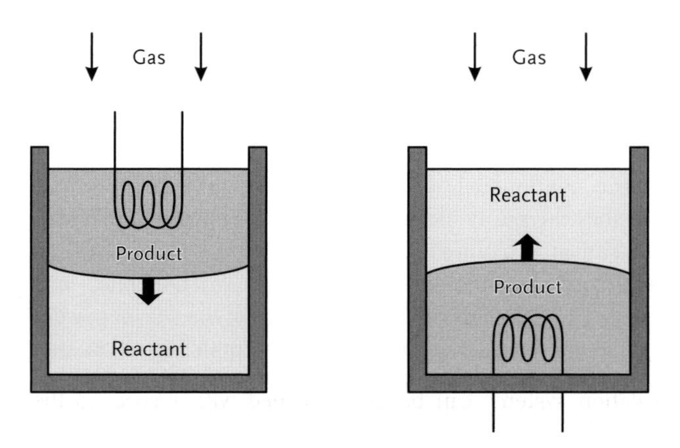

Figure 2.20 Gas flow in self-sustaining gas/solid reactions.

Another option to avoid gas depletion is the addition of solid compounds that cleave the reactant gas upon heating. For example, complete conversion of the metal to the corresponding nitride was achieved in some cases by compacting the metal powder with a stoichiometric portion of NaN_3. An oxidizer (e.g., nitrates) may be added when oxides are prepared. A fuel (e.g., hydrazine, glycine, urea) can also be helpful to obtain spontaneous combustion, especially in combination with oxidizers.

When a pure metal burns in a gas atmosphere, the heat generated during the reaction is so high that melting of the reacting metal or the product and/or decomposition of the product may occur (e.g., transition-metal hydrides are unstable at temperatures between 500 and 1000 °C). Excessive melting of the solid reactant can cause a significant decrease in the gas → permeability, and hence incomplete reaction. Dilution of the metal reactant with an inert compound (usually the product of the reaction) can reduce the reaction temperature and therefore the extent of melting.

2.3 Intercalation Reactions

The most important group of topochemical reactions in inorganic chemistry are intercalation and deintercalation reactions.

The term *intercalation* describes the insertion of guest species (atoms, molecules, or ions), called intercalates, into a crystalline host lattice that retains its general structural features. However, during intercalation reactions the guest and host may undergo perturbations in their geometrical, chemical, and electronic environment. Intercalation results in the modification of the properties of the host material, giving rise to materials with new properties, such as inorganic–organic hybrid materials. Intercalation reactions are usually reversible, and typically occur near room temperature.

After some general remarks on layered solid compounds and a technically important example of the use of intercalation compounds – the lithium ion battery – we will briefly touch on the mechanistic aspects of intercalation reactions, and then turn to methods to prepare intercalation compounds (Section 2.3.2). At the end of this section, two new promising topics of intercalation chemistry will be introduced: the preparation of inorganic–organic hybrid materials and of porous solids by "pillaring" (Section 2.3.3).

2.3.1 General Aspects

Two classes of intercalation systems can be distinguished with respect to the dimensionality of the host lattice:

1. Three-dimensional lattices with parallel channels or interconnected channel systems, as in zeolites (see Section 6.3). The uptake of guests is restricted to those having smaller diameters than the pores ("molecular sieves").
2. Low-dimensional host lattices (layered lattices or chain structures consisting of one-dimensional stacks). The spacings between the layers or chains can adapt to the dimensions of the guest species.

The remainder of this section is restricted to layered compounds, the most important group in terms of chemical reactivity and applications. The ability to adjust the interlayer separation is responsible for the wider occurrence of intercalation compounds for layered structures.

A large variety of guest species can be intercalated, ranging from protons and metal atoms or ions up to large molecular species and polymers. The separation of two host layers can vary over wide ranges, depending on the size and orientation of the host. This is illustrated by some graphite intercalation compounds. Graphite is probably the best investigated host material, and forms intercalation compounds with several hundred chemical species. The graphite layers are only one atom thick, and therefore relatively flexible materials are obtained after intercalation. The interlayer distance (335 pm) is only slightly enlarged (to 371 pm) when lithium is intercalated. Intermediate cases are, for example, potassium (535 pm; Figure 2.21) or AsF_5 (815 pm) as intercalates. Intercalation of KHg leads to an interlayer separation of 1022 pm with potassium and mercury arranged in three layers in between the graphite layers.

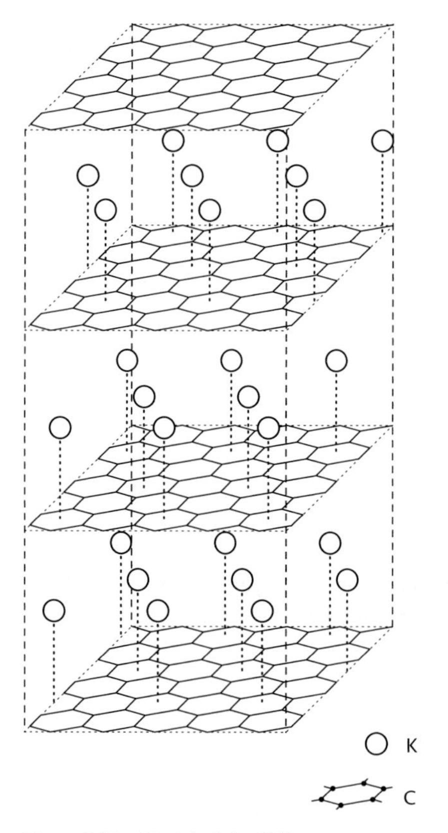

Figure 2.21 Model of the C_8K structure.

In general, layered compounds are characterized by strong *intra*layer bonds and weak *inter*layer interactions. The layers may be electrically neutral or charged:

- **Neutral layers**, held together by van der Waals interactions: for example, graphite, transition-metal oxides (e.g., V_2O_5 or α-MoO_3), transition-metal dichalcogenides (see below), oxychlorides (FeOCl).
- **Negatively charged layers**, separated by mobile (exchangeable) cations: for example, clays, transition metal phosphates.
- **Positively charged layers**, separated by anions: layered double hydroxides (LDH), $[M^{2+}_{1-x}M^{3+}_x(OH)_2]^{x+}$

Examples of the different types of layered compounds will be given below. With respect to the *electronic properties* of the host lattice two qualitatively different types are found:

1. For *insulator host lattices* (layered aluminosilicates, metal phosphates, etc.) the basic physical properties of the host lattice are not affected by intercalation. They are interesting as catalysts or catalyst supports, adsorbents, ion exchangers, and so on.
2. Host *lattices that can reversibly uptake electrons from or* (less often) *donate electrons to the host* (graphite, metal dichalcogenides, metal oxyhalides, etc.) may be reduced or oxidized during intercalation. This results in strong changes of the physical properties of the host matrix, that is, the electric conductivity. Such intercalation compounds are of interest as electrode materials for batteries, as electrochromic systems, sensor materials, and so on.

A prototypical example of the latter case is the reversible reaction of lithium with TiS_2 [Eq. (2.23)].

$$Li + TiS_2 \longrightarrow Li^+[TiS_2]^- \tag{2.23}$$

Many of the metal → dichalcogenides MX_2 (M=Ti, Zr, Hf, V, Nb, Ta, Mo, W, Sn; X=S, Se) possess layered structures. The layers consist of two layers of close-packed chalcogenide ions between which cations reside in octahedral or trigonal prismatic sites. Thus, the sequence of atoms perpendicular to the layers is X-M-X... X-M-X... (Figure 2.22).

During the intercalation of lithium into TiS_2, Ti(IV) is reduced to Ti(III), and lithium cations are inserted between the sulfide layers. The van der Waals forces between the layers of TiS_2 are replaced by Coulomb interactions between the negatively charged layers and the Li^+ cations. The interlayer spacing of the final product, $LiTiS_2$, is expanded relative to the parent TiS_2.

Several other transition-metal oxides or chalcogenides are capable of reversible lithium insertion. Apart from protons, no other cations apart from Li^+ can penetrate so easily into solids. This is the chemical basis for ambient-temperature rechargeable batteries ("lithium batteries") used in cellular phones, portable computers, and so on. Lithium batteries offer several advantages, such as higher voltages, higher energy density, and longer shelf-life compared to other systems. The operating principle of a typical battery is shown in Figure 2.23.

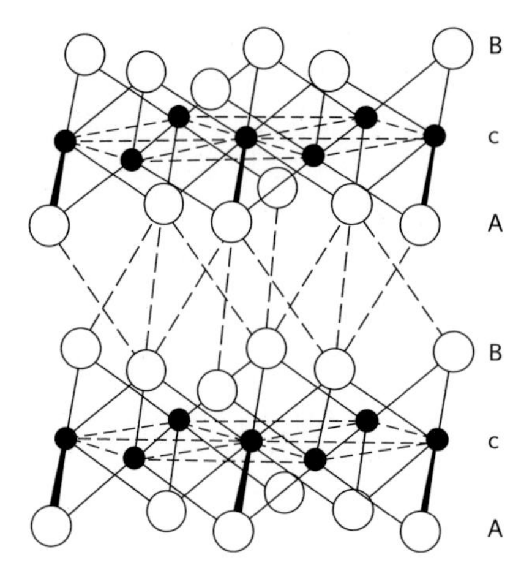

Figure 2.22 The structure of TiS_2. The open circles (A and B layers) represent the sulfur atoms, the full circles (c layers) represent the metal atoms.

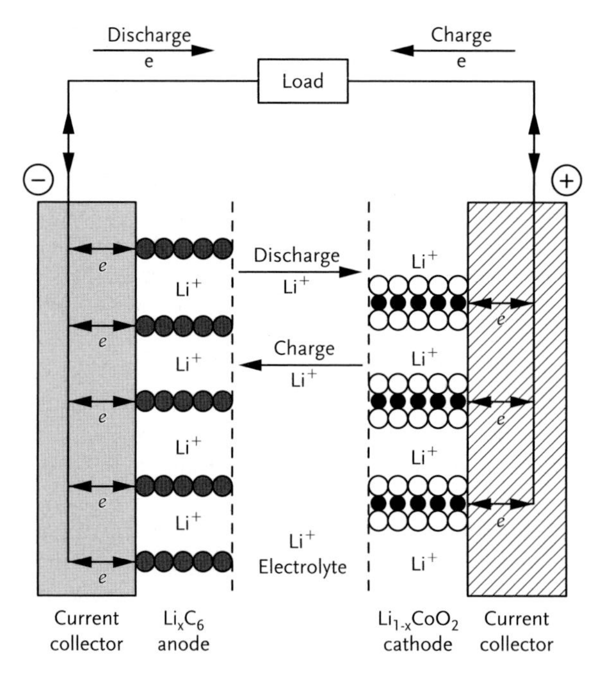

Figure 2.23 Schematic illustration of the charge/discharge process in a rechargeable lithium ion battery.

In lithium ion transfer cells both electrodes are capable of reversible lithium insertion. Because of the difference in chemical potentials of Li in the two electrodes, the transfer of Li^+ from the cathode through the electrolyte to the anode (discharge) delivers energy, while the reverse process (charge) consumes energy

(Figure 2.23). Such cells are also called "rocking-chair" cells, because the Li^+ ions shuttle between the cathode and anode host during the discharge–charge cycles.

The majority of commercially available cells use layered $LiCoO_2$ as the cathode, a graphitic carbon as the anode, and liquid organic (lithium salt dissolved in an aprotic organic solvent) or polymer electrolytes [Eq. (2.24)].

$$\text{cathode:} \quad Li_{0.35}CoO_2 + 0.65\ Li^+ + 0.65\ e^- \underset{\text{charge}}{\overset{\text{discharge}}{\rightleftharpoons}} LiCoO_2$$

$$\text{anode:} \quad C_6Li_x \underset{\text{charge}}{\overset{\text{discharge}}{\rightleftharpoons}} C_{graphite} + x\ Li^+ + x\ e^- \qquad (2.24)$$

$$\text{overall:} \quad \underset{\textit{charged}}{Li_{0.35}CoO_2 + C_6Li} \rightleftharpoons \underset{\textit{discharged}}{LiCoO_2 + \text{graphite}}$$

The carbon material used for the anode is polycrystalline and consists of aggregates of graphite crystals. At ambient pressure, a maximum loading of C_6Li can be reached for highly crystalline graphitic carbon.

The structure of $LiCoO_2$ is derived from the layered metal → dichalcogenide structures MX_2 discussed above (Figure 2.22). The MO_2 equivalents are not stable, due to the higher electronegativity and lower polarizability of oxygen compared to sulfur or selenium. This leads to more ionic M=O bonds, and a stronger repulsion between adjacent oxygen layers. However, alkali metal ions (M′) between the layers stabilize the layered structures (M'_xMO_2), because they screen the anionic repulsion. The Li^+ ions have a high two-dimensional mobility between the CoO_2 layers. Because of the instability of a layered CoO_2 structure, not all Li^+ is electrochemically removed from $LiCoO_2$. Layered materials of the same type ($LiMO_2$) based on mixed Ni, Co, and Mn have been proposed as alternatives to $LiCoO_2$.

2.3.1.1 **Mechanistic Aspects of Intercalation Reactions**

Intercalation reactions in general involve breaking the interlayer interactions in the host lattice and the formation of new interactions between the guest and host. Therefore, layered host lattices with stronger interlayer bonding are more difficult to intercalate. This effect can be observed qualitatively in the intercalation reactions of the layered transition-metal dichalcogenides (MX_2). The metal disulfides intercalate guest molecules under milder conditions than the equivalent isomorphous diselenides (MSe_2), and there is no report of intercalation of a layered metal ditelluride (MTe_2). This trend is attributed to increased covalent bonding between the dichalcogenide layers in the series $S < Se < Te$.

The difficulty of breaking the interlayer interactions can be minimized by the phenomenon of staging. Staging refers to situations in which certain interlayer regions are *totally vacant*, while others are *partially or totally occupied.* Staging can occur readily in compounds with single-atom layers like graphite, but was also observed in other host lattices. In the case of charged layers, a staged compound has different interlayer ions between different pairs of layers. The order of the staging is given by the number of layers between successive filled or partially filled layers (Figure 2.24).

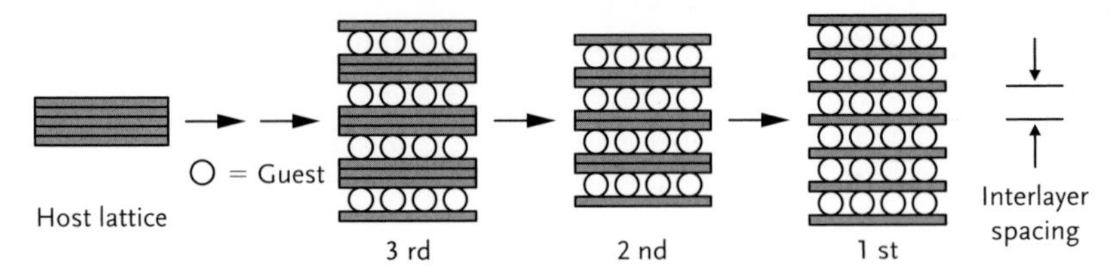

Figure 2.24 Representation of *n*th stage intercalates ($n = 1-3$).

Two extreme staging behaviors are observed in the reaction between guest and host species. A successive sequence of stages, from higher stages down to the first stage is observed, when potassium is reacted with graphite, for example. Each stage corresponds to a specific stoichiometry (e.g., first stage: C_8K, Figure 2.21) and can be converted to the $(n-1)$th stage by reaction with additional potassium.

An example for the other staging behavior is Li_xTiS_2 ($0 \leq x \leq 1$), where a single first stage phase (i.e., a certain amount of Li ions between each pair of successive layers) is formed over the entire composition range. In reality, most intercalation reactions lie somewhere between.

There are two generally accepted models for staging. In the first model, the *n*th stage compound has every *n*th layer space filled (Figure 2.24). The second model explains staging in terms of a domain structure with all interlayers being involved, but with local concentrations of the guest species in certain regions (Figure 2.25).

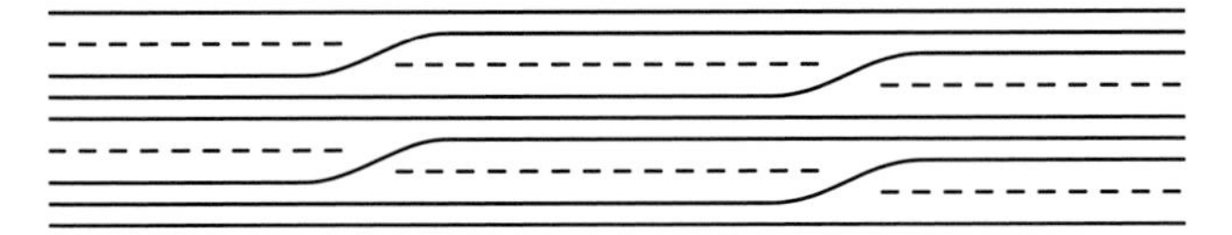

Figure 2.25 Model of a stage 3 compound with the intercalated compound (---) between all layers.

The kinetics of incorporation of guest species into host lattices is complex. The reaction appears to be initiated at defects on the host surface. Therefore, variations in reactivity often arise from one batch to another of the same host due to microcrystalline or particle-size differences.

2.3.2
Preparative Methods

The products of intercalation reactions are insoluble and therefore cannot be purified or separated.

2.3.2.1 Types of Intercalation Reactions

Direct reaction The simplest and most widely used method is the direct reaction of the guest with a potential host. For example, the first alkali metal intercalates of the metal dichalcogenides were prepared by the reaction of the solid host with the metal vapor at temperatures of 600–800 °C as in Eq. (2.23). An alternative is the use of alkaline metals dissolved in liquid ammonia. This offers many advantages, but ammonia is often cointercalated. Heating *in vacuo* to remove the ammonia may result in undesired side reactions.

Butyllithium (LiBu) is a simple and generally applicable reagent for lithium intercalation, giving products of high purity [Eq. (2.25)].

$$x\mathrm{LiBu} + \mathrm{MX}_n \longrightarrow \mathrm{Li}_x\mathrm{MX}_n + x/2\mathrm{C_8H_{18}} \qquad (2.25)$$

Liquid compounds or low-melting solids can be used neat, whereas solid organic and organometallic guests are usually dissolved in polar organic solvents. Although highly polar solvents often accelerate the intercalation process, they can complicate the product distribution by cointercalating with the guest molecules, as mentioned above for ammonia.

Examples for the direct reaction are given in Table 2.4. Note that there are example of reactions proceeding with and without electron transfer.

Almost all organometallic compounds known to form intercalates are good reducing agents, that is, they form stable cations. Metallocenes, Cp_2M, are the most prominent examples. There must be a match between the reducing power of the organometallic guest and its ability to intercalate specific host lattices. For example, cobaltocene, Cp_2Co, with a first ionization potential of 5.5 eV, will intercalate into both TaS_2 and FeOCl, but ferrocene, Cp_2Fe, which is much less reducing (first ionization potential 6.88 eV) will only intercalate into the more oxidizing host lattices of FeOCl or α-$RuCl_3$. The FeOCl structure type (e.g., FeOCl, TiOCl, VOCl) consists of neutral layers of edge-sharing distorted *cis*-$[FeCl_2O_4]$ octahedra, with the chlorine atoms directed towards the interlayer space (Figure 2.26).

Table 2.4 Examples of the direct reaction.

Reactants (host/guest)	Products
TaS_2/pyridine	$TaS_2(C_5H_5N)_{0.5}$
VSe_2/Cp_2Co	$VSe_2(Cp_2Co)_{0.25}$
TiS_2/Na(naphthalene)	$NaTiS_2$
α-$Zr(HPO_4)_2 \cdot H_2O/RNH_2$	$Zr(HPO_4)_2(RNH_2)_2$
$V(O)PO_4$/EtOH	$V(O)PO_4(EtOH)_2$
Graphite/K	C_8K
Graphite/H_2SO_4	$C_{24}{}^{+}HSO_4^{-} \cdot 2H_2SO_4$
Graphite/OsF_6	$C_8(OsF_6)$
Graphite/$AlBr_3/Br_2$	$C_9(AlBr_3)(Br_2)$
FeOCl/pyrrole	$FeOCl(polypyrrole)_{0.34}$
FeOCl/Cp_2Fe	$FeOCl(Cp_2Fe)_{0.16}$
$NiPS_3$/LiBu	$Li_{4.4}NiPS_3$
MoO_3/H^+/reductant	$H_{1.7}MoO_3$
V_2O_5/LiI	$Li_xV_2O_5$

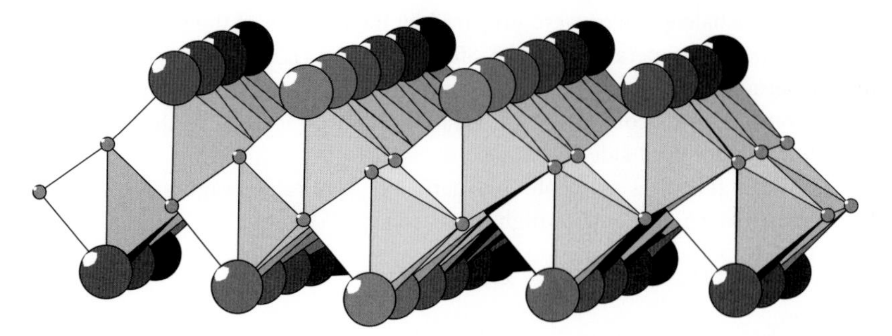

Figure 2.26 One layer of the FeOCl structure. The large spheres represent the chlorine atoms directed towards the interlayer space.

Graphite is the only host lattice known to be able to undergo intercalation reactions by either reduction (e.g., formation of C_8K by reaction with metallic potassium) or oxidation of the layers (e.g., formation of C_8Br by reaction with bromine or of $C_{24}{}^{+}NO_3{}^{-}3 \cdot HNO_3$ by reaction with nitric acid).

Metal phosphorus trisulfides, MPS_3, or more precisely $M_2(P_2S_6)$ ($M{=}Mn^{2+}$, Cd^{2+}, and many other divalent cations), are structural analogs of the metal dichalcogenides (see Figure 2.22), if one considers half of the metal atoms replaced by P_2 pairs. They exhibit two types of intercalation reactions. The first are redox intercalations with reducing guests (see the reaction of $NiPS_3$ with LiBu in Table 2.4 as an example). The second type is unique for the MPS_3 compounds and involves intercalation of cations with concomitant loss of the M^{2+} cations from the layers into solution to maintain charge balance (see the reaction of $MnPS_3$ with KCl in Table 2.5 as an example).

Intercalation reactions sometimes take weeks or even months for completion, depending on the host, and sometimes require elevated temperatures. Sonification of reaction mixtures increases the rates of intercalation in many cases. This is not due to improvement of mass transport, but in a significant decrease in the particle size and surface damage of the host. In fact, the crystallinity of the final product is much inferior compared to the material obtained using conventional methods.

Table 2.5 Examples of ion-exchange reactions.

Reactants	Products	Solvent
$Na_xTiS_2/LiPF_6$	Li_xTiS_2	dioxolane
$K[Al_2(OH)_2(Si_3AlO_{10})]/Na^+$	$Na[Al_2(OH)_2(Si_3AlO_{10})]$	H_2O
$Na_{0.33}(H_2O)_{0.66}TaS_2/[Cp_2Co]I$	$TaS_2(Cp_2Co)_{0.2}$	MeOH
$MnPS_3/KCl$	$Mn_{0.8}PS_3(K)_{0.4}(H_2O)_{0.9}$	H_2O
$Mn_{0.8}PS_3(K)_{0.4}(H_2O)_y/[Ru(bipy)_3]^{2+}$	$Mn_{0.8}PS_3[Ru(bipy)_3]_{0.2}{}^{*}$	H_2O

*bipy = 2,2′-bipyridyl

Ion exchange Intercalated guest ions can often be replaced by immersing the material in a concentrated solution containing another potential guest ion. Some illustrative examples are given in Table 2.5. In some cases, preintercalating with an alkali-metal ion and then ion exchanging the alkali-metal cation with the final guest cation is a useful strategy for intercalating large cations that do not intercalate directly (see intercalation of the tris(2,2′-bipyridyl)ruthenium dication into $MnPS_3$ in Table 2.5).

Intercalation reactions involving silicate hosts (zeolites, clays, pyrochlores, etc.) generally proceed by ion-exchange processes. Long before there was a scientific understanding of their properties, clays enjoyed wide practical us as clarifiers for alcoholic beverages and as decolorizers for edible oils. The practice of mixing raw wool with an aqueous slurry of clays to remove grease dates back to antiquity. *Smectic clays* consist of negatively charged layers that are compensated and bound together by readily exchangeable interlayer cations (normally hydrated Na^+ or Ca^{2+}). A charged layer contains a central sublayer of edge-sharing octahedra sandwiched by two sublayers of linked tetrahedra ("2 : 1 layer"). The cations in the tetrahedral layer are normally Si^{4+}, and those in the octahedral layer divalent or trivalent cations (e.g., Al^{3+}, Fe^{3+}, Mg^{2+}, or Fe^{2+}). Substitution of metal cations with similar size but lower valency (i.e., substitution of Si^{4+} by Al^{3+} in the tetrahedral sublayers, or Al^{3+} by Mg^{2+} or Mg^{2+} by Li^+ in the octahedral sublayer) creates a net negative charge for the layers. This is compensated by the interlayer cations. An example is the mineral muscovite $K[Al_2(OH)_2(AlSi_3O_{10})]$ (see left side of Figure 2.29). One quarter of the silicon atoms in the tetrahedral layer are replaced by Al^{3+} ions. The thus created negative layer charge is balanced by interlayer K^+ cations. The amount and the site of the substitution influences the properties of the clays, since they determine the charge density and the strength of the interaction between the layers and the interlayer cations. Their swelling property makes smectic clays particularly attractive. By dispersion in water, such clays form a stable → thixotropic gel.

Another class of clay minerals is represented by kaolinite, $Al_2(OH)_4(Si_2O_5)$ consisting of one tetrahedral silicate layer and one layer of edge-sharing octahedra with central Al^{3+} cations. The 1 : 1 layers are electrically neutral and are linked head-to-tail via hydrogen bridges. Kaolinite is one of the few clay minerals capable of intercalating neutral guest species. However, the guest species must be able to form strong hydrogen bonds with the host's layers (e.g., urea, acetamide).

Exfoliation and reflocculation Layered lattices are able to disintegrate under appropriate reaction conditions to give → colloidal solutions of the layers. Among the oldest examples of this behavior are smectic clays. Complete delamination (exfoliation) to colloid suspensions is achieved under appropriate conditions. Another example is the hydrated alkali-metal dichalcogenides (e.g., $Na_x(H_2O)_y$-TaS_2) that can be completely dispersed in solvents with high dielectric constants.

An interesting aspect of colloid formation is the possibility of reconstructing the layered structure in the presence of a guest. Large molecules, such as cyanines, substituted ferrocenes, hydrocarbons, or polymers can be inserted in the host lattices by reacting them with the colloid solution. Subsequent precipitation of the intercalated systems (reflocculation) is achieved by removal of the solvent or increasing the electrolyte concentration.

Electrointercalation Electrointercalation has been discussed earlier in the case of lithium batteries. The host lattice serves as the cathode of an electrochemical cell. Some of the advantages of the electrolysis method over conventional techniques are its simplicity, ease of control of stoichiometry, and fast reaction rate at room temperature.

Important examples of electrointercalation are the oxidation and reduction of various compounds $A_xB_yO_z$ whose structures are derived from → perovskite ABO_3 (see Section 2.1.1). Oxidation and reduction can be performed either electrochemically, or by reaction with oxygen or hydrogen. In the reduced state, the compounds belong essentially to the $A_nB_nO_{3n-1}$ and $A_{n+1}B_nO_{3n+1}$ series, with A=La, Nd, Sr, or Ba, and B=Fe, Co, Ni, or Cu. The redox reaction can be summarized as in Eq. (2.26), δ representing the amount of inserted oxygen. Typical examples in the $A_nB_nO_{3n-1}$ series are $Sr_2M_2O_5$ (M=Co and/or Ni), resulting in metallic $SrMO_3$ after oxidation. Examples of the $A_{n+1}B_nO_{3n+1}$ series include La_2CuO_4, M_2NiO_4 (M=Nd or La) and $La_3Ni_2O_7$. Their oxidation gives → superconducting, semiconducting or metallic compounds, respectively. In the $A_{n+1}B_nO_{3n+1}$ series, the compounds are only partially oxidized (δ typically 0.1–0.25).

$$A_xB_yO_z + 2\delta OH^- \longrightarrow A_xB_yO_{z+\delta} + \delta H_2O + 2\delta e^- \qquad (2.26)$$

The structures of the compounds in the reduced state can be considered as oxygen-deficient → perovskites with channels perpendicular to the *z*-axis. Upon oxidation, oxygen diffuses in and either partially or completely fills the vacant sites in these channels. Note that annealing of $YBa_2Cu_3O_6$ (Section 2.1.1.1) is a related topochemical process.

2.3.2.2 **Intercalation of Polymers in Layered Systems**

Considerable interest has focused upon the insertion of macromolecules into layered host lattices. Extended studies have been performed with iron oxychloride (FeOCl), metal phosphates, transition-metal → dichalcogenides, layered double hydroxides and especially, layered silicates. The structure of polyaniline/MoS_2 is shown in Figure 2.27, as an example.

Due to the size of the polymer molecules there is a smooth transition between truly intercalated polymers (not necessarily between each pair of adjacent layers) and exfoliated sheets dispersed in the polymer matrix. The latter is an example of nanocomposites (see Section 7.3.5).

There are several preparative strategies for the synthesis of inorganic–organic hybrid materials from layered solids (Figure 2.28):

- Intercalation of monomer molecules with subsequent *in-situ* → polymerization (Figure 2.28, left).

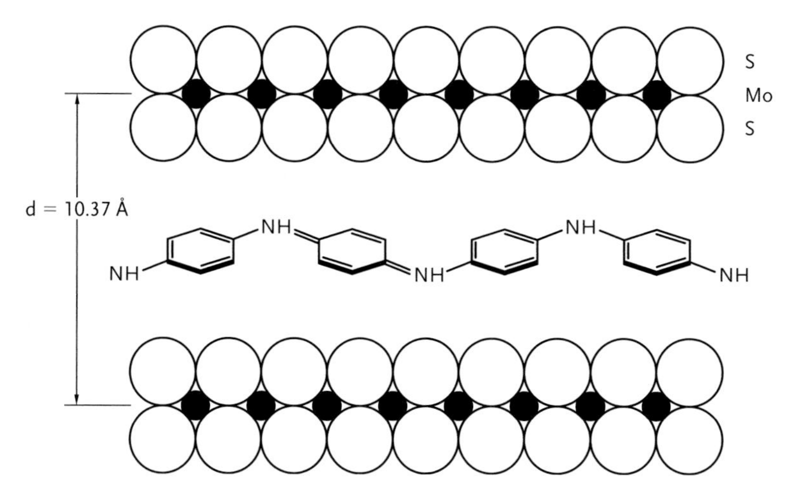

Figure 2.27 Schematic structure of polyaniline/MoS_2.

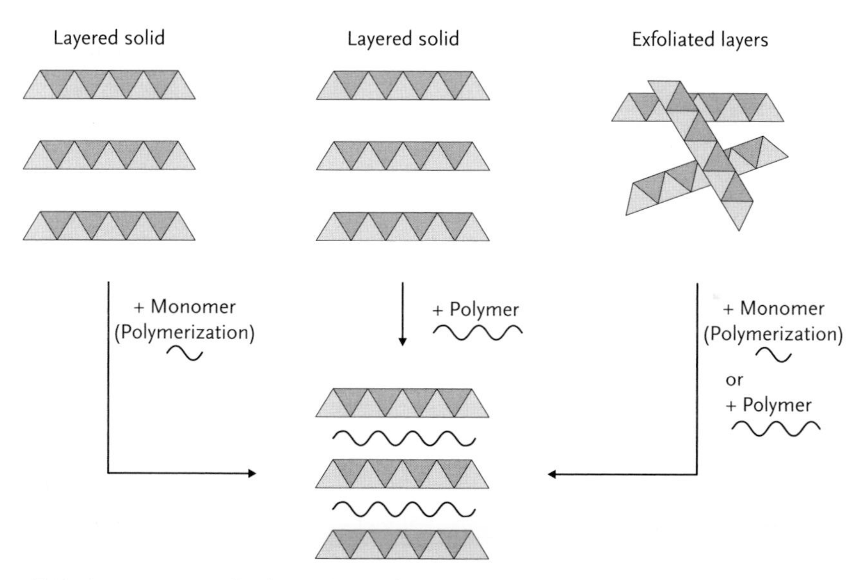

Figure 2.28 Strategies for the synthesis of inorganic–organic hybrid materials from layered solids.

- Direct intercalation of polymer chains into the host lattice (Figure 2.28, center). A disadvantage is the slow kinetics for the transport of macromolecules in the interlayer space. An example is the formation of poly (ethylene oxide) in mica-type sheet silicates by reaction of the polymer melt with Na^+ or alkylammonium-exchanged host lattices. Another approach is first to make the silicates organophilic by exchanging the interlayer cations with organic cations and also swelling the lattice in the presence of a polar solvent. The enlarged interlayer spacing then allows polymer molecules to intercalate.

- Delamination of the host lattice into a → colloid system in appropriate solvents, addition of the polymer and subsequent reprecipitation of one-dimensionally disordered intercalates (see above) (Figure 2.28, right).

2.3.3 Pillaring of Layered Compounds

The development of systems with nanometer-sized rigid pores is of considerable interest in shape-selective catalysis, adsorption, ion exchange, and the design of new structural and functional materials (see Chapter 6). One approach is based on the transformation of *layered* systems into porous framework materials by introducing "pillars" into the interlayer space (Figure 2.29).

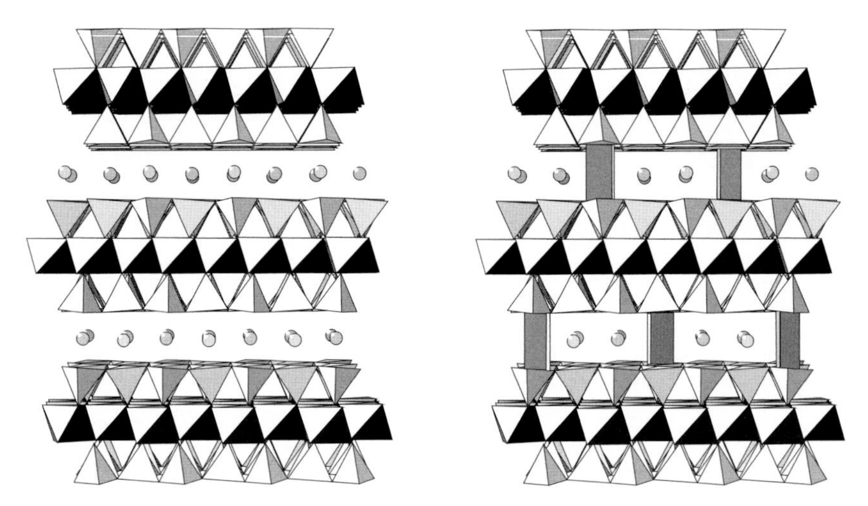

Figure 2.29 The layer structure of muscovite (left) with 2 : 1 layers (see above) and schematic representation of the clay with pillars (right).

The most extensively studied and best understood materials are alumina pillared clays, which are stable in both oxidizing and reducing atmospheres and offer high thermal stability and high surface areas. They may serve as an example of how the pillared materials are prepared.

The cation $[Al_{13}O_4(OH)_{24}(H_2O)_{12}]^{7+}$ (abbreviated Al_{13}) consists of 12 aluminum octahedra surrounding a central aluminum tetrahedron (Figure 2.30) and is approximately 860 pm in diameter. Al_2O_3-pillared clays have been obtained by exchanging the ions in the interlayer space of smectic clays with solutions

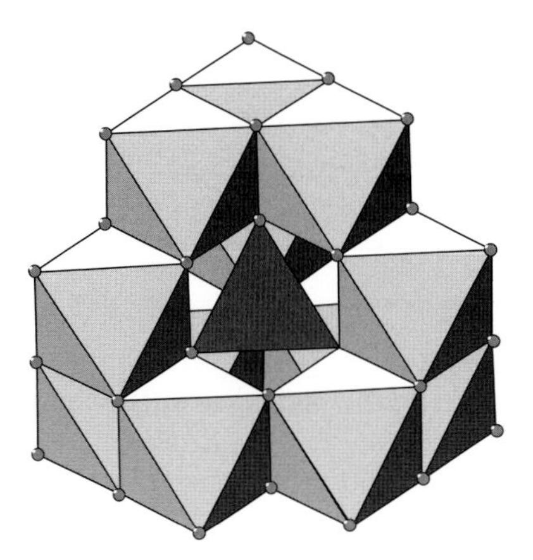

Figure 2.30 The structure of $[Al_{13}O_4(OH)_{24}(H_2O)_{12}]^{7+}$ ("Al_{13}").

containing the Al_{13} ion. Upon calcining the intercalates at about 500 °C, the Al_{13} ions dehydrate and convert to aluminum oxide particles, propping up the layers as pillars and generating interlayer space of molecular dimensions, that is, a two-dimensional porous network.

The alumina pillars are anchored to the tetrahedral aluminosilicate layers of smectic clays by chemical reactions. These pillared clays have interlayer free spacings of 700–800 pm, corresponding to the size of the Al_{13} unit, and surface areas of $\sim 300\ m^2\,g^{-1}$.

In general, the micropore structure is tailored by the nature of the host material and of the pillaring species, and may have pore sizes larger than those of zeolites and related materials.

The ion-exchange technique described for $[Al_{13}O_4(OH)_{24}(H_2O)_{12}]^{7+}$ can similarly be employed for other cationic metal compounds, such as other hydroxo or oxo cluster ions (e.g., $[Fe_{13}O_4(OH)_{24}(H_2O)_{12}]^{7+}$ or $[Zr_4(OH)_8(H_2O)_{16}Cl_z]^{(8-z)+}$), cationic metal–organic complexes, cationic → polynuclear halide compounds or positively charged colloidal particles.

Other layered host materials than smectic clays show limited swellability. Hence, they usually cannot exchange directly with the pillaring species, and preswelling of the layered substance by amine is performed prior to the insertion of pillaring species. Amines intercalate readily into these hosts to form alkylammonium intercalates. The expanded layers permit the subsequent ion-exchange reaction of the interlayer alkylammonium ions with large cationic pillar precursors. This method was applied successfully to the synthesis of Al_2O_3- or Cr_2O_3-pillared phosphates, for example.

The majority of layered host materials consists of neutral or negatively charged layers. The few compounds with positively charged layers, that is, anion-exchange properties, are the so-called layered double hydroxides (LDH). Their structures consist of *positively* charged layers $[M^{2+}_{1-x}M^{3+}_{x}(OH)_2]^{x+}$ (M^{II}=Mg^{2+}, Ni^{2+}, etc.;

$M^{III}=Al^{3+}$, Fe^{3+}, etc.). The structure of the layers is similar to that of the metal → dichalcogenides, with the OH groups instead of the chalcogenide ions. They contain exchangeable anions and water in the interlayer spaces and can be pillared by using *negatively* charged polyoxometalate anions such as $V_{10}O_{28}{}^{6-}$, $[SiV_3W_9O_{40}]^{7-}$ or $PW_{12}O_{40}{}^{3-}$. For LDHs that do not swell in water, the preswelling approach can also be used: the hydroxides are first intercalated with large organic anions prior to polyoxometalate exchange.

Microporous zirconium phosphonate materials with organic pillars were derived from layered metal phosphates or phosphonates. The best known members of this group are zirconium phosphate, $Zr(HPO_4)_2$, and other M(IV) phosphates and arsenates (M(IV)=Zr, Ti, Ge, etc.). The Zr atoms of $Zr(HPO_4)_2$ are octahedrally coordinated by the oxygen atoms of six phosphate anions. In γ-Zr $(HPO_4)_2$ (more precisely written as $Zr(PO_4)(H_2PO_4)$) two layers of Zr octahedra are bridged by the $PO_4{}^{3-}$ anions, while the $H_2PO_4{}^{-}$ tetrahedra terminate the layers on both sides, with the OH groups pointing towards the interlayer space. The "surface" $H_2PO_4{}^{-}$ groups can be exchanged with other phosphorous-containing groups, for example phosphonates. This is a → topotactic reaction because the layered structure is maintained, that is, the resulting metal phosphonate derivatives have the same general structure, part of the OH groups being replaced by organic groups R. Compounds with organic pillars were obtained by using bisphosphonic acids of the type HO_3P-X-PO_3H. While the two PO_3 groups are incorporated into adjacent layers, the interlayer distance can be adjusted by the size of the organic group X that bridges the layers (Figure 2.31).

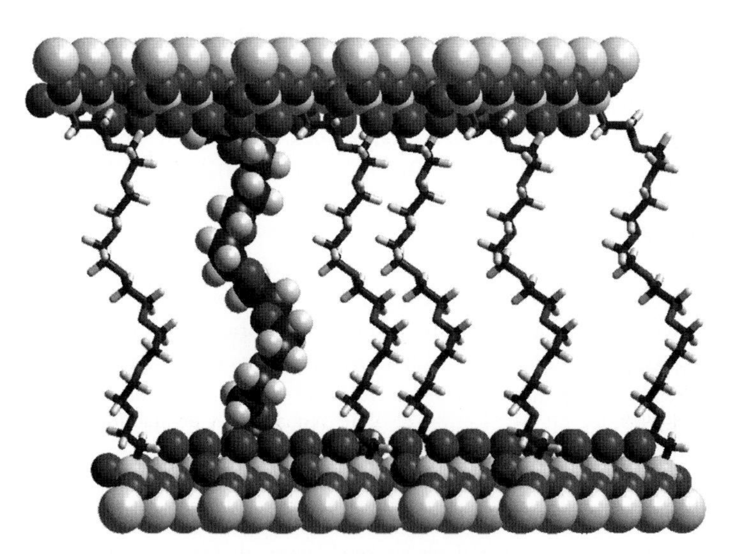

Figure 2.31 Idealized molecular model of pillared γ-$Zr(HPO_4)_2$ prepared by partial exchange of the $H_2PO_4{}^{-}$ groups with oligo(ethyleneglycol) diphosphonate. Only part of the layer atoms are shown (half of the Zr atoms and $H_2PO_4{}^{-}$/ $RPO_4{}^{-}$ groups, respectively).

Further Reading

1 Bhaduri, S.B. and Bhaduri, S. (1999) Combustion synthesis, in *Non-Equilibrium Processing of Materials* (ed. Suryanarayana, C.), Pergamon, Amsterdam, pp. 289–309.
2 Bowen, C.R. and Derby, B. (1997) Self-propagating high temperature synthesis of ceramic materials. *Br. Ceram. Trans.*, **96**, 25–31.
3 Dresselhaus, M.S. (ed.) (1987) *Intercalation in Layered Materials*, Plenum, New York.
4 Dresselhaus, M.S. and Dresselhaus, G. (1981) Intercalation compounds of graphite. *Adv. Phys.*, **30**, 139–326.
5 Gillan, E.G. and Kaner, R.B. (1996) Synthesis of refractory ceramics via rapid metathesis reactions between solid-state precursors. *Chem. Mater.*, **8**, 333–343.
6 Grigorieva, T.F., Barinova, A.P., and Lyakhov, N.Z. (2001) Mechanochemical synthesis of intermetallic compounds. *Russ. Chem. Rev.*, **79**, 45–63.
7 Khan, A.I. and O'Haare, D. (2002) Intercalation chemistry of layered double hydroxides: recent developments and applications. *J. Mater. Chem.*, **12**, 3191–3198.
8 Lange, F.F. (1989) Powder processing science and technology for increased reliability. *J. Am. Ceram. Soc.*, **72**, 3–15.
9 Lange, H., Wötting, G., and Winter, G. (1991) Silicon nitride – from powder to ceramic materials. *Angew. Chem. Int. Ed. Engl.*, **30**, 1579–1597.
10 Manthiram, A. and Kim, J. (1998) Low temperature synthesis of insertion oxides for lithium batteries. *Chem. Mater.*, **10**, 2895–2909.
11 Moore, J.J. and Feng, H.J. (1995) Combustion synthesis of advanced materials. *Prog. Mater. Sci.*, **39**, 243–273 (part I), 275–316 (part II).
12 Munir, Z.A. and Holt, J.B. (eds) (1990) *Combustion and Plasma Synthesis of High Temperature Materials*, Wiley-VCH, Weinheim.
13 O'Hare, D. (1992) Inorganic intercalation compounds, in *Inorganic Materials* (eds Bruce, D.W. and O'Hare, D.), John Wiley & Sons Ltd, Chichester, pp. 165–235.
14 Ohtsuka, K. (1997) Preparation and properties of two-dimensional microporous pillared interlayered solids. *Chem. Mater.*, **9**, 2039–2050.
15 Parkin, I.P. (2002) Solvent free reactions in the solid state: solid state metathesis. *Transition Met. Chem.*, **27**, 569–573.
16 Rao, C.N.R. (1993) Chemical synthesis of solid inorganic materials. *Mater. Sci. Eng.*, **B18**, 1–21.
17 Salvador, P.A., Mason, T.O., Hagerman, M.E., and Poeppelmeier, K. (1998) Layered transition-metal oxides and chalcogenides, in *Chemistry of Advanced Materials – an Overview* (eds L.V. Interrante and M. Hampden-Smith), Wiley-VCH, New York, pp. 449–498.
18 Schöllhorn, R. (1996) Intercalation systems as nanostructured functional materials. *Chem. Mater.*, **8**, 1747–1757.
19 Segal, D. (1997) Chemical synthesis of ceramic materials. *J. Mater. Chem.*, **7**, 1297–1305.
20 Subrahmanyam, J. and Vijayakumar, M. (1992) Self-propagating high-temperature synthesis. *J. Mater. Sci.*, **27**, 6249–6273.
21 Weimer, A.W. (ed.) (1997) *Carbide, Nitride and Boride Materials – Synthesis and Processing*, Chapman & Hall, London.
22 Whitt ingham, M.S. and Jacobson, A.J. (eds) (1982) *Intercalation Chemistry*, Academic Press, New York.
23 Winter, M., Besenhard, J.O., Spahr, M.E., and Novak, P. (1998) Insertion electrode materials for rechargeable lithium batteries. *Adv. Mater.*, **10**, 725–763.

24 Yanagida, H., Koumoto, K., and Miyayama, M. (1996) *The Chemistry of Ceramics*, John Wiley & Sons Ltd, Chichester.

25 Yi, H.C. and Moore, J.J. (1990) Self-propagating high-temperature (combustion) synthesis (SHS) of powder-compacted materials. *J. Mater. Sci.*, **25**, 1159–1168.

26 Zyryanov, V.V. (2008) Mechanochemical synthesis of complex oxides. *Russ. Chem. Rev.*, 77, 105–135.

3
Formation of Solids from the Gas Phase

This chapter deals with reactions in which solid products are obtained from gaseous precursors or intermediates. The product phases may be formed by gas-phase reactions (Section 3.3) or by thermal reaction of the gaseous precursors after adsorption to surfaces in chemical vapor deposition (CVD) processes (Section 3.2). In chemical transport reactions (Section 3.1), gaseous intermediates are in thermodynamic equilibrium with the solid phase. Their diffusion results in an overall mass transport of the solid.

3.1
Chemical Vapor Transport

In this section we will first deal with the general principle of chemical vapor transport phenomena. We will then demonstrate their important role in halogen lamps, before discussing some of the chemical reactions involved.

The general principle behind transport reactions is that a non-volatile solid (A) *reversibly* reacts with a gaseous agent (B) to form a gaseous product (AB). The equilibrium constant associated with this reversible reaction is temperature dependent, and therefore the equilibrium concentration of AB changes when the temperature is changed. Thus, a concentration gradient of AB develops when the system is exposed to a temperature gradient. This provides the driving force for an overall mass transport by diffusion from one site at the temperature T_1 (the source), where AB is formed, to another site at the temperature T_2 (the sink), where the chemical equilibrium position is shifted towards the side of A and B (Figure 3.1).

The equilibrium positions at the two temperatures should not be too far on either side for an efficient transport reaction. Exothermic reactions cause migration into the region of higher temperature ($T_1 < T_2$), and endothermic reactions in the opposite direction ($T_1 > T_2$). The intermediate formation of gaseous compounds distinguishes chemical transport from sublimation.

Chemical transport is most often used for the purification of solids and particularly for the growth of single crystals. This is possible because only one chemical species is transported and thus the very pure compound is deposited at the sink.

Synthesis of Inorganic Materials, Third Edition. Ulrich Schubert and Nicola Hüsing.

Published 2012 by WILEY-VCH Verlag GmbH & Co. KGaA

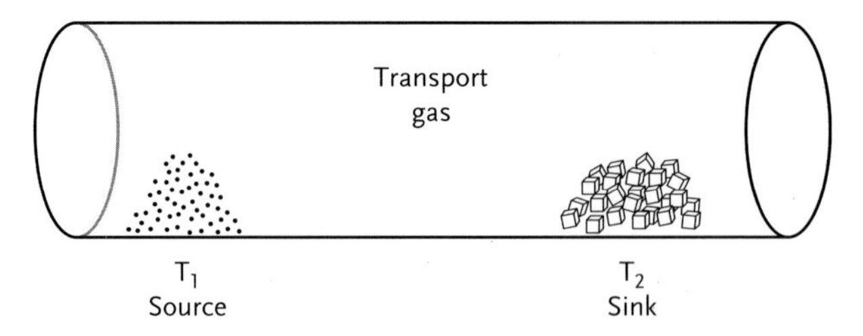

Figure 3.1 The principle of chemical transport reactions. The reaction vessel can be open or closed (see text).

However, chemical transport can also be used for the preparation of new compounds. In this case the transport reaction is coupled with a subsequent reaction at the sink (see below).

Transport reactions can be performed in open systems with a continuous stream of the transport agent, or in closed systems. An example of the first is found in nature. In the presence of volcanic gases, Fe_2O_3 is volatilized and deposited at another site. At hot sites in a volcano, Fe_2O_3 reacts with HCl (the transport agent), and gaseous $FeCl_3$ and water vapor are formed. $FeCl_3$ is transported with the gases and deposited at cooler sites by backreaction [Eq. (3.1)].

$$3Fe_2O_3(s) + 6HCl(g) \rightleftharpoons 2FeCl_3(g) + 3H_2O(g) \tag{3.1}$$

A technical example of the chemical transport in a continuous stream of the transport agent is the Mond process in which elemental nickel in high purity (99.9–99.99%) is produced [Eq. (3.2)]. Nickel is purified by reversible reaction with CO. $Ni(CO)_4$ is formed in an exothermic reaction at 50–80 °C and atmospheric pressure, transported out of the solid mixture formed in the metallurgical process and then decomposed at 230 °C.

$$Ni + 4CO \underset{230\,^{\circ}C}{\overset{50\,^{\circ}C}{\rightleftharpoons}} Ni(CO)_4 \tag{3.2}$$

3.1.1 Halogen Lamps

The most important technical application of chemical transport reactions in closed systems is in halogen lamps. Since Edison made the first incandescent lamp in 1879, with a carbon filament, the principle has not changed very much. When an electric current is passed through a solid body, it is heated to a high temperature and electromagnetic radiation is emitted, mostly as infrared radiation (heat). The wavelength of the radiation decreases with increasing temperature. In incandescent lamps, only a few per cent of the electrical energy are converted into visible light, and most of the energy is lost as heat (Figure 3.2). For example, the yield of light of a heated tungsten filament is only 1.4% at 2400 °C.

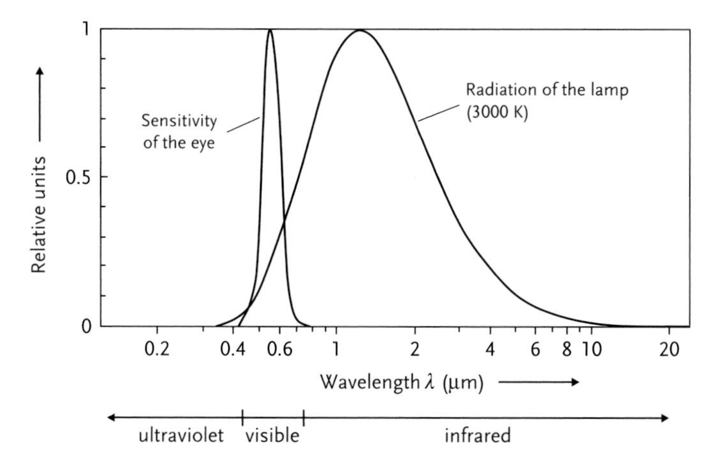

Figure 3.2 The spectrum of incandescent lamps.

The major improvements since Edison's invention have been the replacement of carbon by tungsten, the use of an inert-gas filling instead of evacuating the bulb, the coiling of the filament, and finally the addition of halogens.

Tungsten filaments The great advantage of tungsten compared with other filament materials is that its vapor pressure is low – so that loss of the filament material by evaporation is slow – and the melting point is high (3410 °C). Other advantages of tungsten are its selective emissivity and its great strength at elevated temperatures. Improving the yield of visible light by a higher temperature of the tungsten filament is limited, because the higher the temperature, the greater the rate at which the filament evaporates, resulting in a shorter lifetime. For example, in a lamp whose life is completely determined by vapor transport of tungsten, a 70% increase in efficacy, achieved by raising the filament temperature from 2500 to 2900 °C, will shorten the lifetime by a factor of about 100.

By increasing the temperature of the tungsten, the vapor pressure of the filament material is increased. The vaporized metal condenses at the colder regions of the bulb. This has two consequences: the tungsten wire becomes thinner and eventually "blows." Furthermore, the bulb blackens and thus the yield of light is reduced. In ordinary incandescent lamps, the size of the bulb therefore is large to keep blackening of the bulb, and hence light losses, within reasonable bounds.

Inert-gas filling The inert-gas filling reduces the rate of tungsten evaporation by inhibiting the transport of tungsten. The rate of diffusion is inverse proportional to the total pressure. The gas should preferentially have a high molar mass because thermal conductivity is inversely proportional to the molar mass of the gas. An economic compromise is the use of argon with a low portion (7%) of nitrogen in conventional light bulbs.

Halogens With the addition of a small amount of iodine (about 0.1 mg per ml bulb volume) as transporting agent (halogen lamps) the vaporized tungsten is transported back to the hot filament (Figure 3.3).

Figure 3.3 Deposition of tungsten crystals on the tungsten filament of a halogen lamp (130-fold magnification).

A simplified picture is that the evaporated tungsten from the hot filament diffuses through the inert gas towards the bulb wall, reacting in the cooler zones with the transport medium to form a volatile compound. The reaction products diffuse backwards, that is, in the direction of the filament. The volatile compounds dissociate at the filament, and the tungsten is released. Iodine as the transport medium is then available again for renewed reaction in cooler zones.

The "backreaction" of the deposited tungsten with iodine requires a certain minimal temperature of the glass wall. For this reason, the size of halogen bulbs is much smaller than that of conventional light bulbs. The filament heats the glass wall to the required temperature (about 600 °C). The lower volume of the halogen bulbs allows the use of the more expensive xenon or krypton as gas filling instead of argon. Furthermore, because of their greater mechanical strength, these bulbs can be filled with inert gas at a higher pressure. Both measures reduce the transport of tungsten and thus permit a higher filament temperature.

The chemical processes in halogen lamps are less straightforward than it may appear. For example, it was found that low concentrations of oxygen are necessary for a good performance of the lamps. The compound responsible for the transport of tungsten is WO_2I_2 (instead of WI_2, as originally thought). This is formed at about 600 °C at the glass wall by reaction of the deposited tungsten with iodine and oxygen (or water) in an exothermic reaction [Eq. (3.3)].

$$W + O_2 + I_2 \underset{}{\overset{600\,^{\circ}C}{\rightleftharpoons}} WO_2I_2 \tag{3.3}$$

When WO_2I_2 migrates towards the hot filament (about 3000 °C), it dissociates sequentially into WO_2, WO, and eventually tungsten atoms (close to the filament) [Eq. (3.4)]. The tungsten atoms are then deposited at the hot filament. Thus, the process at the filament surface is not a chemical reaction, but a condensation process.

$$WO_2I_2 \xrightleftharpoons{-I_2} WO_2 \xrightleftharpoons{-1/2O_2} WO \rightleftharpoons W + 1/2\,O_2 \tag{3.4}$$

increasing temperature →

As a consequence, the tungsten atoms are deposited at the coldest (i.e., thickest) parts of the filament. The transport reaction therefore is not a self-healing process, since tungsten preferentially evaporates from the hottest parts of the filament but is redeposited at the coldest parts. Tungsten evaporates from so-called “hot spots” at a faster rate until the lamp fails through localized melting or fracture of the filament. The “hot spots” are small regions in the filament where the temperature is higher because of local inhomogeneities. The development of a truly regenerative cyclic process, which would increase the life of bulbs considerably, is therefore still a challenging task.

Another type of lamp where transport reactions play an important role is the metal-halide lamp. A typical construction of such a lamp is shown in Figure 3.4. The arc tube of a metal halide lamp is made from quartz glass or polycrystalline alumina with two tungsten electrodes. The arc tube contains mixtures of metal iodides and/or bromides that vaporize during operation of the lamp.

Of particular interest are alkali-halide/lanthanide-halide systems. DyI_3, TmI_3, and HoI_3 are important for light generation, because these metals display many lines in the visible spectrum when excited in the hot plasma of the lamp. The intensity of the light emission increases if the concentration of the rare-earth metal in the arc column is raised. This is possible by the formation of gaseous hetero-metallic complexes by chemical vapor transport reactions. For example, $NaDyI_4$ and Na_2DyI_5 have been identified in the NaI/DyI_3 system. In the high-temperature arc core, the metal complexes dissociate into the elements or ionized species, which are then transported outward by convection and diffusion to the lower-temperature mantle where they recombine again.

3.1.2 Transport Reactions

Under laboratory conditions, transport reactions are mostly carried out in closed tubes typically having a length of 10–20 cm and a diameter of 1–2 cm. The sample

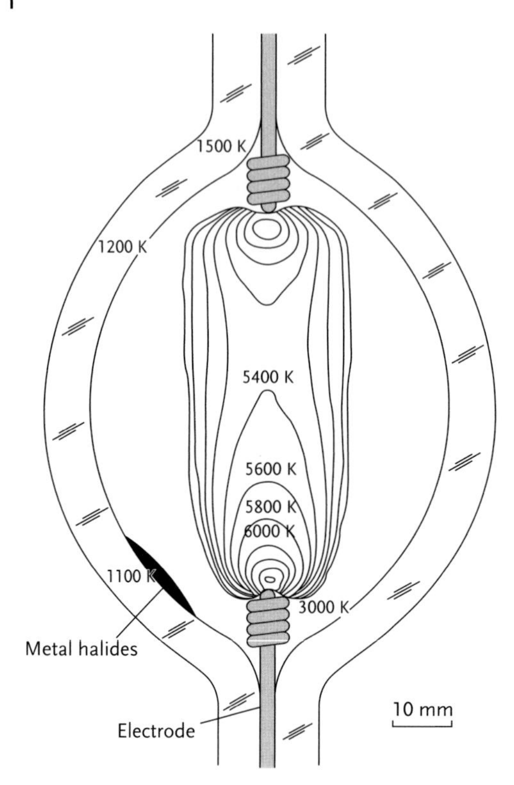

Figure 3.4 Arc tube of a metal halide lamp with temperature distribution under operating conditions.

is placed in the ampoule, and the transport agent is added. The tube is then evacuated at different pressures, depending on the kind of transport agent, sealed and placed in a tubular furnace to which a temperature gradient (Figure 3.1) is applied. The transport agent reacts at the source, but is liberated during the backreaction and diffuses back to the source. It therefore remains in the system and is available for further reactions. Thus, small amounts of the agent can transport large amounts of the solid.

In general, transport reactions in closed systems can be described in four steps:

1. The chemical reaction at the source, leading to equilibrium between the condensed phase and the gas phase.
2. Mass transport by diffusion of the gaseous compounds from the source to the sink.
3. Deposition of the condensed phase at the sink.
4. Diffusion of the gaseous transport agent back to the source.

The overall reaction is in most cases diffusion limited. When a gaseous compound is produced at one site of the reaction vessel and consumed at another, the different partial pressures of the gas give rise to a mass flux between the two temperature regions. It is obvious that there is faster transport if the partial pressure gradient is large. On the other hand, a high total pressure slows down the balancing of the partial pressures. Therefore, the total pressure in the ampoules is usually kept at around 1 bar at the reaction temperature by addition of the appropriate amount of the transport agent.

A large number of chemical transport reactions are known and some typical examples are given for chromium compounds in Table 3.1. In most cases, the reactions are not as simple as the chemical equations may indicate, because very often different gaseous species are involved. In principle, it is possible to calculate the best choice of the transport agent from the thermodynamic data of all species involved in the reaction.

Halogens are very often used as the transport agent for elemental metals, as has been discussed earlier for halogen lamps. Another example is the van Arkel–de Boer process, the first deliberately applied transport reaction. This process is used for the purification of metals (such as Ti, Cr, or V), and makes use of the exothermic reaction between metal and iodine to form a volatile iodide (e.g., CrI_2 from Cr and I_2). The metal is redeposited at a hot filament, thus allowing the metals to be extracted from their oxides, nitrides, carbides, and so on. The volatile metal halide can also be formed by reaction with other halogen sources, such as hydrogen halides or metal halides.

Transport of metal oxides is often achieved with hydrogen halides as transport agents. The formation of water as the byproduct shifts the equilibrium for the formation of the metal halide to the right (e.g., Eq. (3.1)). Elemental halogens are less often used for the transport of oxides, because the equilibrium concentration of binary metal halides is usually very low. However, more volatile oxo halides may be formed, as shown in Table 3.1 for chromium oxides.

Many metal halides can be transported with $AlCl_3$, which forms volatile complexes (e.g., Eq. (3.5)). Some metal halides migrate without an additional transport

Table 3.1 Transport reactions of chromium compounds.

Transported compound	Transport agent	Gas-phase species
Cr	I_2	CrI_2
CrX_3 (X–Cl, Br, I)	X_2	CrX_4
$CrCl_3$	$AlCl_3$	$CrAl_3Cl_{12}$
Cr_2O_3	O_2/Cl_2	CrO_2Cl_2
Cr_2O_3	H_2O/Cl_2	$CrO_2Cl_2 + HCl$
Cr_2O_3	$H_2O/HgCl_2$	$CrO_2Cl_2 + Hg + HCl$
$CoCr_2O_4$ (spinel)	Cl_2	$CrO_2Cl_2 + CoCl_2$
$Cr_2(SO_4)_3$	HCl	$CrO_2Cl_2 + SO_2 + H_2O$
$Cr_2P_2O_7$	I_2/P_4	$CrI_2 + P_4O_6$
$CrSi_2$	I_2	$CrI_2 + SiI_4$

agent ("autotransport"). Volatile metal halides acting as transporting agent are formed by partial disproportionation.

$$CoCl_2(s) + Al_2Cl_6 \rightleftharpoons Cl_2Al(\mu\text{-}Cl)_2Co(\mu\text{-}Cl)_2AlCl_2\ (g) \tag{3.5}$$

Two substances can be transported in different directions along the temperature gradient if one reaction is exothermic and the other endothermic. An example is given in Eq. (3.6). CuCl forms exothermically from Cu_2O with HCl and endothermically from Cu. This allows the separation of Cu and Cu_2O, as Cu_2O is redeposited at higher temperature and Cu at a lower temperature.

$$\begin{aligned} Cu_2O(s) + 2HCl(g) &\underset{900\,^\circ C}{\overset{500\,^\circ C}{\rightleftharpoons}} 2CuCl(g) + H_2O(g) \\ Cu(s) + HCl(g) &\underset{500\,^\circ C}{\overset{600\,^\circ C}{\rightleftharpoons}} CuCl(g) + 1/2H_2(g) \end{aligned} \tag{3.6}$$

For use as a preparative method, the transport reaction is coupled with a subsequent reaction at the sink. For example, metallic niobium and silica do not react with each other if heated together at 1100 °C in vacuum. However, in the presence of small amounts of H_2, gaseous silicon monoxide, SiO, is formed that migrates to the niobium and gives the silicide [Eq. (3.7)]. The role of SiO for the transport of silicon has already been discussed in Section 2.1.3.

$$\begin{aligned} SiO_2(s) + H_2 &\rightleftharpoons SiO(g) + H_2O \text{ (transport reaction)} \\ 3SiO(g) + 8Nb &\longrightarrow Nb_5Si_3 + 3NbO \end{aligned} \tag{3.7}$$

3.2 Chemical Vapor Deposition

Chemical vapor deposition (CVD) is one of the most important methods to prepare thin films and coatings of inorganic materials of various compositions. Thin films are of increasing technological importance because they can add new chemical and physical properties to a substrate material. Unique combinations of properties can thus be obtained through particular combinations of substrate and coating.

After an introduction into general aspects of CVD, such as precursor chemistry, the different steps occurring during CVD, equipment and techniques, we will discuss the deposition of metals (Section 3.2.2), diamond (Section 3.2.3), metal oxides (Section 3.2.4), metal nitrides (Section 3.2.5), and compound → semiconductors (Section 3.2.6). These classes of compounds were selected because they allow an elaboration to be made on the different chemical processes that occur during the thermal decomposition of different precursor compounds on the substrate surface.

3.2.1 General Aspects

Chemical vapor deposition (CVD) is a process where one or more volatile precursors are transported via the vapor phase to the reaction chamber, where they decompose on a heated substrate. This results in the deposition of a solid thin film. Materials deposited by CVD include metals and different multielement materials, such as oxides, sulfides, nitrides, phosphides, arsenides, carbides, borides, silicides, and so on.

A first example is given in Eq. (3.8). TiB_2, which has outstanding properties as an armor material, melts at 3325 °C. Therefore, the preparation of thin films by melting TiB_2 is practically impossible. However, CVD of TiB_2 films according to Eq. (3.8) already occurs at about 1000 °C, that is, substantially below the material's melting point.

$$TiCl_4 + 2BCl_3 + 5H_2 \longrightarrow TiB_2 + 10HCl \quad (3.8)$$

Initial applications of CVD were largely hard coatings to improve the life and performance of cutting tools or turbine components. This is still a large market. More recently, CVD is used heavily in the microelectronics industry, as intricate three-dimensional structures can be built on substrates to perform complex tasks in electronic components (see Section 3.2.2). In the glass industry, CVD is used to coat large panels of glass with SnO_2, TiN, or SiO_2. Typically, the glass panels are formed on a float line in which the hot glass is transported on top of molten tin. The panels pass through several CVD reactors in which reactants are directed toward the hot substrate, where they form coatings. Other applications are layers in solar cells, coatings for catalysis, membranes, or optical layers in waveguides. "Synthetic gold" coatings (non-stoichiometric TiN) are deposited on large volumes of personal jewelry.

CVD processes involve a series of steps (Figure 3.5):

- Transport of the reagents ($TiCl_4$, BCl_3, and H_2 in the above example) in the gas phase, often in a carrier gas, to the deposition zone.
- Diffusion or convection of gaseous precursors through the boundary layer (hot layer of gas immediately adjacent to the substrate).
- Adsorption of film precursors onto the growth surface.
- Surface diffusion of precursors to growth sites. The probability that a precursor molecule reacts directly at the first point of contact with the surface should not approach unity, because this may result in rough surface topologies, that is, some surface mobility of the precursors to growth sites is desirable.
- Surface chemical reactions leading to deposition of a solid film (TiB_2 in the above example) and the formation of byproducts (HCl in the above example).
- Desorption of byproducts.
- Transport of gaseous byproducts out of the reactor.

In the majority of CVD processes, gas-phase reactions (mostly in the hot boundary layer) prior to adsorption of the precursor on the surface are undesirable,

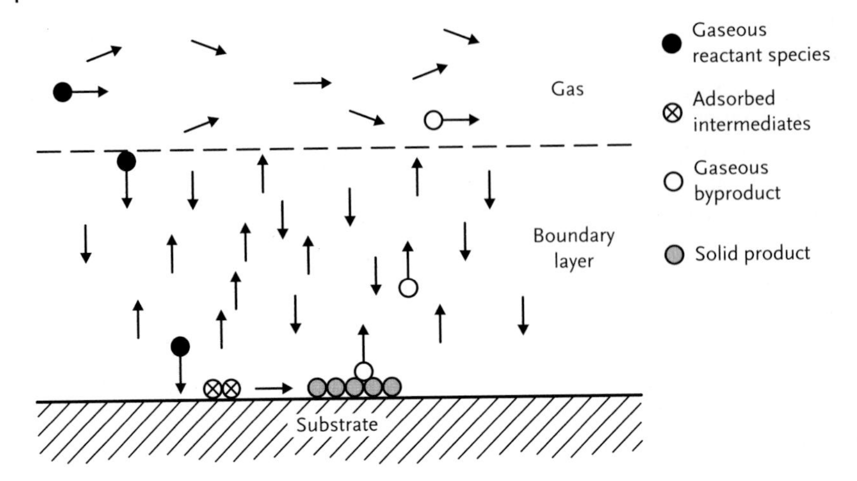

Figure 3.5 Schematic representation of the steps in CVD processes.

because they can result in particle formation, incomplete reaction of the precursor, and depletion of the precursor concentration at the surface. However, there are examples (see Section 3.2.6 for CVD of GaAs) where gas-phase reactions are beneficial for the film quality. The total pressure in the reactor controls the degree of gas-phase reactions, as when it is lowered the probability of reactant collisions in the gas phase becomes smaller.

Once the precursor has been adsorbed onto the substrate and has diffused to growth sites it should undergo surface-initiated chemical reactions to deposit the desired material. The nature of the → ligand and the type of metal–ligand bonds determine and control the decomposition pathway.

An ideal precursor should:

- be liquid rather than solid or gaseous;
- have a good volatility;
- have good thermal stability in the delivery system, and during its evaporation and transport in the gas phase;
- decompose cleanly and in a controlled manner on the substrate, without incorporation of contaminations;
- give byproducts that are stable and readily removed from the reaction zone;
- have a high purity or should be readily purified;
- be readily available in consistent quality and quantity at low cost; and
- be non-toxic and non-pyrophoric.

In reality, precursors rarely meet all criteria, and compromises have to be made. In particular, the requirement of long-term thermal stability at ordinary temperatures (shelf-life) on the one hand, and a high reactivity at higher temperatures (to achieve high deposition rates) on the other hand, often results in a narrow thermal stability window.

Most industrially important CVD processes utilize simple precursors such as metal hydrides (SiH_4, AsH_3), metal alkyls (Al^iBu_3, $GaEt_3$), or volatile metal halides (WF_6, $TiCl_4$). Metal halides generally require the presence of a reducing agent, and are often corrosive or liberate corrosive byproducts (HX, X_2). Metal–organic and organometallic compounds are usually more volatile than most inorganic compounds. However, they often thermally decompose at relatively low temperatures, which can lead to carbon or other contaminations of the substrate (see below).

Volatility is enhanced by minimizing all types of interactions between the precursor molecules in the condensed state (e.g., hydrogen bonds, dipole–dipole interactions, van der Waals interactions). Fluorination of ligands or substituents helps to reduce van der Waals interactions between the molecules. Since smaller non-polar compounds have a higher vapor pressure, oligomerization or aggregation of the precursors should be prevented. For this reason, β-diketonate → ligands are often used in metal-containing precursors. They are generally → chelating and thus occupy two coordination sites at the metal center. Therefore, β-diketonate complexes of metals with low coordination numbers are often monomeric and volatile.

Volatility problems arise with metals that prefer higher coordination numbers. This results in oligomerization to satisfy the metal's coordination number, mostly by the formation of ligand bridges. There are two strategies to prevent oligomerization:

1. Employment of sterically demanding → ligands. They limit the steric accessibility of ligands of an adjacent molecule to the metal center. However, this approach may result in two problems:
   a. The metal center is generally very reactive, due to its low coordination number. Therefore, small molecules (such as O_2 or H_2O) can penetrate the ligand shell and thus can make the compounds difficult to handle from a practical viewpoint.
   b. The large ligands result in a relatively high molecular weight of the compound, which tends to decrease the volatility.
2. Introduction of → chelating multidentate ligands into the coordination sphere of the metal, which can satisfy the preferred higher coordination number of the metal while preventing oligomerization. The problems associated with this approach are:
   a. The multidentate ligands may prefer to bridge two or more metal centers rather than chelate; or
   b. They may dissociate prior to or during transport, which can lead to oligomerization.

Chemical solutions for this problem will be discussed in Section 3.2.4 for alkaline-earth precursors for CVD of high-temperature → superconductors.

A general problem associated with the use of organometallic compounds is the carbon incorporation into the deposited film. To avoid this, precursors must be designed that completely eliminate the supporting organic ligands during the surface reactions. The volatile products must additionally be removed rapidly from the deposition zone in order to prevent contamination of the film by unwanted decomposition reactions. This is possible when the ligand is itself a stable molecule, for example, CO, alkenes, or when a thermal reaction results in a stable byproduct by a facile reaction, such as β-hydrogen elimination. In many cases, additional reagents are used to oxidize the ligands or the metal, to reduce the metal, or to protonate the ligands. Examples will be given in the following sections.

There are two approaches for the CVD of multielement materials, exemplarily shown in Eqs. (3.9) and (3.10) for the formation of GaAs (see also Section 3.2.6):

- Two (or more) individual precursors (*multiple-source precursors*) may be used that decompose individually on the substrate and eventually give a film of the desired multielement composition. In the case of GaAs, the most commonly employed precursors are $GaMe_3$ and AsH_3 [Eq. (3.9)].

$$GaMe_3 + AsH_3 \longrightarrow GaAs + 3CH_4 \tag{3.9}$$

- Single-source precursors contain the elements desired in the final film in one component, preferably in the ratio also required in the film. An example for GaAs is the cyclic compound $[Et_2Ga\text{-}As^tBu_2]_2$ [Eq. (3.10)]. In single-source precursors, the bonds between the film-forming elements must be stronger than those to the supporting ligands or substituents. The latter can then be removed cleanly by selective bond breaking. Otherwise, the advantages of a single source are lost.

$$[Et_2Ga\text{-}As^tBu_2]_2 \longrightarrow GaAs + 4C_2H_4 + 4CH_2{=}CMe_2 + 4H_2 \tag{3.10}$$

One problem associated with the multiple-source precursor approach is the reproducible control of film stoichiometry. This problem arises because the precursors or the intermediates formed during film deposition have different volatilities and reactivities. A typical example is the deposition of lead-containing films such as $PbTiO_3$, which can result in formation of lead-deficient materials due to desorption of volatile PbO.

Some of the problems associated with multiple-source precursors are circumvented by single-source precursors. Their use additionally allows a simplification of the precursor delivery system. However, single-source precursors have higher molecular weights than multiple-source precursors. They therefore are generally less volatile. It is also difficult to deposit films with non-integral stoichiometry, doped films, or materials with two or more metals. Furthermore, the stoichiometry of the precursor is not always retained in the film.

3.2.1.1 **Equipment**

A schematic of a typical CVD reactor is shown in Figure 3.6. Features common to all CVD reactors include a precursor delivery system, the reactor, and an exhaust system to remove byproducts. In conventional CVD processes, the substrate temperature typically varies between 200 and 800 °C, depending on the nature of the layer, and the pressure in the reactor cell varies between 0.1 mbar and 1 bar.

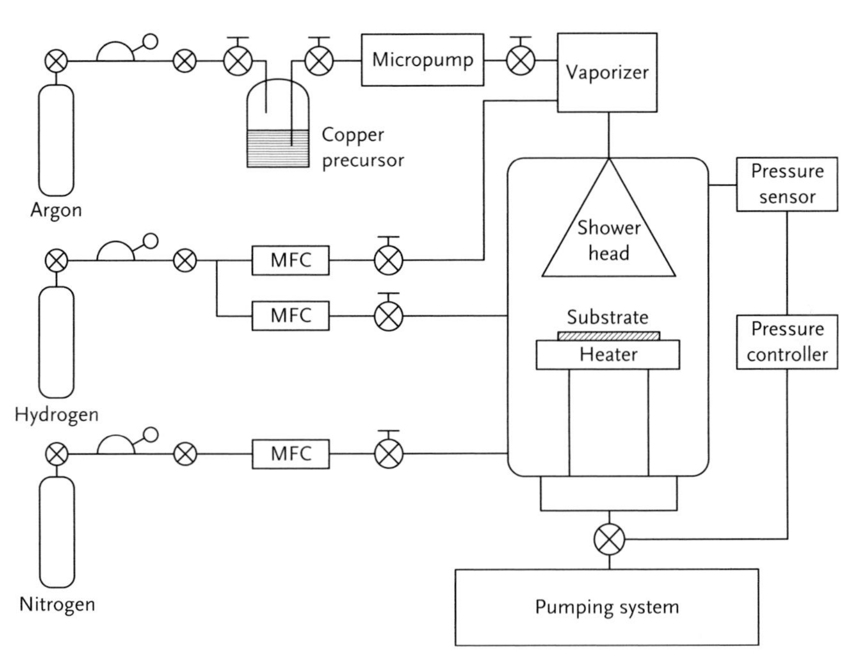

Figure 3.6 Schematic of a thermal CVD reactor for copper CVD from a Cu(II) precursor (see Eq. (3.19)) (MFC = mass flow controller).

The *precursor delivery method* plays a critical role because the overall deposition rate can be limited by the feed rate of precursor delivery into the reactor (see below). The transport of liquid or solid precursors with relatively low vapor pressures is controlled by bubbling a carrier gas through the precursor. The vapor pressure of liquids should be >0.01 bar at 25 °C for efficient industrial utilization. If the volatility of the precursors at room temperature is not sufficient, the storage vessel must be heated. Conventional delivery methods are unsuitable if the precursors are thermally unstable. Alternative methods of delivery must then be used such as liquid delivery, → aerosol delivery, spray pyrolysis (see Section 3.3), and supercritical fluid delivery. In the direct liquid-injection delivery method, the precursors are in liquid form (liquid or solid dissolved in a solvent) and are injected in a vaporization chamber. Then, the precursor vapors are transported to the substrate as in classical CVD processes. This technique is suitable for use on liquid or solid precursors. High growth rates can be reached using this technique.

Two main *reactor types* can be distinguished in conventional CVD systems: hot- and cold-wall reactors (Figures 3.7 and 3.8).

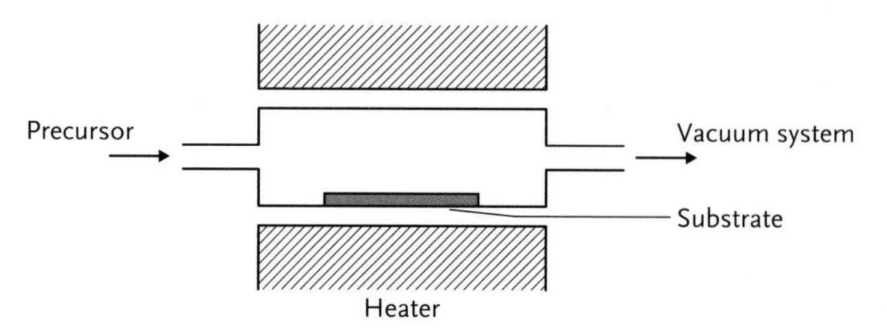

Figure 3.7 Schematic drawing of a hot-wall reactor.

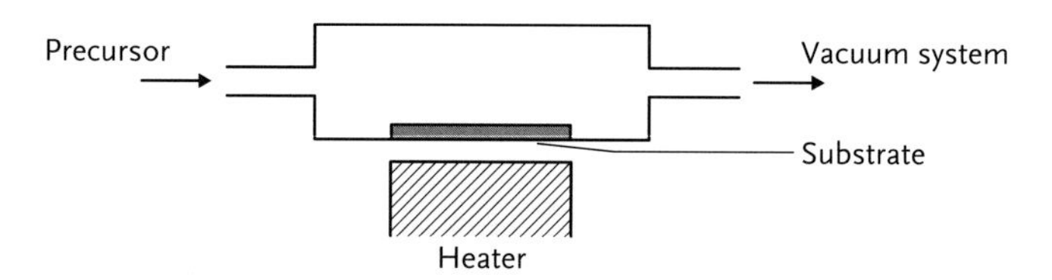

Figure 3.8 Schematic drawing of a cold-wall reactor.

In *hot-wall reactors* (Figure 3.7), the substrate and the chamber walls are maintained at the same temperature. The advantages are that:

- they are simple to operate,
- they can accommodate several substrates,
- obtaining a uniform substrate temperature is easy,
- they can be operated under a range of pressures and temperatures, and
- different orientations of the substrate relative to the gas flow are possible.

The major problems are that:

- deposition occurs not only on the substrate but also on the reactor walls (these deposits can eventually fall off the reactor walls and contaminate the substrate);
- the large consumption of the precursor makes control of the gas composition more difficult and can result in feed-rate-limited deposition (see below); and
- gas-phase reactions in the heated gas can occur.

For these reasons, hot-wall reactors are used primarily at the laboratory scale, and in industry for CVD of → semiconductors and oxides, with precursors having high vapor pressures. They are also often used to determine product distributions because the large heated surface area can consume the precursor completely and provide high yields of the reaction products.

Most metal CVD in industry is carried out in *cold-wall reactors* (Figure 3.8). The substrate is maintained at higher temperature then the reactor walls. The advantages are that:

- pressure and temperature can be controlled,
- plasmas can be used,
- deposition does not occur on the reactor walls,
- gas-phase reactions are suppressed, and
- higher deposition rates can be obtained because deposition occurs only on the heated substrate (higher precursor efficiency).

A disadvantage is that the steep temperature gradients near the substrate surface may lead to severe convection. This can result in non-uniform coatings.

CVD is a complex process with a large number of variables that influence the film properties: reactor geometry, reactant delivery, total pressure, gas and substrate temperature, substrate environment, gas composition and homogeneity, flow rate, substrate surface, time of deposition, gas flow behavior, deposition rate, → nucleation density, thermodynamic and kinetic properties of all involved species, particularly gas and surface chemistry, composition of the products, reaction mechanism, and so on.

Because of the many parameters, different results are often obtained for the same precursor, making comparison of results from different systems difficult. Coatings properties influenced by the process variables are composition, thickness, morphology, density, uniformity, adhesion to the substrate, crystallinity, → stress, and so on.

3.2.1.2 **Growth Rates**

The growth rate of the product layer must often be greater than 0.1 μm min^{-1} for CVD processes used in microelectronics processing, and higher for large-area coating processes such as those in the glass industry.

The typical dependence of the film growth rate on the substrate temperature is shown in Figure 3.9.

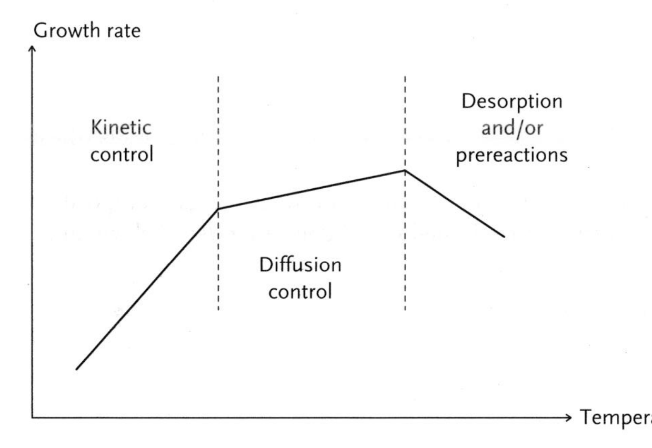

Figure 3.9 Dependence of the film growth rate on the substrate temperature.

There are three regimes:

1. At low temperatures, the growth rate is *surface-reaction (kinetically) limited*. This is the case when the feed rate is sufficiently high and diffusion limitations do not occur.
2. In the intermediate temperature region, the growth rate is *diffusion* or *mass-transport limited*. Essentially all the reactants that reach the substrate decompose. The reaction proceeds more rapidly than the rate at which the reactant is supplied to the surface by diffusion through the boundary layer.
3. At high temperatures, the growth rate tends to decrease because of an increased desorption rate of film precursors or film components, together with depletion of reactants by reaction at the reactor walls.

If both the mass transport and diffusion are fast, the deposition rate also may be limited by the rate at which the reactant(s) are fed to the chamber (*feed-rate-limited* process). This is often encountered for precursors with low vapor pressures.

The pressure of the CVD reactor determines the relative importance of each of these regimes in the growth process. For instance, there is a significant boundary layer at atmospheric to intermediate pressures (e.g., 0.2 bar). Therefore, growth may be characterized by the intermediate and high-temperature regimes described above. However, as the pressure in the reactor is lowered (low-pressure CVD, <1 mbar) layer growth is essentially surface-reaction controlled, as in the low-temperature region.

The growth in bimolecular systems (molecules A and B) can follow two different mechanisms (Figure 3.10):

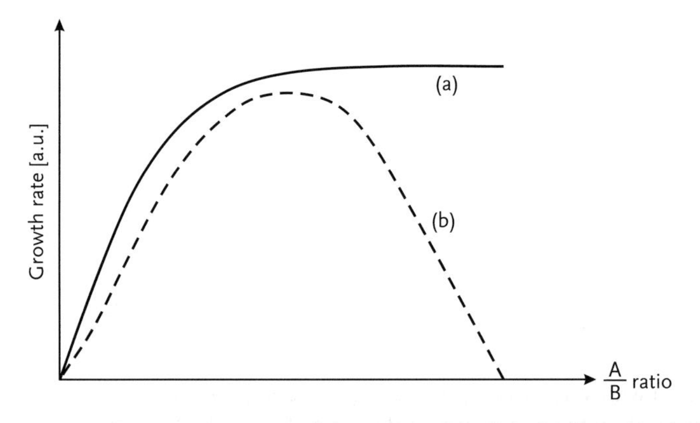

Figure 3.10 Growth rates in bimolecular systems: (a) Eley–Rideal mechanism, (b) Langmuir–Hinshelwood mechanism.

- **The Eley–Rideal mechanism**: Only one kind of molecule (A) is adsorbed. The adsorbed molecule reacts directly with a molecule of the other kind (B) from the gas phase. The growth rate shows a saturation behavior for high A/B ratios, that is, the limiting growth rate is determined by the complete coverage of the surface by molecules A.

- **The Langmuir–Hinshelwood mechanism**: Both A and B are adsorbed onto the substrate, and reaction takes place between the adsorbed molecules. When the A/B ratio is increased, the growth rate increases to a certain limit and then drops again. The species present in excess occupies the free adsorption sites on the substrate, and therefore there are not enough sites for the other species.

3.2.1.3 Selective Deposition

In the fabrication of multilevel electronic devices, the patterning of surfaces is essential. Patterned films can be prepared by covering the whole surface with the deposited material ("blanket deposition") followed by selective-area etching. Alternatively, selective CVD deposits a material only on one substrate (the growth surface) in the presence of another substrate (the non-growth surface) (Figure 3.11). Typical examples are metals or silicon as growth surfaces and silica as the non-growth surface. The possibility of selective growth of films is unique to CVD, and not possible with other methods.

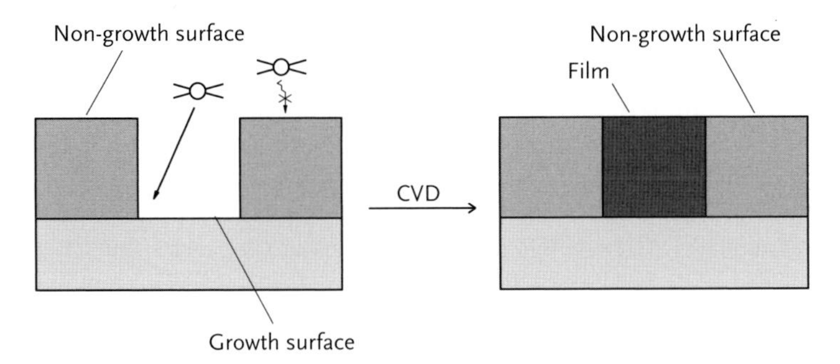

Figure 3.11 Selective deposition by CVD.

There are several strategies for selective deposition that rely on either inhibiting the adsorption and reaction of the precursor on the non-growth surface or promoting its reaction on the growth surface:

- The reaction rate of the precursor on the non-growth surface is intrinsically slower than its reaction rate on the growth surface. For example, copper CVD using (hfac)Cu(PMe_3) (see Section 3.2.2.3) occurs on Cu, Pt and other metal surfaces, but not on SiO_2 and other non-metal surfaces.
- The growth surface acts as a coreactant and is selectively consumed by a precursor (e.g., reaction of Si with WF_6 or MoF_6, see Eqs. (3.15) and (3.16), Section 3.2.2.2), while the reaction rate at the non-growth surface (e.g., SiO_2 or Si_3N_4) is slower.
- A chemical reaction of a gaseous coreactant occurs on the growth surface (e.g., dissociation of H_2 on a metal), but not on the non-growth surface (SiO_2 or metal oxides).
- The rate is increased on the growth surface by radiation while the thermal reaction at the non-irradiated non-growth surface is slow.
- Selective passivation of the non-growth surface prevents adsorption and reaction of the precursor, while adsorption and reaction occur readily on the

growth surface. For example, adsorption on silica can be prevented in many cases by conversion of the surface Si–OH groups in Si–OSiMe$_3$ groups.
- A free-energy barrier exists for → nucleation on the non-growth surface. It is assumed that the surface reaction is rapid and that the nucleation rate is only limited by the rate of formation of thermodynamically stable clusters on the growth surface.

3.2.1.4 **CVD and CVD-Related Techniques**

CVD processes resulting in the → epitaxial growth of a crystalline film are called *vapor phase epitaxy* (VPE). Epitaxial films have the same crystallographic orientation as the single-crystal substrate with similar lattice constants. VPE is generally used to prepare → semiconductor films (see Section 3.2.6) and is utilized in large-scale commercial operations.

Several prefixes are used to either specify the kind of precursors or variations of the CVD technique:

- The prefix "OM" (OMCVD, OMVPE, and so on) is used when the precursor is an organometallic compound, that is, a compound in which one or more organic → ligands are bonded to the metal by a metal–carbon bond. "MO" (MOCVD, MOVPE, and so on) is used for metal-organic precursors, that is, metal compounds with ligands containing organic substituents that are not bonded to the metal by a metal–carbon bond, such as amides, alkoxides, and so on.
- The basic CVD process is often modified, such as low-pressure CVD (LPCVD), ultrahigh vacuum CVD (UHVCVD), plasma-enhanced or plasma-assisted CVD (PECVD or PACVD), laser-assisted (laser-induced) CVD (LCVD), remote-plasma CVD (RPCVD) or microwave plasma-assisted CVD (MPCVD).
- Alternative methods of delivery must sometimes be used, such as aerosol-assisted chemical vapor deposition (AACVD, see Section 3.3) or direct liquid injection CVD (DLICVD).
- Chemical vapor infiltration (CVI) is CVD on internal surfaces of porous preforms (Section 3.2.7).

There are numerous CVD-related deposition methods with widespread applications. These include atomic layer deposition (ALD, also called atomic layer epitaxy, ALE), or chemical-beam epitaxy (CBE). The latter is also called metal-organic molecular-beam epitaxy (MOMBE). Some of the more important acronyms are summarized in Table 3.2.

Plasma-enhanced chemical vapor deposition (PECVD) This involves generation of reactive species in a glow discharge. The working pressure is usually in the range 0.1–1.0 mbar. Inelastic collisions between high-energy electrons and the gaseous precursors generate excited or ionized molecules. As a result, reactions proceed at lower temperatures. This is the major advantage of PECVD over a thermal CVD, along with potential high growth rates. PECVD is the established commercial technique for low-temperature deposition of a number of important materials, especially insulating films such as Si_3N_4 and SiO_2. The low gas and substrate temperatures in PECVD processes reduce the mobility of the reactants on the

Table 3.2 Survey of frequently used acronyms in CVD and related methods.

CVD	Chemical vapor deposition
PVD	Physical vapor deposition
VPE	Vapor phase epitaxy
OMCVD	Organometallic CVD
MOCVD	Metal-organic CVD
LPCVD	Low-pressure CVD
UHVCVD	Ultrahigh vacuum CVD
PECVD	Plasma-enhanced CVD
PACVD	Plasma-assisted CVD
LCVD	Laser-induced CVD
RPCVD	Remote-plasma CVD
MPCVD	Microwave plasma-assisted CVD
CVI	Chemical vapor infiltration
ALD	Atomic layer deposition
CBE	Chemical-beam epitaxy
MOMBE	Metal-organic molecular-beam epitaxy
AACVD	Aerosol-assisted CVD
DLICVD	Direct liquid injection CVD

substrate surface. Therefore, PECVD layers are often amorphous, and the → step coverage in microelectronic devices is poorer. Other potential disadvantages are the lack of substrate selectivity and plasma-induced substrate damage. In *remote plasma-enhanced chemical vapor deposition* (RPECVD), the potential of plasma damage to the substrate and film is reduced by separating the plasma-excitation region from the growth region.

Laser-assisted chemical vapor deposition (LCVD) In thermal LCVD, a laser is used to heat the substrate, analogous to local heating in a cold-wall reactor. In LCVD, and photoassisted CVD processes in general, the precursor molecule is activated by photochemical processes. This allows the deposition temperature to be significantly lower. Furthermore, deposition processes in selected areas are possible. For example, three-dimensional patterns or structures can be created in microelectronic devices without using masks by scanning of the laser over the surface.

Chemical beam epitaxy (CBE) (= MOMBE) This technique is mainly used in semiconductor technology. The molecular beams are effusive jets of the gaseous precursors. The method combines the advantages of MOCVD with those of molecular beam epitaxy (MBE; see below). In contrast to MBE, the precursor is supplied from outside the system to the deposition zone. In contrast to MOCVD, CBE film growth is performed in a ultrahigh vacuum (UHV) chamber (pressure $< 10^{-10}$ mbar). At these low pressures, gas transport proceeds via molecular flow, with no collisions in the gas phase. Furthermore, a hot boundary layer of gas close to the substrate is absent under these conditions, and layer growth is controlled by the substrate temperature and by the kinetics of the surface reactions. The major drawbacks of CBE are the very high cost of UHV equipment and the low deposition rates and throughput.

Atomic layer deposition (ALD) This method is used for the controlled deposition of monolayers. The film is grown one atomic layer at a time, and the deposition process is based on alternating chemisorption of the different precursors and their surface reaction. This is shown in Figure 3.12 for the formation of a ZnS film from $ZnCl_2$ and H_2S, as an example. Repetition of the growth cycle produces controlled layer-by-layer film growth.

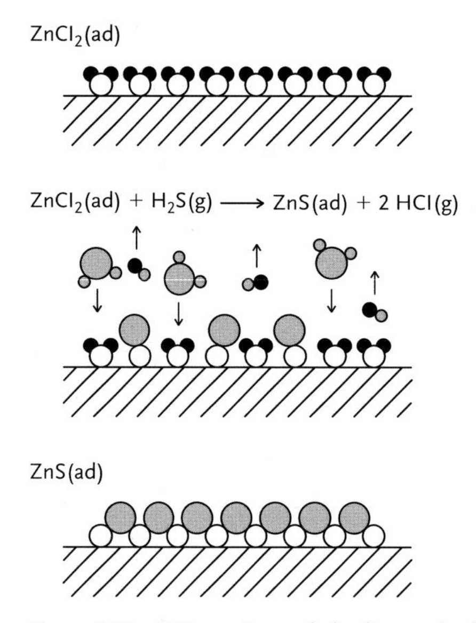

Figure 3.12 ALD reaction cycle leading to the formation of a ZnS thin film.

Ideally, the precursor chemisorbs only to active sites on the substrate until all sites are occupied. When carried out within certain temperature limits ("ALD-window"), ALD is a self-limiting deposition process. It has the potential to grow extremely homogeneous crystalline films over large areas and offers unprecedented control at the monolayer level. One major disadvantage of ALD is that film growth is much slower than conventional CVD, because the growth rate is typically limited by the time required for switching from one precursor to the other, which usually requires several seconds.

3.2.1.5 Non-CVD Processes for the Gas-Phase Deposition of Thin Films

These processes include PVD (*physical vapor deposition*) by evaporation or sputtering of a target material onto a substrate, and MBE. They are discussed briefly at this point for comparison with the CVD methods.

Today, most metallization for microelectronics is performed using PVD processes. These involve three steps:

1. vaporizing a solid;
2. transporting the gaseous compound from the source to the substrate; and
3. condensing the gaseous compound on the substrate surface, followed by nucleation and growth of a new layer.

The main difference between CVD and physical methods is that in CVD, films are formed by reaction of some precursor on the substrate (e.g., deposition of metallic copper by thermal decomposition of metal-organic precursors, see Section 3.2.2.3), while in PVD (and related processes) some material is volatilized and deposited as such on the substrate (e.g., elemental copper itself).

PVD by *thermal evaporation* is inherently a simple method, and is widely used for depositing metallic films for integrated circuit fabrication. An electron or laser beam is directed at the target material under high vacuum conditions (10^{-5}–10^{-8} mbar) to vaporize the material. Alternatively, a crucible containing the material is heated by induction or by resistive heating. The rate of evaporation is controlled by the temperature of the source. The evaporated material deposits on a substrate that is maintained at a lower temperature than that of the vapor. At the low pressure, the mean free path of the evaporated species is very long (5–5000 m) compared to the source-to-substrate distance. This allows an essentially collisionless transport of the evaporated material to the substrate surface.

PVD by *sputtering* is a process where the surface atoms of a target material are liberated by bombardment with energetic ions that are generated in a glow discharge or plasma. The sputtered atoms are then ballistically transported to the substrate surface, where they condense. Sputter-deposited films are widely used in integrated circuits. Sputtering processes are generally applicable to all inorganic materials. The deposition of multielement materials by sputtering is possible because the composition of the film is the same as the composition of the target. An alternative for the deposition of ceramic films is "reactive sputtering." Chemical compounds are then formed by reaction of the metal vapor with a reactive gas. Examples of reactive gases are methane, ammonia, or nitrogen, and diborane to deposit carbide, nitride, and boride materials, respectively.

One of the biggest advantages of PVD methods is their lower substrate temperature that allows the deposition of films also on thermolabile substrates such as organic polymers. However, in contrast to CVD, the coating of surfaces not facing the vapor source is a serious problem. This also includes → step coverage in electronic devices. In CVD processes, diffusion or convection can transport reactants to hidden areas (Figure 3.13).

In PVD processes, the particle stream from the evaporating or sputtering sources is directed. Furthermore, the deposited species are not sufficiently mobile on the substrate surface to migrate to areas not facing the source. Other advantages of CVD over PVD include the possibility of → epitaxial growth and selective deposition, the capability of large-scale production, and the ability to produce → metastable materials.

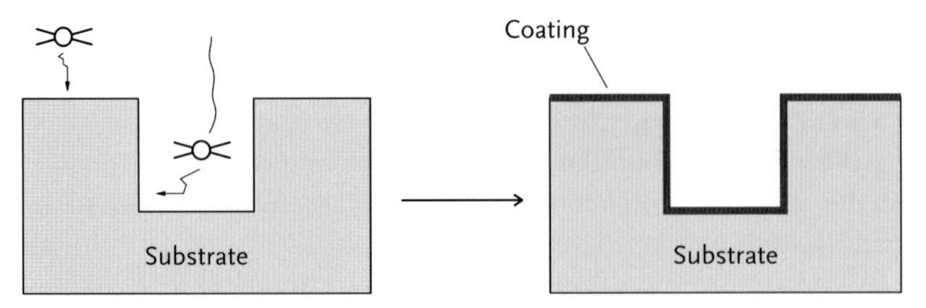

Figure 3.13 Step coverage by CVD methods.

Molecular-beam epitaxy (MBE) This is a relatively simple process in which elemental sources are independently evaporated at a controlled rate, forming molecular beams that intercept at the heated substrate (Figure 3.14). MBE is usually carried out under UHV conditions (10^{-10} mbar) and at low growth rates. Films with high purity and very complex layer structures (e.g., nanostructures) can be deposited with a precise control of doping of the deposited layers. The low processing temperature is also important for microelectronic production.

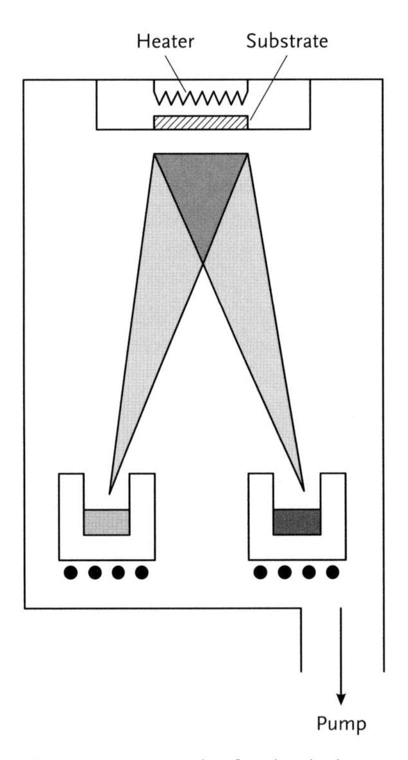

Figure 3.14 Principle of molecular-beam epitaxy (MBE).

3.2.2
Metal CVD

Metal coatings on various substrates are needed for a variety of applications. Typical examples include oxidation, corrosion, or abrasion-protection films, reflective or conducting coatings, or electrodes. Particularly important applications are in microelectronics. The increasing demand for more and more sophisticated electronic devices requires the speed of the circuits to be increased and their performance to be improved. Therefore, the size of the devices must be decreased and the number of levels in complex multilevel structures must be increased. Figure 3.15 shows an example for the complexity of such a device.

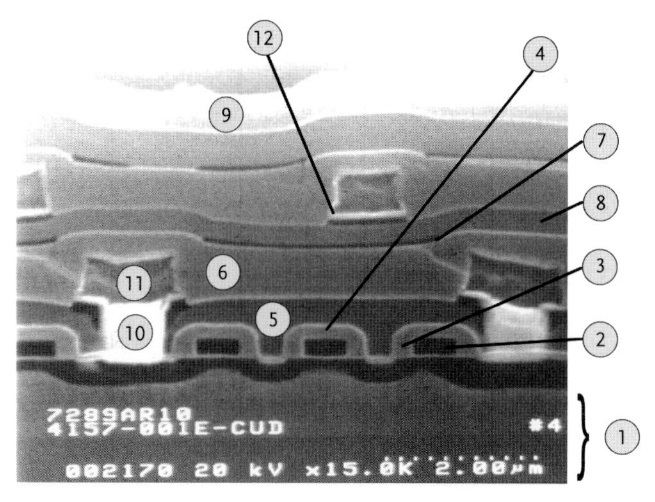

Figure 3.15 A cross-section through a multilevel structure of an electronic device. 1: silicon substrate; 2: gate; 3: spacer (CVD oxide); 4: covering oxide; 5: borate-phosphate-silicate glass (PACVD); 6: silica (PACVD from $Si(OEt)_4$); 7: glass layer (spin-on); 8: phosphate-doped silica (plasma CVD from $Si(OEt)_4$); 9: oxynitride/nitride passivation layer (PACVD); 10: tungsten plug (thermal CVD); 11: AlCu (PVD); 12: TiN barrier (reactive PVD).

The primary reason for metallization is to provide interconnections between the various circuit components. In addition to having good electrical properties, it is essential that these interconnection materials do not react with the materials which they contact. This is a problem for many metal–silicon contacts. Silicon may diffuse into the metal film to form → solid solutions or metal silicides during thermal processing or by → electromigration. This becomes an increasingly important topic as the size of the interconnect lines decreases. Therefore, diffusion barrier layers are often introduced between the metal and the → semiconductor. Common materials that suppress this migration for Si/Al contacts are TiN, Ti, TiW, ZrN, or RuO_2.

Most commercial metallic films are produced by PVD methods. In this section, the CVD of Al, Cu, and W is discussed as an example, because this allows an elaboration to be made on the chemistry of different types of precursors.

3.2.2.1 **Aluminum**

The main uses for aluminum films are:

- Metallized polymers in food packaging (where the Al film provides a gas diffusion barrier) and as reflective films (mirrors, CDs, and so on).
- Interconnects in microelectronics. The → resistivity of bulk Al (2.74 μΩ cm) is only slightly greater than that of Cu (1.70 μΩ cm) or Ag (1.61 μΩ cm). An important aspect of its use in device technology is that Al as an amphoteric element is easily etched both by strong bases or acids. On the negative side, Al is prone to → electromigration and is soluble in or reacts with semiconductor materials. One possibility of reducing electromigration is to alloy Al with small amounts of Cu and Si.

The best investigated precursor for depositing high-quality Al films by CVD is triisobutylaluminum, Al^iBu_3, a colorless, pyrophoric liquid, which at room temperature has a vapor pressure of about 0.1 mbar. It is monomeric in the gas phase.

The first isobutyl ligand of Al^iBu_3 is degraded by β-hydride elimination [Eq. (3.11)] already at temperatures above 50 °C. The diisobutylaluminum hydride formed is a trimer, and therefore has a substantially lower vapor pressure (0.01 mbar at 40 °C). The formation of $[HAl^iBu_2]_3$ in the precursor vessel and the gas tubings can be suppressed by addition of isobutene to the carrier gas, because the β-elimination process is reversible.

$$\underset{(Al^iBu_3)}{3\ Al(CH_2CHMe_2)_3} \rightleftharpoons 3\ CH_2{=}CMe_2 + \underset{[HAl^iBu_2]_3}{\text{cyclo-}(^iBu_2Al\text{–}H)_3} \quad (3.11)$$

Typical conditions for the deposition of Al from Al^iBu_3 are temperatures of 200–300 °C in hot-wall reactors. Growth rates of 20–80 nm min^{-1} are achieved. The decomposition chemistry of Al^iBu_3 on aluminum surfaces was very well studied, and sheds light on the processes leading to film growth. The overall reaction is given in Eq. (3.12).

$$Al^iBu_3 \longrightarrow Al + 3CH_2{=}CMe_2 + 3/2H_2 \quad (3.12)$$

The three isobutyl groups become equivalent after adsorption. They diffuse over the Al surface and are no longer attached to just one aluminum atom (roughly speaking, once a layer of aluminum is formed on a particular substrate, further decomposition of Al^iBu_3 occurs on an aluminum surface). Therefore, each isobutyl group equally participates in the rate-determining β-hydrogen elimination (Figure 3.16).

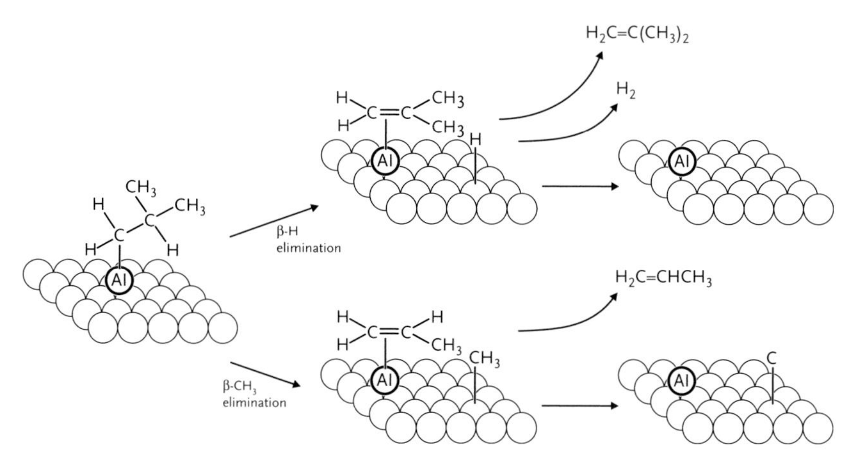

Figure 3.16 Schematic representation of the thermal decomposition of Al^iBu_3. The upper pathway (at low temperatures) results in the clean deposition. At higher temperatures (lower pathway) carbon impurities may be incorporated into the film.

As a consequence, the hydrogen atoms created by β-elimination are also spread over the Al surface. Both H_2 and isobutene readily desorb from Al at temperatures below the decomposition temperature.

Carbon is only incorporated in the films at temperatures above 330 °C. β-Methyl elimination, leading to propene and surface-bonded methyl groups, has a higher activation energy than β-hydrogen elimination, and therefore only plays a role at higher temperatures. The surface-bonded methyl groups eventually decompose to give carbon (see Figure 3.16).

Trimethylaluminum, Al_2Me_6, is a less suitable precursor for the same reason. β-Hydrogen elimination is not possible; the decomposition path of Al_2Me_6 is complex, and depends strongly on the conditions. Al films deposited between 350 and 550 °C from Al_2Me_6 contain high levels of carbon. Thermal decomposition under N_2 even results in the formation of thermodynamically favored Al_4C_3 [Eq. (3.13)]. However, Al films with little carbon incorporation are formed when the atmosphere is changed to hydrogen [Eq. (3.14)]. Thermodynamic calculations predict that the reaction of Al_2Me_6 at 200 °C in the presence of H_2 results in the formation of AlH_3 and methane, which readily desorbs.

$$2Al_2Me_6 \longrightarrow Al_4C_3 + 9CH_4 \tag{3.13}$$

$$Al_2Me_6 + 3H_2 \longrightarrow 2Al + 6CH_4 \tag{3.14}$$

With Al^iBu_3 as the precursor, the selective deposition of Al is possible. Film growth occurs on Si or Al, but not on silica. When Al^iBu_3 is adsorbed on silica, two of the three isobutyl groups are readily lost already at room temperature. Due to the inductive effect of the two oxygen atoms then bonded to the aluminum atom (Figure 3.17), elimination of the third isobutyl group is suppressed. The surface-bonded $^iBuAl{=}$ species inhibit film growth on silica by preventing further adsorption.

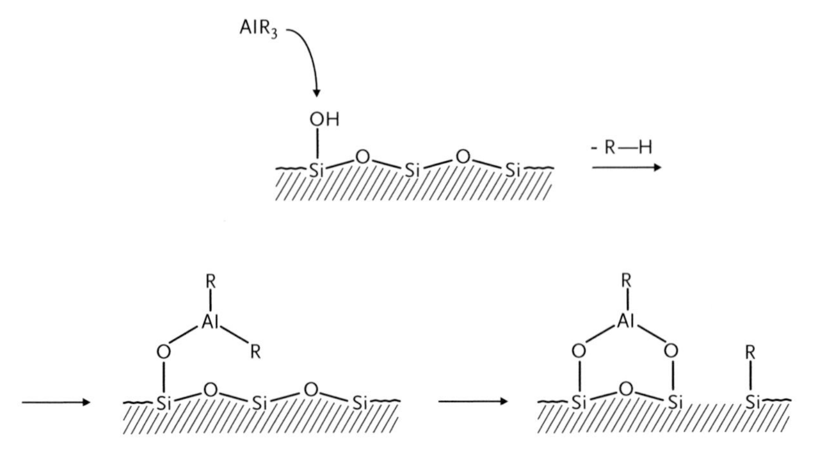

Figure 3.17 Inhibition of Al deposition from AlR_3 on a silica surface.

Al^iBu_3 can be considered as a source of AlH_3 formed by β-elimination of isobutene. However, AlH_3 itself is a solid not suited for CVD. AlH_3 is also contained in donor–acceptor complexes of the type L-AlH_3, where L is a Lewis base. For CVD purposes, adducts with L = tertiary amine are mainly used, their stability and volatility being influenced by the electronic properties and the steric bulk of L. They are easily prepared and are significantly less air-sensitive than are aluminum alkyls. Trimethylamine forms a solid 1 : 1 and a 2 : 1 adduct, $Me_3N–AlH_3$ and $(Me_3N)_2AlH_3$. Liquids are obtained for L = $EtMe_2N$ or Et_3N.

There are only a few problems with carbon incorporation with these precursors, because the compounds contain no Al–C bond, and the Al–N bond is easily cleaved. The decomposition mechanism of $Me_3N–AlH_3$ on Al surfaces is shown schematically in Figure 3.18. The first step after adsorption is dissociation of NMe_3 and its fast desorption. Similar to the dispersion of the butyl groups after adsorption of Al^iBu_3, the hydrogen atoms of the AlH_3 group also disperse over the Al surface.

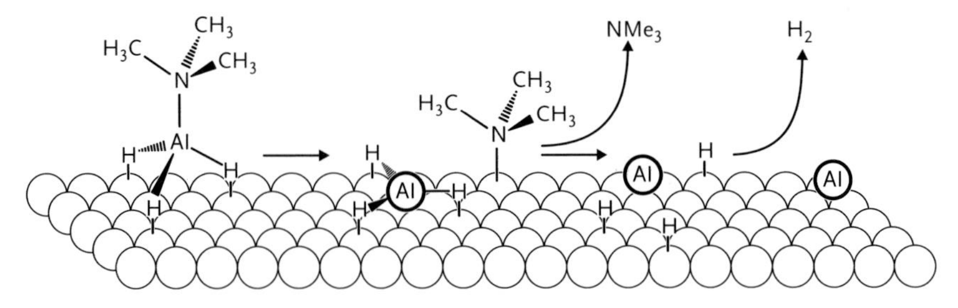

Figure 3.18 Thermal decomposition mechanism of $Me_3N–AlH_3$ on Al surfaces.

3.2.2.2 Tungsten

There are two important areas of application for tungsten films:

- Wear protection for cutting and grinding tools and corrosion protection for various materials, even at high temperatures. The → hardness and chemical inertness can be enhanced by adding small amounts of carbon, oxygen, or nitrogen.
- Metallization in integrated circuits. In addition to favorable electrical, mechanical, and chemical properties, tungsten has a high resistance to → electromigration, a low thermal expansion coefficient, and does not react with major → semiconductor materials below 600 °C. A disadvantage is the low adhesion to the SiO_2 layer on Si, which necessitates the use of adhesion promoters such as TiN.

The dominant tungsten CVD precursor is tungsten hexafluoride (WF_6) despite some disadvantages (see below). This is used routinely in the microelectronics industry. WF_6 has a boiling point of 17 °C, a high vapor pressure (1.3 bar at 25 °C), and is commercially available at low cost. Although thermal decomposition can be achieved above 750 °C, most methods of tungsten CVD involve reducing agents (mostly H_2, Si, or SiH_4) to lower the deposition temperatures.

The reduction of WF_6 by a silicon substrate allows the selective deposition of W. SiF_4 is formed below 400 °C [Eq. (3.15)], and SiF_2 above 400 °C [Eq. (3.16)]. The high-temperature reaction [Eq. (3.16)] consumes twice the amount of Si than does the low-temperature reaction.

$$2WF_6 + 3Si \xrightarrow{<400\,^\circ C} 2W + 3SiF_4 \tag{3.15}$$

$$WF_6 + 3Si \xrightarrow{>400\,^\circ C} W + 3SiF_2 \tag{3.16}$$

The reactions are very fast, but stop when a certain film thickness is reached (mostly within seconds). For film growth, either silicon has to diffuse through the W layer, or WF_6 has to diffuse to the W/Si interface. Both become increasingly difficult with increasing film thickness, particularly since tungsten is an excellent diffusion barrier for silicon. The presence of water has to be carefully avoided, because WF_6 is readily hydrolyzed and the thus generated HF attacks the silica layer on silicon and the tungsten layer.

WF_6 is reduced by hydrogen [Eq. (3.17)] at temperatures between 300 and 800 °C.

$$WF_6 + 3H_2 \longrightarrow W + 6HF \tag{3.17}$$

Because of the formation of HF and the problems associated with that, tungsten is not deposited directly on silicon [Eq. (3.18)]. This problem is avoided when SiH_4 is used as the reductant. Furthermore, no silicon is consumed (as in the WF_6/Si combination), and the deposition rates are high (up to ~1 μm min^{-1} in cold-wall LPCVD). In principle, two deposition processes compete with each other: reduction of WF_6 (giving W) and thermal decomposition of SiH_4 (giving Si that then reacts with elemental W). The deposition reactions are less straightforward as in the Si or H_2 reduction. Equation (3.18) gives the most probable reactions. There is no reduction of WF_6 by the produced H_2, because the reduction by SiH_4 is much faster.

$$\begin{aligned} 2WF_6 + 3SiH_4 &\xrightarrow{250\,^\circ C} 2W + 6H_2 + 3SiF_4 \\ 4WF_6 + 3SiH_4 &\xrightarrow{T>600\,^\circ C} 4W + 12HF + 3SiF_4 \end{aligned} \quad (3.18)$$

The advantages of the SiH_4 reduction are somewhat impaired by the very difficult handling of SiH_4/WF_6 mixtures. The two compounds react vigorously with each other. Therefore, silane reduction is only used to initiate tungsten deposition, followed by the hydrogen reduction process.

Their volatility and thermolability makes organometallic compounds potentially interesting precursors for tungsten deposition under less corrosive conditions in the electronics industries. A variety of such compounds has been investigated, for example, tungsten hexacarbonyl [$W(CO)_6$], or bis(cyclopentadienyl)dihydridotungsten [$(\eta^5\text{-}C_5H_5)_2WH_2$]. However, the major drawback of tungsten CVD using organometallic compounds is contamination of the films with sometimes high levels of carbon and/or oxygen.

3.2.2.3 Copper

Copper may replace Al in part of the interconnections in multilevel integrated circuits, because it is more resistant to → electromigration. Furthermore, it has a better thermal expansion coefficient, a higher melting point, and a lower electric → resistivity. Thus, Cu interconnects could allow an increase of the operation frequency of devices and higher current densities compared to Al. Disadvantages of Cu are its fast diffusion in Si and drift in SiO_2-based insulating materials (→ dielectrics), the poor adhesion to SiO_2, and the lack of suitable etch processes.

Since copper does not form organometallic compounds suitable for CVD, and copper halides are not volatile enough, metal-organic compounds are mainly used. There are two chemically different approaches for the CVD of Cu films. In the first approach, volatile Cu(II) compounds are decomposed at the substrate surface in the presence of hydrogen. The second approach makes use of the well-known property of Cu(I) compounds to disproportionate into elemental Cu and Cu(II) compounds.

Copper(II) precursors Attention has mostly focused on the use of Cu(II) bis(β-diketonate) complexes and, to a limited extent, on bis(β-ketoiminate) complexes. Two of the best investigated complexes are shown in Figure 3.19.

An important feature, also characteristic of the Cu(I) compounds discussed below, is the use of fluorinated ligands, such as 1,1,1,5,5,5-hexafluoroacetylacetonate (hfac) (Figure 3.19). The volatility of β-diketonate complexes increases with the number of fluorine atoms due to the reduction of van der Waals interactions between the molecules. The high electronegativity of the fluorine atoms may also lead to a weakening of the Cu–ligand bonds.

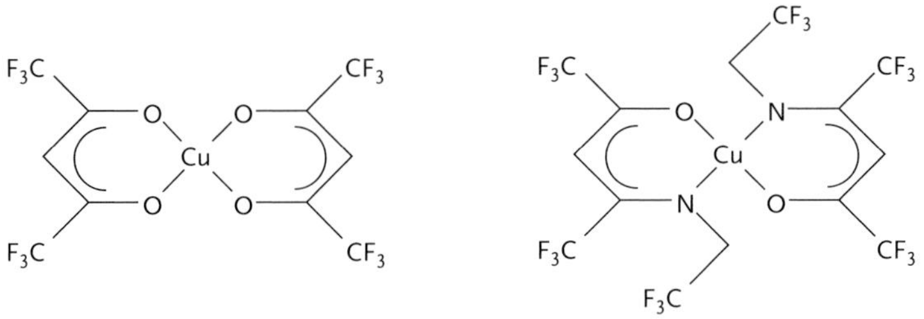

$Cu(hfac)_2$ $Cu(nona\text{-}F)_2$

Figure 3.19 $Cu(hfac)_2$ and $Cu(nona\text{-}F)_2$ as representative Cu(II) precursors with β-diketonate (left) and β-ketoiminate ligands (right).

Equation (3.19) gives the stoichiometry of the CVD process for $Cu(hfac)_2$ in the presence of hydrogen. The ligands are liberated in their protonated forms.

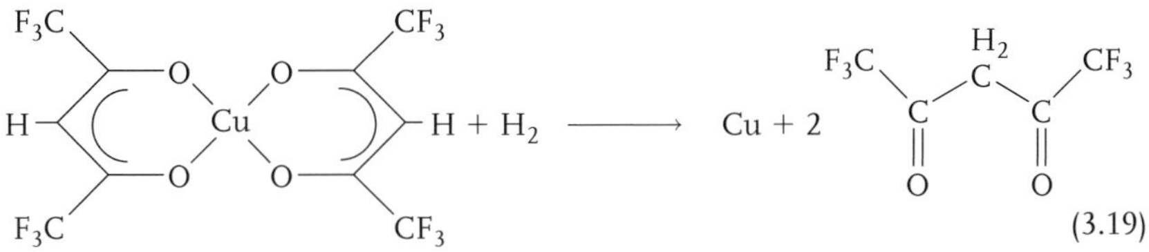

$$Cu(hfac)_2 + H_2 \longrightarrow Cu + 2\ F_3C\text{-}C(=O)\text{-}CH_2\text{-}C(=O)\text{-}CF_3 \qquad (3.19)$$

$Cu(hfac)_2$ adsorbs dissociatively under CVD conditions [Eq. (3.20)]. If the film grows on a copper surface (i.e., after the first Cu layer has been formed), the Cu atoms become part of the surface, which corresponds to an electron transfer from the surface to the incoming Cu atom. The hfac ligands spread over the Cu surface, as discussed above for the butyl ligands of Al^iBu_3. When $Cu(hfac)_2$ is adsorbed onto non-metallic surfaces, electron transfer is no longer facile and the intact molecule is adsorbed.

$$\begin{aligned} Cu(hfac)_2(g) &\rightleftharpoons Cu(s) + 2(hfac)_{ads} \\ H_2(g) &\rightleftharpoons 2H_{ads} \\ (hfac)_{ads} + H_{ads} &\rightleftharpoons H\text{-}hfac(g) \end{aligned} \qquad (3.20)$$

Dissociative adsorption of H_2 occurs readily and reversibly at room temperature and above on most metals. The heat of adsorption is rather low, which means that the surface will not be blocked by hydrogen under reaction conditions. There is a significant kinetic barrier for hydrogen adsorption (about 40–60 $kJ\,mol^{-1}$). Therefore, when the coverage of the surface by hfac ligands is high and the partial pressure of H_2 in the gas phase is low, there may be conditions where H_2 adsorption becomes the rate-limiting step for the deposition of Cu films.

Kinetic measurements suggest that the rate-limiting step in the overall Cu deposition is the desorption of H-hfac formed by bimolecular surface reaction between adsorbed H and hfac. Instead of desorbing as H-hfac, the hfac ligands may undergo decomposition reactions. Therefore, the surface coverage by hfac also affects film purity.

Copper(I) precursors Although the Cu(II) precursors have high thermal stability and high vapor pressures, deposition rates are usually low (0.1–0.5 nm min^{-1}) and high substrate temperatures ($\sim 400\,°C$) are required. They only result in pure Cu films in the presence of a reducing agent, such as H_2. Cu(I) precursors are mostly liquids and have lower vapor pressures and lower decomposition temperatures ($< 200\,°C$). Typical deposition rates are around 0.1–1 μm min^{-1}, and no reducing agent is needed.

The most widely studied family is Lewis base-stabilized Cu(I) β-diketonates. The → ligand L may be phosphines (PMe_3), olefins (mostly 1,5-cyclooctadiene [COD] or vinyltrimethylsilane [VTMS]), alkynes (3-hexyne), and so on.

These complexes have the following advantages:

- the β-diketonate ligand provides volatility to the complex, particularly if it contains fluorinated substituents;
- the neutral ligands L do not decompose at temperatures where Cu deposition occurs; and
- both the β-diketonate ligand and the ligand L allow easy modifications to tailor the physical state of the complex and its volatility and reactivity.

The overall deposition process is a thermally induced disproportionation reaction according to Eq. (3.21). The proposed mechanism for deposition is shown in Eqs. (3.22) and (3.23) for (hfac)CuL. Since the ligand L is only weakly bound, the thermal decomposition of the Cu(I) complex occurs at relatively low temperatures once the precursor has been adsorbed to the surface. The neutral ligand L desorbs readily, while at the same time adsorbed Cu(hfac) disproportionates into $Cu(hfac)_2$ and elemental Cu, [Eq. (3.23)]. $Cu(hfac)_2$ desorbs from the surface, because the deposition temperature of copper is below the decomposition temperature of $Cu(hfac)_2$.

$$2\ (\text{R}_2\text{-diketonate-O,O})\text{Cu–L} \longrightarrow \text{Cu} + (\text{R}_2\text{-diketonate-O,O})_2\text{Cu} + 2\,\text{L} \quad (3.21)$$

$$[(\text{hfac})\text{CuL}]_{ads} \longrightarrow [\text{Cu}(\text{hfac})]_{ads} + \text{L}(g) \quad (3.22)$$

$$2[\text{Cu}(\text{hfac})]_{ads} \longrightarrow \text{Cu}(s) + \text{Cu}(\text{hfac})_2(g) \quad (3.23)$$

3.2.3 Diamond CVD

Diamond has outstanding properties, most of which are relatively insensitive to lattice defects:

- mechanical properties: extreme hardness (~ 90 GPa), very high bulk modulus ($1.2 \times 10^{12}\,N\,m^{-2}$), very low compressibility ($8.3 \times 10^{-13}\,m^2\,N^{-1}$);
- acoustic properties: high sound velocity ($18.2\,km\,s^{-1}$);

- thermal properties: very high thermal conductivity at room temperature ($>2 \times 10^3$ W m^{-1} K^{-1}), low thermal expansion coefficient at room temperature (0.8×10^{-6} K^{-1});
- optical properties: transparency from the deep UV to the far IR;
- electrical properties: good insulator (→ resistivity $\sim 10^{16}$ Ω cm at room temperature), → semiconductor (10–10^6 Ω cm), with a bandgap of 5.47 eV (for comparison: Si 1.1 eV or GaN 3.44 eV);
- chemical properties: resistant to chemical corrosion and radiation.

These properties result in several highly interesting applications for diamond thin films. The films can be polycrystalline for most applications (Figure 3.20). The most developed are heat sinks in electronic devices and wear-resistant hard coatings for tools. Applications also include precision optical components for high-power lasers, radiation detectors and surface-acoustic-wave (SAW) devices. In the future, diamond films could be used as high-temperature semiconducting devices, high-frequency field-effect transistors, or X-ray-lithography masks.

Figure 3.20 Diamond films (on a SIALON surface) deposited by microwave activation of 1.7% methane in hydrogen. The different morphologies originate from different gas pressures and different microwave powers.

The synthesis of polycrystalline diamond has been achieved by several CVD methods (Figure 3.21).

The most common methods for diamond synthesis are the activation of methane in an excess of hydrogen (typically 1–2 vol% CH_4 in H_2) by using microwave, r.f. or d.c. discharges (Figure 3.21(b)) and the decomposition of the gas mixture by hot ($\sim 2300\,^\circ C$) filaments (Figure 3.21(a)). These variations have in

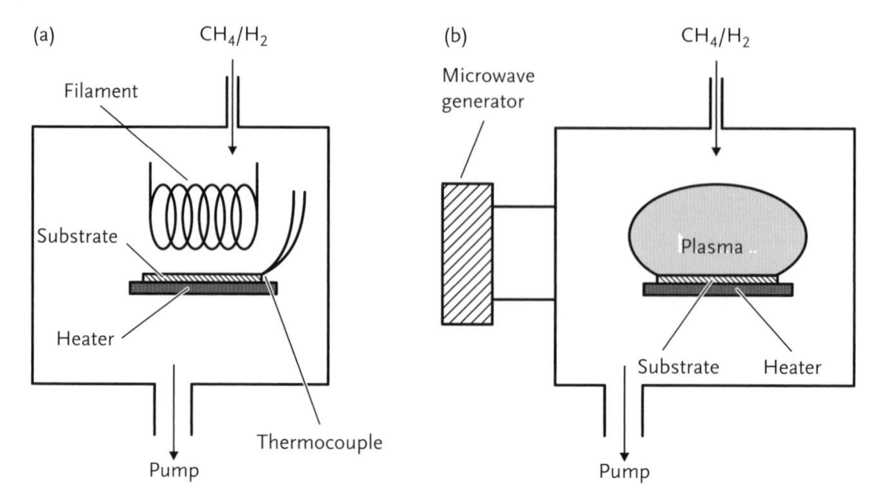

Figure 3.21 Schematic diagrams of the two most commonly used apparatus for diamond deposition. (a) Hot-filament reactor. (b) Microwave plasma-enhanced CVD reactor.

common that atomic hydrogen is produced from H_2 near the surface during the decomposition process. Gas-phase hydrogen-abstraction reactions lead to the formation of hydrocarbon radicals that probably are the main precursors for diamond. However, the chemistry at the growth surface is very complex. The choice of appropriate deposition parameters is important for obtaining specific diamond morphologies.

Diamond is → metastable with respect to graphite at low pressure. The growth of metastable diamond is possible because of the presence of atomic hydrogen ($H^{\bullet}$) that selectively etches codeposited graphitic nuclei and leaves diamond. The dissolution rate of graphite and amorphous carbon by hydrogen is about 50 times that of diamond. Thus, the basic chemistry of diamond deposition can be written in two equations [Eq. (3.24)]

$$\begin{aligned} &\text{etching of graphite:}\ y\text{H} + x\text{C}_{\text{graphite}} \longrightarrow \text{C}_x\text{H}_y \\ &\text{diamond deposition:}\ m\text{H} + \text{C}_x\text{H}_y \longrightarrow m + y/2\text{H}_2 + x\text{C}_{\text{diamond}} \end{aligned} \tag{3.24}$$

Furthermore, hydrogen prevents the formation of sp^2-hybridized carbon atoms at the growth front, and thus stabilizes the growing diamond surface. Hence, the growth surface is usually hydrogen terminated.

Diamond films are also obtained when a hydrocarbon (typically acetylene or methane) is burnt with oxygen in a welding torch. If this flame is directed to a water-cooled substrate at temperatures of 700–1000 °C, diamond growth occurs at the intersection between the substrate and the primary combustion zone. The method is inherently simple, but mostly not economical. The yield of diamond and the area of deposition are relatively small, the deposits are rough and have varying crystal quality.

Diamond → nucleation is very sensitive to the substrate material itself as well as to its surface condition (surface composition and morphology, dislocations, pretreatment, and so on). Diamond nucleation on highly perfect substrate surfaces, for example silicon wafers, is sluggish and has a long initiation period. Nucleation is also difficult if the surface carbon concentration is lowered rapidly by diffusion into the substrate (e.g., Fe, Co, Ni). It only starts when the substrate is saturated. Chemical reactions occurring between the substrate and reaction-gas components (e.g., atomic hydrogen) can also have an important influence on the nucleation and on the adhesion of the films. To enhance the diamond nucleation density, pretreatment of the substrate surface by polishing, abrading, and so on, is generally very effective.

Freestanding diamond sheets in sizes of several hundreds cm^2 are commercially produced by diamond deposition on dummy substrates, which are dissolved after formation of the film.

3.2.4 CVD of Metal Oxides

Thin films of metal or semimetal oxides are employed in numerous technological applications. Although there is major competition from sol–gel techniques (Section 4.5), CVD is widely used, particularly in the microelectronics industry. A variety of transition-metal oxide films has been prepared by CVD, mainly for optical purposes or as materials with high → dielectric constant. Hafnium oxide (HfO_2) is a promising material for the latter application. Since this book concentrates on chemical aspects rather than applications, only SiO_2 and $YBa_2Cu_3O_{7-x}$ (YBCO) will be discussed as case studies.

The precursors for metal oxide CVD are mostly the same as those for the deposition of the metals together with an oxygen source. The most common oxygen sources are O_2, N_2O, or water. Interesting alternatives are often metal alkoxides that combine a (mostly) sufficient and tailorable volatility with the advantage of being single-source precursors, and their use as precursors will be exemplarily discussed for SiO_2.

3.2.4.1 SiO_2 and Silicate Glasses

The formation of SiO_2 thin films is one of the most important processes in industrial microelectronic device manufacturing. SiO_2 thin films are used for gate insulation layers, surface passivation, planarization, and packaging (see Figure 3.15).

Each device within a microcircuit needs to be *isolated* from adjacent devices. This is achieved by the deposition of an insulator material. For this application, → step coverage is an important issue. Borosilicate, phosphosilicate, and borophosphosilicate glasses are also frequently used. They have lower intrinsic stress, lower melting temperatures, and better → dielectric properties than SiO_2 itself. *Passivation* of a → semiconductor is a process by which a film is deposited to protect it from the environment and/or to provide electronic stabilization of the surface by saturating all bonds of the surface atoms. During processing of multilevel electronic devices, the surface becomes increasingly non-planar. In order to allow the subsequent deposition of conducting layers without breaking metal lines, the surface must be flat and smooth. This is achieved by the *planarization* process in which a → dielectric material such as silica or silicate glass is deposited and then etched to give a smooth surface.

The silica layer covering the surface of silicon under ambient conditions is of no technological value. Thicker films can be prepared by heating the silicon in either dry oxygen or water vapor, or by plasma oxidation. However, these methods are not suitable during multistep fabrication, and CVD is used instead.

Silane (SiH_4) as precursor Silane is used widely as a precursor for SiO_2. It is a highly reactive gas that thermally decomposes around 1000 °C to give Si and H_2. Therefore, SiH_4 is also used for the CVD of elemental silicon. SiO_2 is obtained by oxidation of SiH_4 with O_2 or N_2O as most common oxygen sources. H_2O_2 is gaining importance as an oxidant. Growth rates up to $3\,\mu m\,min^{-1}$ have been achieved by rapid thermal LPCVD, which also provides a highly uniform → step coverage. Equation (3.25) oversimplifies the reaction with O_2; the detailed mechanism involves complex branching-chain reactions. Water is formed at high partial pressures of oxygen [Eq. (3.26)].

$$SiH_4 + O_2 \longrightarrow SiO_2 + 2H_2 \tag{3.25}$$

$$SiH_4 + 2O_2 \longrightarrow SiO_2 + 2H_2O \tag{3.26}$$

N_2O is an alternative source of oxygen in the reaction with SiH_4. The overall reaction is given in Eq. (3.27). The reaction is probably initiated by decomposition of N_2O generating N_2 and atomic oxygen and proceeds in a complex sequence of radical reactions.

$$SiH_4 + 2N_2O \longrightarrow SiO_2 + 2H_2 + 2N_2 \tag{3.27}$$

Silicon halides as precursor The most widely used high-temperature LPCVD process for the growth of SiO_2 (~900 °C) is the N_2O oxidation of H_2SiCl_2 [Eq. (3.28)], which is a liquid at room temperature.

$$H_2SiCl_2 + 2N_2O \longrightarrow SiO_2 + 2HCl + 2N_2 \tag{3.28}$$

Tetraalkoxysilane ($Si(OEt)_4$, TEOS) as single-source precursor The high-temperature growth of SiO_2 (~750 °C at ambient pressure and ~600 °C under LPCVD conditions) from TEOS requires no external oxygen source, [Eq. (3.29)],

$$Si(OEt)_4 \longrightarrow SiO_2 + 2C_2H_4 + 2EtOH \tag{3.29}$$

Surface-bound alkoxysilyl species apparently play a very important role and are the immediate source of ethylene, [Eq. (3.30)]. The thus created surface silanol groups (Si–OH) can react with gaseous TEOS by elimination of ethanol to form new surface-bound alkoxysilyl species [Eq. (3.31)]. Note that this reaction is totally different from that of SiO_2 formation from TEOS by hydrolysis and condensation, which will be discussed in Section 4.5.

$$\equiv Si-OEt \longrightarrow \equiv Si-OH + C_2H_4 \tag{3.30}$$

$$\equiv Si-OH + Si(OEt)_4 \longrightarrow \equiv Si-O-Si(OEt)_3 + EtOH \tag{3.31}$$

The deposition temperature in atmospheric pressure CVD is not influenced by the addition of O_2, which, however, removes carbon contaminations from the films. The deposition temperature can be lowered to 300 °C when ozone is added. Furthermore, the quality of the films is improved. For these reasons, the $TEOS/O_3$ system has become widely used.

Boron-containing glass films are grown by using $B(OR)_3$ ($R = Me$, Et, $SiMe_3$)/TEOS mixtures, and $P(OMe)_3$ or $PO(OMe)_3$ are preferred as phosphorus sources. The use of the alkoxides allows for deposition at lower temperatures (500–650 °C) at comparable growth rates than when using the hydrides.

3.2.4.2 **Yttrium Barium Copper Oxide (YBCO)**

Since the discovery of high-temperature → superconductors, major efforts have been made to grow high-quality films by CVD. The discussion in this section is restricted to YBCO (see also Section 2.1.1), but the strategies are similar for other high-temperature superconductors.

For reasons already discussed in Section 3.2.2, β-diketonates are very useful precursors: they are available in high purity for nearly all metal ions, and their volatility can be maximized by preventing molecular association. Saturating the metal coordination sphere with sterically demanding, non-polar or fluorinated groups is an attractive strategy for that purpose. Thus, β-diketonate derivatives of Y, Ba, and Cu are nearly exclusively employed as precursors for YBCO, mostly the dipivaloylmethanate (dpm) derivatives (dpm is also named tetramethylheptanedionate [tmhd or thd]) (Figure 3.22). Other derivatives used for CVD include $Cu(acac)_2$ and β-diketonates with fluorinated substituents.

For the delivery of the precursors in the reaction zone, a non-reactive gas (usually argon) is passed over or through the Ba, Y, and Cu complexes heated to temperatures at which they are sufficiently volatile for transport (e.g., source temperatures for $Y(dpm)_3$ 100–170 °C, $Ba(dpm)_2$ (Figure 3.22) 230–300 °C,

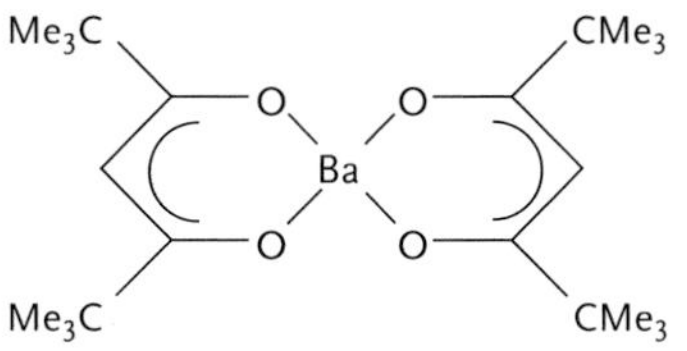

Me3C CMe3 O O Ba O O Me3C CMe3

Figure 3.22 Barium dipivaloylmethanate, $Ba(dpm)_2$.

$Cu(dpm)_2$ 120–160 °C). The metal complex vapors are mixed, and the oxygen source is then added just before entry into the reaction chamber.

With $Y(dpm)_3$, $Ba(dpm)_2$ and $Cu(dpm)_2$ or $Cu(acac)_2$ as precursors, high-temperature decomposition (800–900 °C) in the presence of O_2 or N_2O presumably involves both pyrolysis and oxidation, while hydrolysis is probably also involved when H_2O/O_2 is used.

While Y and Cu precursors fulfill the requirements for CVD precursors such as stability and no degradation during delivery to the deposition zone, significant problems exist with the stability of Ba precursors. This is a consequence of the large size-to-charge ratio of the Ba^{2+} ion. Coordination numbers of up to nine have been observed for Ba compounds. Two β-diketonate ligands balance the charge of the cation, but occupy only four coordination sites. The empty coordination sites must be blocked by Lewis bases in order to prevent intermolecular aggregation. For example, tetraene, triethylenglycoldimethylether (triglyme) or phenanthroline adducts of $Ba(dpm)_2$ have been used to stabilize the $Ba(dpm)_2$ precursor. Another means of solving this problem is to add Lewis bases to the carrier gas, such as ammonia, amines, or ethers.

An important development of YBCO films was the transition to liquid-source delivery systems. Advantages of DLICVD include easier temperature control, better control of vapor pressure, better stability of the precursors, and higher deposition rates. In the liquid delivery system, the $M(dpm)_n$ precursors (M = Y, Ba, Cu) are dissolved in an organic solvent. The solution is then vaporized as described in Section 3.2.1.1.

3.2.5
CVD of Metal Nitrides

Many metal and semimetal nitrides exhibit interesting properties such as high → hardness, high melting points, and high chemical inertness. The electrical properties vary from insulating (e.g., silicon nitride, Si_3N_4) to conducting (e.g., titanium nitride, TiN). Owing to these properties, nitride layers are applied for a variety of purposes. Some aspects of the CVD of Si_3N_4 and TiN are discussed here as case studies.

The deposition of *silicon nitride*, Si_3N_4, is a broadly applied industrial process. Si_3N_4 layers are often used in the microelectronics industry for passivation and encapsulation, because they are a very good barrier to diffusion of water, oxygen, and sodium ions. Additionally, Si_3N_4 is very hard, and chemically resistant.

Titanium nitride (TiN) has a unique combination of properties, including high → hardness, good electrical conductivity, a high melting point (3300 °C), and chemical inertness. Thin films of TiN have found practical uses as wear-resistant and friction-reducing coatings for machine tools. The gold-like color of TiN makes it useful for decorative purposes, for example coatings on jewelry and watches. More recently, TiN has found applications in microelectronics applications as low-resistance contact material (~22 μΩ cm) and as a metal diffusion barrier.

Elemental nitrogen is only rarely used as the nitrogen source for the CVD of nitrides, because it is too unreactive. The most common nitrogen source for nitride films is ammonia, NH_3. This is a cheap gas and easily purified, but it is only sufficiently reactive at high temperatures. Approaches to lower the reaction temperatures in nitride CVD include the use of hydrazine (N_2H_4) as a more reactive nitrogen source and precursors containing nitrogen–element bonds.

An inherent disadvantage of the often-used metal halide/ammonia combinations is the formation of ammonium halides (NH_4X) by reaction of ammonia (which is often used in large excess) and the hydrogen halide (HX) generated as a byproduct [Eq. (3.32)]. This secondary reaction is not explicitly stated in the following examples. Although NH_4X is sublimable, it can lead to chlorine contamination of the deposited films. Since the equilibrium in Eq. (3.32) is shifted to the left side at higher temperatures, and HX are gaseous compounds, the level of chlorine contaminations is lowered at higher temperatures.

$$NH_3 + HX \rightleftharpoons NH_4X \tag{3.32}$$

As in the CVD of SiO_2, the most often employed silicon sources for the CVD of Si_3N_4 are SiH_4 or H_2SiCl_2 [Eq. (3.33)], and, more recently, Si_2Cl_6. In commercial systems, a large excess of NH_3 is used to obtain films with a stoichiometric composition.

$$3H_2SiCl_2 + 4NH_3 \longrightarrow Si_3N_4 + 6HCl + 6H_2 \tag{3.33}$$

Only Ti(IV) compounds are sufficiently volatile to be suitable as CVD precursors for TiN. The metal atom in TiN is in the formal oxidation state +III, Thus, other than in the analogous reactions leading to Si_3N_4 (e.g. Eq. (3.33)), the overall process of TiN deposition also involves the reduction of the metal atom from +IV to +III.

While the combination $TiCl_4/N_2/H_2$ [Eq. (3.34)] requires temperatures >700 °C for high-quality films, the use of NH_3 [Eq. (3.35)] reduces the necessary temperature to between 320 and 700 °C. In the latter reaction, ammonia has a dual role: it delivers the nitrogen for TiN, and it acts as a reductant. Therefore, nitrogen is formed as a byproduct.

$$2TiCl_4 + 4H_2 + N_2 \longrightarrow 2TiN + 8HCl \tag{3.34}$$

$$6TiCl_4 + 8NH_3 \longrightarrow 6TiN + 24HCl + N_2 \tag{3.35}$$

Thermodynamic calculations suggest that $TiCl_3$ may be the actual precursor in the $TiCl_4/N_2/H_2$ system. $TiCl_3$ is probably formed by gas-phase reaction between $TiCl_4$ and H_2, and is then adsorbed to the substrate surface where it reacts with chemisorbed hydrogen and nitrogen.

A very important development with respect to the nitride deposition are transamination reactions between metal dialkylamides and ammonia. Metal dialkylamides are relatively easy to prepare and to handle. Transamination reactions provide a means of producing the corresponding amides (with M–NH_2 groups) *in situ*, which readily undergo condensation reactions to give M–N–M linkages (Eq. (3.36), see also Section 5.6). Examples are given in Eqs. (3.37) and (3.38).

$Si(NMe_2)_4$ (b.p. 196 °C), $HSi(NMe_2)_3$ (b.p. 142 °C) or $H_2Si(NMe_2)_2$ (b.p. 93 °C) are good precursors for transamination reactions with ammonia [Eq. (3.37)]. In TiN CVD, $Ti(NMe_2)_4$ and $Ti(NEt_2)_4$ are primarily employed because of their commercial availability. $Ti(NMe_2)_4$ is a liquid at room temperature with a vapor pressure of 1 mbar at 60 °C, and $Ti(NEt_2)_4$ has a vapor pressure of about 0.1 mbar at 100 °C. TiN films have been deposited from both precursors by transamination reactions with NH_3 [Eq. (3.38)] at temperatures as low as 450 °C with low impurity levels.

$$\equiv M{-}NR_2 + NH_3 \longrightarrow \equiv M{-}NH_2 + HNR_2$$

$$3 \equiv M{-}NH_2 \longrightarrow \begin{array}{c} M{-}N{-}M \\ | \\ M \end{array} + 2NH_3 \qquad (3.36)$$

$$3Si(NMe_2)_4 + 4NH_3 \longrightarrow Si_3N_4 + 12HNMe_2 \qquad (3.37)$$

$$6Ti(NR_2)_4 + 8NH_3 \longrightarrow 6TiN + 24HNR_2 + N_2 \qquad (3.38)$$

3.2.6 CVD of Compound Semiconductors

The majority of important → semiconducting compounds are isoelectronic with elemental silicon, and they have related structures. These include combinations of groups III and V such as GaN, GaAs, or InP ("III–V compounds"), or combinations of groups II and VI such as CdS or ZnSe ("II–VI compounds"). The bandgaps of common semiconductors are shown in Figure 3.23. They are in "useful" regions of the electromagnetic spectrum. For example, those in the visible region can be used for displays and solar cells, and those in the infrared region for thermal imaging technologies.

The bandgap energies of II–VI compounds are larger than those of the III–V compounds. Thus, these materials are potentially important for the emission,

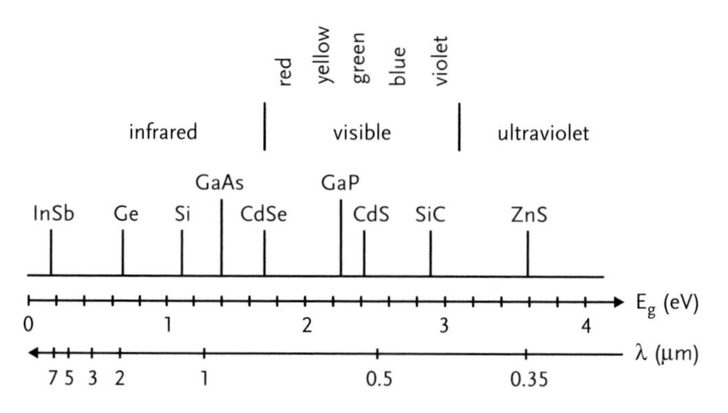

Figure 3.23 The bandgaps of common semiconductors.

detection or modulation of light in the green to near-UV. Progress in II–VI semiconductor applications has been hampered by inherent disadvantages, which include the difficulty to grow high-quality single crystals, problems associated with polytypism (cubic and hexagonal modification), and the reproducible control of the conductivity. However, new technologies are being developed and may open new applications. Thin-film electroluminescent displays based on ZnS are already commercially available. Applications for Cd → chalcogenides (CdS, CdSe, CdTe) include solar cells, photoconductors, sensors, and transducers.

Commercial applications of III–V materials are very well developed. Examples include solar cells, light-emitting diodes (LED), solid-state lasers and electronic devices (electrons can move with a higher velocity through GaAs than through Si, which allows the manufacturing of faster electronic devices).

The energy gap (and the lattice parameters) can be engineered by variation of the composition (→"semiconductor alloys"). A typical example is $In_{1-x}Ga_xN$ used for blue and green LEDs.

Two devices for light generation have already been discussed in Section 3.1: incandescent lamps and metal halide lamps. The use of LEDs is a technology of increasing importance. Light is created in a semiconductor layer of these devices by the recombination of positive and negative charge carriers. A high crystallinity and a low density of defects are required for a high quantum yield. The color of the LED is determined by the bandgap energy of the semiconductor material. The bandgaps of phosphides and arsenides are suitable for red and yellow LEDs, but they are too small to generate blue or green emissions. The latter are also required for white light, where a blue LED is coated with a phosphor, for example, Ce-doped yttrium aluminum garnet (YAG, $Y_3Al_5O_{12}$), that emits yellow light. The break-through was achieved in 1994 by the development of stable blue $In_{1-x}Ga_xN$ on a sapphire substrate. The emission can be shifted between 0.37 and 0.58 μm by varying the In proportion; an increasing In proportion shifts the emission maximum to higher wavelengths.

Figure 3.24 shows the structure of a blue-green LED device. The structure consists of a 30-nm GaN buffer layer grown at low temperatures (550 °C) on a sapphire substrate. This buffer layer is required to match the lattice parameters of sapphire and GaN. This is followed by a 4-μm layer of Si-doped GaN (n-type GaN) grown at 1010 °C and the active layer of $In_{0.45}Ga_{0.55}N$ (3 nm). On top of the active layer are a 100-nm p-type $Al_{0.2}Ga_{0.8}N$ barrier layer and a 500-nm layer of Mg-doped GaN (p-type GaN). The layers are grown by MOCVD using Al_2Me_6, $GaMe_3$, $InMe_3$, ammonia, SiH_4, and Cp_2Mg, respectively.

The chemistry of compound semiconductor CVD is exemplarily discussed for GaAs. The principles of the precursor and deposition chemistry for other III–V and II–VI combinations (the nitrides have been discussed in Section 3.2.5) are very similar.

The most commonly employed precursors for GaAs OMVPE or CBE are the commercially available trimethylgallium ($GaMe_3$) and arsine AsH_3. Typical growth temperatures are 600–750 °C. The deposited GaAs layers have remarkably low levels of carbon, considering that organometallic compounds are used.

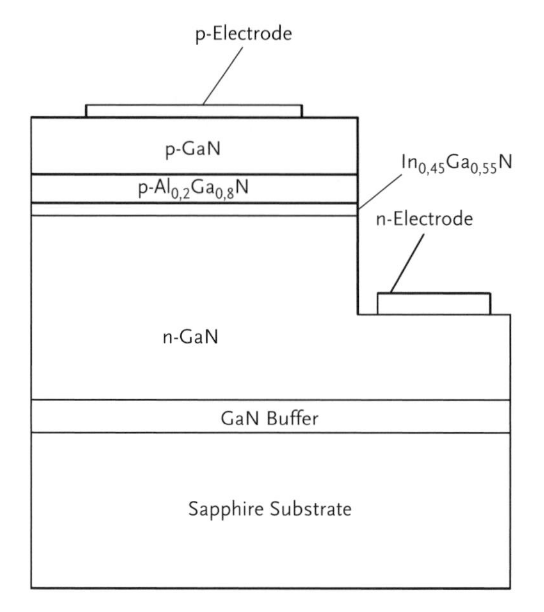

Figure 3.24 Structure of a blue-green LED device. See text for explanations.

A mechanism has been proposed that involves both gas-phase and surface reactions. Pyrolysis reactions occur in the boundary layer which primarily lead to $GaMe_x$ species ($x < 3$, mainly $x = 1$) by abstraction of methyl radicals. The methyl radicals can react with AsH_3 in the gas phase to give methane and AsH_2 [Eq. (3.39)]. The formation of these species lowers the decomposition temperature of AsH_3. The gas-phase formation of the adduct $Me_3Ga–AsH_3$ has no significant contribution to the GaAs growth mechanism due to the weakness of the Ga–As interaction. After adsorption of AsH_2 and the $GaMe_x$ species, the remaining methyl group(s) are removed by surface reactions. Hydrogen is transferred from the AsH_x species to surface-adsorbed methyl groups, and the formed methane is desorbed. This clean removal of carbon-containing groups from the surface leads to layers with very little carbon incorporation. Similarly, high-purity InP can readily be grown from $InMe_3$ and PH_3.

$$GaMe_x + AsH_y \longrightarrow GaMe_{x-1} + AsH_{y-1} + CH_4 \quad (3.39)$$

$GaEt_3$ is a less convenient precursor than $GaMe_3$, because of its lower vapor pressure (4 mbar at 20 °C compared with 243 mbar for $GaMe_3$). However, it has been used successfully in combination with AsH_3 at low pressures. The pyrolysis chemistry of $GaEt_3$ is different from that of $GaMe_3$. As expected, the major decomposition route is by β-elimination which yields C_2H_4 together with Et_2GaH and $EtGaH_2$. These species are thermally unstable and, when formed in the boundary layer, can lead to the premature formation of Ga and GaAs. Therefore, GaAs layers deposited by using $GaEt_3$ are generally less uniform than those obtained from $GaMe_3$. However, they contain even less residual carbon due to the "cleaner" decomposition mechanism (see also discussion in Section 3.2.2.1, the related chemistry of Al compounds).

Efforts are being made to replace the highly toxic and gaseous AsH_3 with less hazardous liquid precursors, which may also be more easily purified and pyrolyze at lower temperatures. Since hydrogen is needed to remove the methyl radicals from the growth surface, the most promising layer properties were obtained with precursors containing both alkyl and hydrogen substituents. The most successful alternative to AsH_3 to date is tert-butylarsine ($^tBuAsH_2$). This is a liquid with a vapor pressure of 108 mbar at 10 °C, and it is considerably less hazardous than AsH_3. Furthermore, the pyrolysis of $^tBuAsH_2$ creates more active AsH_2 or AsH species near the surface. Dominant decomposition reactions of $^tBuAsH_2$ in the gas phase are dissociation in $Bu^•$ and $AsH_2^•$ radicals, and elimination of butane with the concomitant formation of AsH species.

A wide variety of III–V semiconducting compounds have been deposited from single-source precursors. Potential advantages include a higher air and moisture stability, a lower toxicity and less prereactions in the gas phase. The advantages are somewhat balanced by a lower volatility, a more difficult control of the stoichiometry, particularly for semiconductor alloys (e.g., $Al_xGa_{1-x}As$), and a lower surface mobility of the → polynuclear decomposition fragments. The most widely investigated III–V single source precursors are dimeric compounds $[R_2GaAs^tBu_2]_2$ (R=Et, ^{t}Bu) as in Eq. (3.10). An important feature is that the 1 : 1 stoichiometry is retained during pyrolysis.

A particular problem in semiconductor CVD is the purity of the precursors. Since the semiconducting properties of a material are strongly influenced by impurities, very pure precursors must be applied. Special purification techniques have been developed for this purpose. These include not only classical techniques such as distillation, sublimation, or chromatography, but also chemical techniques such as "adduct purification". Group II and III organometallic compounds are Lewis acids, and therefore form adducts with Lewis bases such as ethers, amines, or phosphines. The donor–acceptor bond is usually the most labile in the molecule. Although the adducts can be purified intact by distillation or recrystallization, the donor–acceptor bond can be thermally cleaved at moderate temperatures, liberating the purified organometallic precursor. For example, the bis-adducts between the trialkyl compounds of Al, In or Ga (MR_3) and 4,4′-methylene-bis(N,N′-dimethylaniline) or 1,2-bis(diphenylphosphino)ethane (diphos) (Figure 3.25) are already formed at room

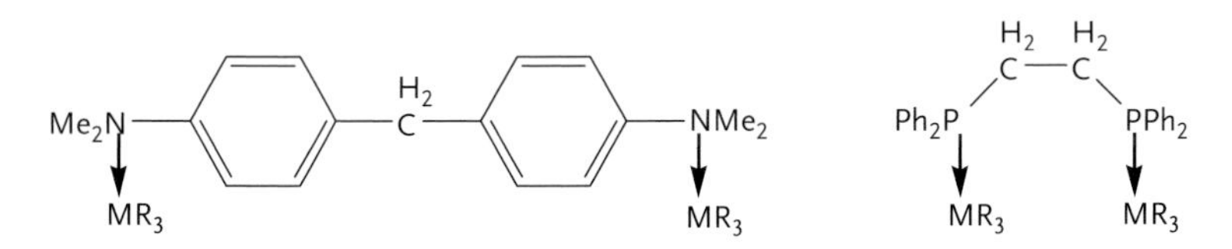

Figure 3.25 4,4′-Methylene-bis(N,N′-dimethylaniline) (left) and 1,2-bis(diphenylphosphino) ethane (right) as Lewis bases in donor–acceptor complexes.

temperature. Traces of Si, Sn, Zn, or Mg compounds do not form such adducts and are thus removed by purification of the adducts. The diphos adducts liberate the pure MR_3 on heating to $>80\,°C$ at 10^{-2} mbar.

3.2.7 Chemical Vapor Infiltration

In CVD processes, solid films are deposited onto a surface. In contrast, in chemical vapor infiltration (CVI) the source gases are flowed through a porous preform at high temperature, during which material is deposited as a matrix in the empty spaces. An example for the latter was already given in Section 2.1.3 (CVI of porous carbon with gaseous SiO, see Figure 2.11). The experimental conditions must be chosen to favor in-depth deposition, that is, to fill up the pores totally. Thus, clogging of the pore entrances must be avoided. CVI offers all the advantages already discussed for CVD processes.

Chemical vapor infiltration is widely used for fabricating fiber- or particle-reinforced → composites, especially ceramic matrix composites (CMC) or carbon–carbon (C/C) composites. Such thermostructural composites have excellent mechanical properties, such as hardness or stiffness, at high or very high temperatures. C/C composites are mainly applied to rocket nozzles and aircraft brake-disks, while CMCs such as C/SiC and SiC/SiC composites are developed for aircraft engines and related applications. C/SiC composites with carbon-fiber reinforcement offers the high-temperature strength and toughness for structural and heat shield components required for aerospace structures.

Any macro- or mesoporous solid, such as solid foams or aerogels, as well as particle agglomerates can also be infiltrated. Fibrous preforms may be cloths, felts or stacked fabric layers of ceramic or carbon fibers, but also carbonized wood, cotton, or paper.

When gaseous precursors are passed through the preforms at high temperatures, the same chemical reactions as have been discussed in the previous sections result in the deposition of solid material in the empty space of the preform. For example, TiN/C composites were obtained by CVI of highly porous carbon substrates with a mixture of gaseous $TiCl_4$, N_2, and H_2 at about 850 °C [Eq. (3.34)]. SiC matrix composites are made on an industrial scale from CH_3SiCl_3/H_2 mixtures at about 1000 °C, and carbon matrix composites by pyrolysis of kerosene (mixture of liquid hydrocarbons).

The standard CVI process proceeds at isothermal and isobaric conditions. To this end, the preforms are put in a hot-wall pressure CVD reactor (see Figure 3.7 and Section 3.2.1.1) fed with the gaseous precursor(s) of the desired matrix. This is a relatively slow process. The obtained materials have some residual porosity and spatial density gradients, the highest density being close to the surface. A compromise must be achieved between deposit uniformity and infiltration rate. Despite some drawbacks industry uses this process because of the following advantages:

- It requires a rather simple technology.
- The reinforcing capability of fibers is retained owing to the relatively low processing temperatures.

- The nature of the matrix can be easily modified by changing the precursors. For example, first coating a fibrous preform with one kind of ceramic material and then depositing a ceramic matrix of another material is feasible.
- Different preforms, which may have different sizes and complex shapes, can be densified in the same run.

Nevertheless, several derived techniques were developed to overcome the drawbacks of isothermal isobaric CVI.

In the *forced CVI* technique, the precursor gas is injected at a pressure p_1 through one side of the porous preform, whereas the exhaust gas is pumped off at a pressure p_2 ($p_2 < p_1$) at the opposite side. The infiltration time is lowered from several hundred hours to a few tens of hours by applying this pressure gradient. However, the deposit may have a density gradient. A disadvantage of this technique is that it is only suited for simple shapes, such as disks or tubes, where a pressure gradient can be easily applied.

In the *pulse CVI* technique, a total pressure cycling is applied in order to periodically regenerate the entire gas phase. To this end, the gaseous precursor is injected rapidly. After holding at the desired pressure and residence time, the gas is pumped off again. This cycle is repeated several times.

3.3 Aerosol Processes

CVD processes (Section 3.2) are controlled in a way that the gaseous precursors are adsorbed to the substrate surface and there form the solid product by thermal reactions. Gas-phase reactions and particle formation in the gas phase are usually undesirable. Contrary to that, in gas-phase powder syntheses – also called → aerosol processes – particles are produced in the gas phase by chemical or physical processes. The advantages of aerosol processes, compared to other processes, are that:

- they do not involve the large volumes of liquids as in wet processes;
- their time scales are much shorter than those for solid–solid reactions;
- they can produce materials of high purity at high yields and with a high throughput; and
- multicomponent or nanophase materials (see Chapter 7) can be produced.

The example we will start with is the Aerosil® process for the production of fine powders of inorganic compounds. We will then treat some general aspects of the two major routes for aerosol processes: the gas-to-particle conversion route, and spray pyrolysis. After a short excursion to reactor types we will then turn to products that can be produced by aerosol processes.

The use of gas-phase reactions to produce titania, silica, or carbon black (soot) with nanometer-size particles is well established. Several million tons of products per year are produced worldwide. A typical example is highly dispersed amorphous silica. The so-called Aerosil® process was patented in 1942 by Degussa.

The chemical reactions [Eq. (3.40)] explain why this process is also called *flame hydrolysis.*

$$\begin{array}{c} 2H_2 + O_2 \longrightarrow 2H_2O \\ SiCl_4 + 2H_2O \longrightarrow SiO_2 + 4HCl \\ \hline \text{overall: } SiCl_4 + 2H_2 + O_2 \longrightarrow SiO_2 + 4HCl \end{array} \tag{3.40}$$

Silicon tetrachloride (obtained by reaction of elemental silicon with HCl) is volatilized and fed into an oxygen–hydrogen flame (Figure 3.26). The water formed by reaction of hydrogen and oxygen serves for the very fast and quantitative hydrolysis of $SiCl_4$ at about 1000 °C.

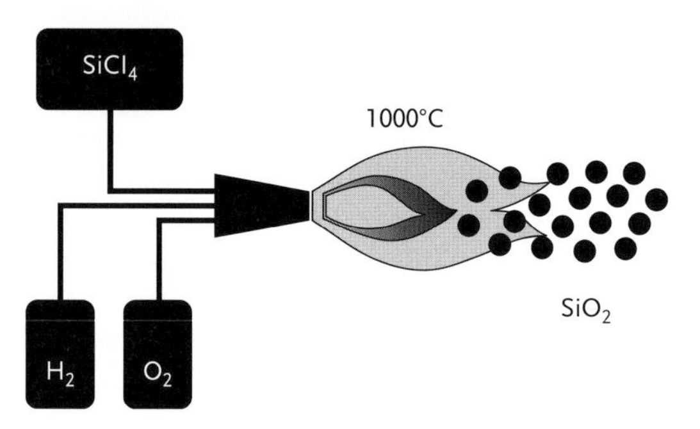

Figure 3.26 Formation of silica by the Aerosil process.

The only byproduct is HCl, which is separated and recycled. The formed fumed silica is cooled, collected by conventional means (e.g., by cyclones, electrostatic precipitators, or baghouse filters) and deacidified to remove adsorbed residual HCl. The surface properties of the silica can be modified by post-treatment with various silanes.

Highly dispersed titania, which is mainly used as pigments, is prepared by the same procedure. Fumed titania and silica are major chemical products. They are very light powders – a 200-liter bag filled with fumed silica can be easily lifted, because it weighs only about 10 kg! Its low weight is paralleled only by aerogels (see Section 4.5.6).

Fumed silica consists of agglomerated spherical, amorphous primary particles of only 7–40 nm diameter (Figures 3.27 and 3.28). The primary particle size can be influenced by the reaction parameters. Owing to the small primary particles, the specific surface area (S) is high (50–400 $m^2\ g^{-1}$). For spherical, unagglomerated particles of diameter d and density ρ, the specific surface area is given by

$$S = 6/(d \cdot \rho) \tag{3.41}$$

A material composed of discrete small particles will possess a higher specific surface area than the same material in a coarser form (see Figure 2.5), because diameter and surface area are inversely related. In contrast to aerogels, the high

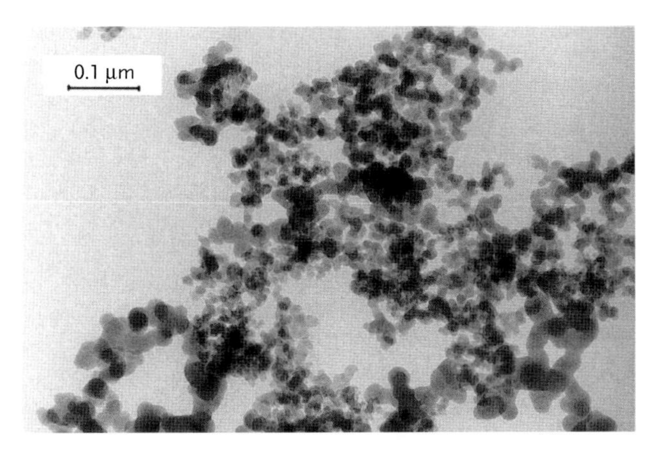

Figure 3.27 Transmission electron micrograph of fumed silica with a primary particle size of 16 nm and a specific surface area of $130 \pm 25\,m^2\,g^{-1}$ (Aerosil® 130 of Evonik Degussa).

surface area of pyrogenic silica originates only from the outer surface, that is, it is not caused by the presence of pores.

There are many uses for pyrogenic silica. It was originally developed as "white soot" as a reinforcing →filler for rubber, but now has found manifold use for

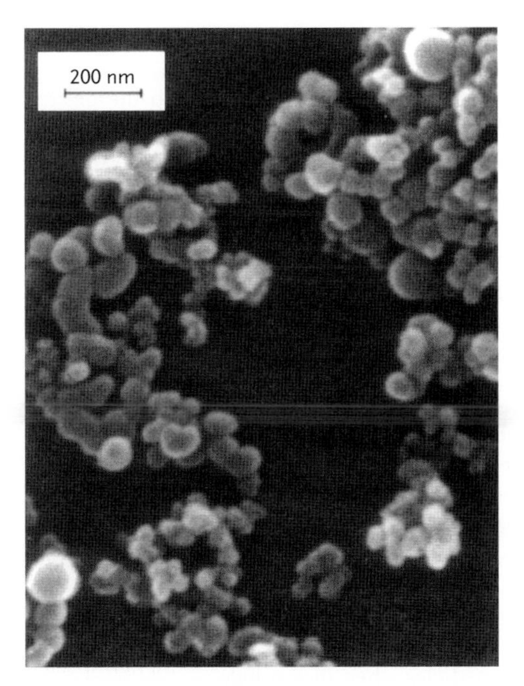

Figure 3.28 Scanning electron micrograph of fumed silica with a primary particle size of 40 nm and a specific surface area of $50 \pm 15\,m^2\,g^{-1}$ (Aerosil® OX 50 of Evonik Degussa).

various applications. For example, all types of liquids are thickened by the addition of pyrogenic silica. Many of these formulations show → thixotropic behavior. This behavior is explained by the formation of a three-dimensional network by hydrogen bonding between the surface OH groups of the silica particles, either directly or via molecules of the liquid. This causes an increase of viscosity. When mechanical → stress is applied, the network is degraded and the viscosity decreases. Influencing the → rheology of a system by addition of pyrogenic silica is used, for example, in drilling fluids, lacquers, paints, plastics, adhesives, greases, creams, ointments, or toothpastes. An additional effect is that the addition of pyrogenic silica prevents or at least slows down the sedimentation of solids in dispersions, such as pigments in lacquers. Pyrogenic silica is also used as a → filler in silicones (see Section 5.2) and other elastomers, as adsorbent or support, to improve the storage stability and free flow of powders (fire-extinguisher powders, tablets, cosmetic powders, toners, table salt, and so on), or to control the → triboelectric properties of powders. It has excellent heat-insulation properties, and is therefore used for thermal insulation of ovens, furnaces, pipelines, heating elements, or turbines.

3.3.1 Aerosol Process Routes

There are two major routes for aerosol processes:

1. **Gas-to-particle conversion route**: Powders are made at high temperatures by reacting gases or vapors.
2. **Spray pyrolysis**: Precursor particles or droplets are converted to the desired product powder either by reaction with gaseous species or by pyrolysis. Such processes are also called particle-to-particle conversion, evaporative decomposition, spray roasting, or aerosol decomposition. The characteristics of the product powders made by this route are distinctly different from those made by gas-to-particle conversion. Although the solid powders are actually not formed from the gas phase, this route is nevertheless dealt with in this section, because of the similarity of the *physical* processes and the reactor operation with that of the gas-to-particle conversion route. The *chemical* processes converting the droplets or particles to the products are dealt with in the relevant sections.

The precursors used for aerosol processes are often the same as in CVD and PVD processes. Examples will be given below. Since mainly commodities are produced, the precursors for aerosol synthesis must be inexpensive and convenient to use. In CVD processes, relatively low partial pressures can yield adequate film deposition rates, whereas powder synthesis reactors require substantial precursor partial pressures to be economical.

Some technical terms associated with aerosol processes include:

Coagulation: attachment of two particles when they collide.
Coalescence: fusion (sintering, condensation) of two particles.

Agglomerates: assemblies of primary particles physically held together by weak interactions. They are also called "soft agglomerates," because they are easier to break up.

Aggregates: assemblies of primary particles connected by the stronger chemical bonds. They are also called "hard agglomerates", because they make powder sintering and consolidation more difficult.

Gas-to-particle conversion In this route (Figure 3.29), mixtures of gaseous precursors are fed into the reactor. They react at high temperatures to form molecular clusters and eventually ultrafine particles of the product. The particles then form aggregates and agglomerates of solid powder. Having passed the reaction zone, the powders are collected. Additional postprocessing is sometimes required to produce high-purity powders.

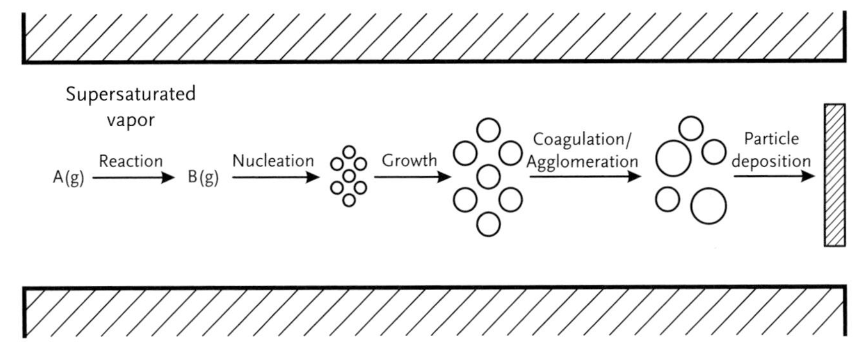

Figure 3.29 Particle generation by gas-to-particle conversion.

The formation of the powders proceeds in several steps (see also Sections 4.1.2 and 4.2 for analogous processes in melts and solutions):

1. **Homogeneous gas-phase reactions.** between precursor gases result in the formation of molecular or cluster (oligomeric) compounds.
2. **Nucleation.** A supersaturated vapor is inherently unstable and tends to form the condensed phase. The initially formed molecules and clusters will form particles by → homogeneous nucleation. If the clusters are thermodynamically stable, nuclei are formed by coagulation. In the case of unstable clusters, the nuclei are formed by balanced condensation and evaporation of molecules or atoms to and from the clusters until a thermodynamically stable particle is formed. Homogeneous nucleation will be discussed in detail in Section 4.1.2.
3. **Particle growth.** The nuclei grow by several mechanisms, including condensation of precursor molecules on the particle surface and coagulation. These processes strongly influence the properties of the product particles. If the rate of particle collision is faster than that of their coalescence, then non-spherical (agglomerate, aggregate) particles are formed. If the sintering rate is faster than the particle collision rate, then regularly shaped, monolithic (dense) particles are formed. The sintering rate depends strongly on the material, primary particle size, and temperature.

When the particles grow by condensation of precursor molecules on the particle surface, the involved processes are similar to CVD processes. The precursor must first be adsorbed to the surface, and the byproducts must be desorbed after formation of the product, and so on. If the rate of reaction is faster than the rate of vapor transport to the particle surface, the latter controls the rate of particle growth (similar to diffusion or mass-transport-limited processes in CVD). The reverse situation (rate of chemical reaction is slower than that of vapor transport) is similar to reaction-limited processes in CVD.

Powders made by the gas-to-particle conversion route are usually aggregates or agglomerates of fine, non-porous primary particles. Temperature, reactor residence time and chemical additives affect the particle sizes and size distributions, extent of agglomeration or aggregation and, consequently, powder morphology. Short reactor residence times (higher flow rates) lead to smaller primary particles, since growth takes place for a shorter time in the reaction zone. Higher reaction temperatures tend to have the same effect, perhaps by increasing the nucleation rate, and hence decreasing the amount of reactant available for each particle to grow.

The difference between soft and hard agglomerates is frequently due to the temperature at which the particles aggregate. If the aggregation takes place at the high temperatures during the synthesis, interparticle diffusion will lead to strong bonds between the primary particles. If aggregation can be suppressed until the temperature has decreased sufficiently, then readily dispersed agglomerates may be formed. Other mechanisms such as vapor deposition into the necks between aggregated primary particles may also contribute to the formation of hard agglomerates. This process also takes place primarily in the high-temperature region of the reactor.

The gas-to-particle conversion route is most suitable for the synthesis of single-component, high-purity powders of small particle size, high specific surface area, and controlled particle-size distribution. Its major disadvantage is that it results in aggregates or agglomerates that can lead to problems in consolidation and sintering during fabrication of large ceramic parts.

Multicomponent powders are more difficult to synthesize by the gas-to-particle conversion route because of differences in vapor pressure, → nucleation and growth rates of the various components that can lead to non-uniform product composition. For this reason, coated particles or particles with varying composition from particle to particle are often obtained. As in CVD, single-source precursors have been used. The advantages and disadvantages of such precursors were discussed in Section 3.2.

Spray pyrolysis In this route (Figure 3.30), a solution or slurry is atomized, or solid precursor powders are suspended in a carrier gas. The → aerosol is passed through a heated region. Inside the furnace the solvent evaporates and the particles are pyrolyzed or reacted with a gas to yield the product powder, which is then removed from the process stream by conventional means. The size of the product particles is proportional to that of the aerosol droplets or particles.

Most work has been done for the aerosol synthesis of oxide powders. With improved syntheses of polymeric precursors for ceramic materials (Section 5.5.2),

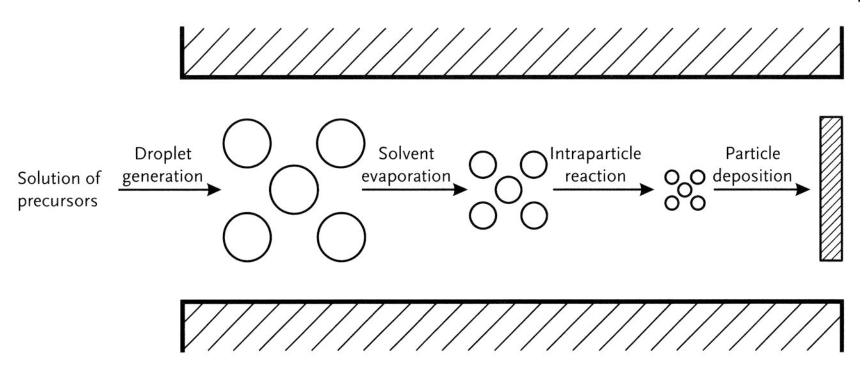

Figure 3.30 Particle generation by spray pyrolysis.

spray pyrolysis (in the presence or absence of a reactant gas) could become a viable means for production of non-oxide powders. Multicomponent powders are easier made by spray pyrolysis than by the gas-to-particle conversion route. Each droplet contains precursors in the same stoichiometry as in the product.

The first step in spray pyrolysis is the suspension or atomization process. This can be carried out for liquid precursors using a variety of atomizers depending on the rheological characteristics of the liquid or the required size of the atomized droplets. In general, atomization results in broad droplet-size distributions. There are limited options for suspension of solids without assistance of a liquid-phase carrier (atomization of slurries).

Particles obtained by this route have a high purity, and they are generally amorphous, unagglomerated, and have a monolithic, spherical morphology. A major advantage of spray pyrolysis is that it can be easily scaled up.

A limitation of this route is that hollow and porous material is easily formed. It is possible to control the porosity of powders by changing the precursor concentration in the droplets and the reactor temperature profile.

Hollow particles can be formed when a solute concentration gradient is created during evaporation (Figure 3.31). The solute precipitates first at the more highly supersaturated surface if there is not enough time for solute diffusion in the droplet. If the crust thus formed is impermeable to solvent, exploded particles may result when the pressure within the particle builds upon further heating.

3.3.2 Reactors

One of the most important aspects of gas-phase powder synthesis is the design and operation of the units that provide the high temperatures needed for gas- or particle-phase reactions.

Flame reactors These employ the combustion of hydrocarbons or hydrogen (see Aerosil process above), or the reaction of hydrogen and chlorine.

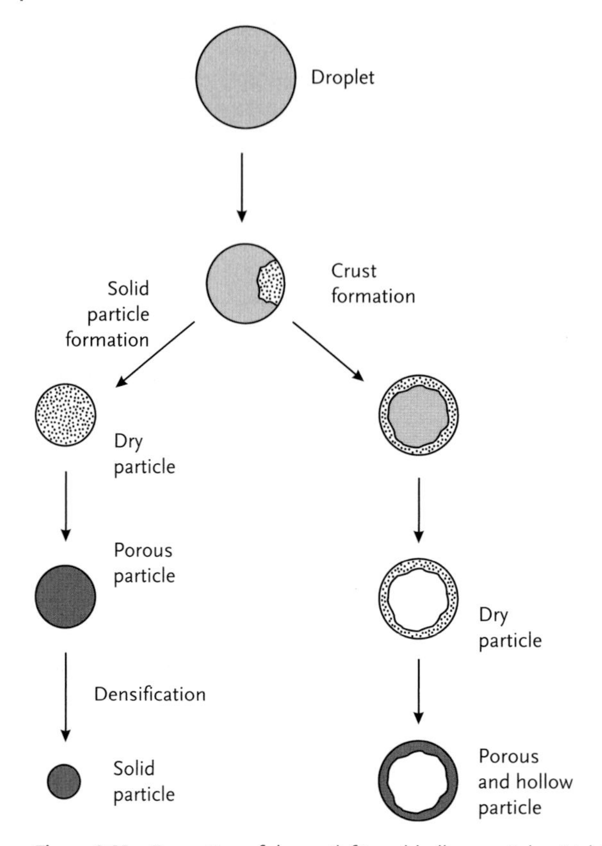

Figure 3.31 Formation of dense (left) and hollow particles (right) by spray pyrolysis.

They are attractive for particle generation as they are easy to construct and operate. They only need to bring fuel and oxidant into contact, and make the maximum use of the heating energy. However, the product powder may be contaminated because of its direct contact with the combustion reactants and flue gases. A unique characteristic of flame reactors is the short residence time in the high-temperature region. As a result, the flame-produced powders are typically aggregates of fine, non-porous primary particles with a relatively narrow size distribution. A disadvantage of powders made in flames is that the agglomerates are often hard to break up and have a broad size distribution. The temperature profile of the flame, additives (to modify powder phase composition, morphology and size) and residence time determine the crystallinity, phase composition and extent of aggregation.

Furnace reactors These are externally heated metallic or ceramic tubes through which the precursor gases are flowing. The closed environment offers significant advantages, because these reactors provide excellent control over temperatures and residence times, and hence over the product particle characteristics. Two of the key problems associated with externally heated reactors are the loss of product powder by deposition on the reactor walls and the formation of hard agglomerates. To prevent extensive agglomeration and further growth of the agglomerates, the submicrometer powders formed in the reaction zone must be removed from the hot and possibly still reacting gases and must be effectively quenched. One method is to expose the product particles to a large volume of inert gas, or to cool the gases containing the particles by sudden expansion. Another effective technique is to limit the length of the reaction zone and rapidly withdraw the submicrometer particles from the reactor.

Laser reactors These utilize the energy of lasers. The aerosol synthesis of powders using high-temperature furnaces involves relatively slow heating of the reactant gases by convection or radiation. Lasers improve the efficiency of the energy transfer to gaseous reactants. Nucleation at the walls is avoided, because the premixed gases react in a small, well-defined zone where the focused laser beam intersects the reactant jet and induces a rapid increase in temperature in the gas stream. Reactants are dilute, and hence give fine, loosely agglomerated powders. In contrast to the thermal method, the steep temperature gradients at the reaction zone allow precise control of particle nucleation and growth rates. It is necessary for one of the reactants to absorb the laser radiation strongly. For example, SiH_4 or BCl_3 strongly absorbs CO_2 laser radiation. If none of the reactants is absorbing, a sensitizer may have to be added to the gas mixture, but these are a potential source of impurities in the product material. Laser-heated reactors have been used primarily for non-oxide ceramics. Powders produced via laser reactions tend to be of high purity, controlled stoichiometry, and uniform particle size.

Plasma reactors The energy of highly ionized gases is transferred to molecules participating in chemical reactions. In plasmas, a significant proportion of the particles are ionized, although overall positive and negative charges compensate each other. The plasmas are generated by ionizing a flowing gas such as argon, either using DC or low-frequency AC, with the electrodes in direct contact with the gas, or by inductively coupling a radio-frequency (r.f.) source to it. The r.f. plasma technique avoids the use of electrodes, which may be sources of product contamination. Solid reactants are used in most plasma syntheses. They are injected into the discharge zone, where vaporization occurs. Since the reactants are heated quickly to high temperatures and quenched rapidly, plasma methods produce very finely divided materials with high specific surface areas.

3.3.3 Products

Large-scale industrial aerosol processes are mainly used for the preparation of highly dispersed silica, titania, and alumina powders from the corresponding chlorides ($SiCl_4$, $TiCl_4$, $AlCl_3$) by flame hydrolysis, as described earlier in this section. On a smaller scale, other oxides are also produced by the Aerosil process, such as Bi_2O_3 from $BiCl_3$, Cr_2O_3 from CrO_2Cl_2, Fe_2O_3 from $FeCl_3$ or $Fe(CO)_5$, GeO_2 from $GeCl_4$, NiO from $Ni(CO)_4$, MoO_2 from $MoCl_5$, SnO_2 from $SnCl_4$ or $SnMe_4$, V_2O_5 from $VOCl_3$, WO_3 from WCl_6 or $WOCl_4$, ZrO_2 from $ZrCl_4$, $AlBO_3$ from $AlCl_3$ and BCl_3, Al_2TiO_5 from $AlCl_3$ and $TiCl_4$, and $AlPO_4$ from $AlCl_3$ and PCl_5. Other reactor types – and particularly also the spray pyrolysis method – have been used, particularly for metal oxides containing two or more metallic species (→ spinels, high-temperature → superconductors, magnetic oxides).

The increased demand for high-purity, non-oxide powders has led to the development of novel routes for their manufacture. Aerosol methods have therefore been employed in the synthesis of various non-oxides such as nitrides, carbides, borides, or silicides (Table 3.3). Even metal powders can be prepared, when the gaseous halides are reacted with the vapors of reactive metals [Eqs. (3.42) and (3.43)]. Many of these processes are variations of similar ones developed for the

Table 3.3 Examples of non-oxide powders prepared by aerosol methods.

Product	Method	Reactants
Carbides		
B_4C	Plasma/Laser	$BCl_3 + CH_4$
B_4C	Thermal	$B_2O_3 + C$
SiC	Laser/Plasma/Thermal	$SiH_4 + CH_4$
SiC	Plasma	$SiCl_4$ or SiO_2 or $SiO + CH_4$
SiC	Plasma/Thermal	$SiMe_4$
SiC	Laser/Plasma	$SiO_2 + C$
SiC	Laser	$H_2SiCl_2 + C_2H_4$
TiC	Plasma/Thermal	$TiCl_4 + CH_4$
Mo_2C	Thermal	$MoCl_5$ or $MoO_3 + CH_4$
WC, W_2C	Plasma/Thermal	WCl_6 or $W + CH_4$
Nitrides		
BN	Laser/Thermal	$BCl_3 + NH_3 + N_2$
AlN	Plasma/Thermal	$Al + N_2$ [$+NH_3$]
AlN	Thermal	$AlCl_3 + NH_3 + H_2$
AlN	Thermal	$Al_2Et_6 + NH_3$
Si_3N_4	Laser/Plasma/Thermal	$SiH_4 + NH_3$
Si_3N_4	Laser/Plasma/Thermal	$SiCl_4 + NH_3 + H_2$
Si_3N_4	Plasma	$Si + NH_3$ or N_2
SiAlON	Plasma	$Si + Al + NH_3 + O_2$
TiN	Plasma	$Ti + N_2$ [$+H_2$]
TiN, ZrN	Thermal	$TiCl_4/ZrCl_4 + NH_3 + N_2 + H_2$
VN_x	Thermal	$VCl_5 + NH_3 + N_2 + H_2$
Borides/Silicides		
B_4Si	Plasma	$B_2H_6 + SiH_4$
$TiSi_2$	Laser	$TiCl_4 + SiH_4$
TiB_2	Laser	$TiCl_4 + B_2H_6$
TiB_2	Thermal	$TiCl_4 + BCl_3 + Na$ or H_2
WSi_2	Plasma	$WF_6 + SiH_4$

synthesis of oxide particles. However, oxygen contamination must be prevented in all process stages.

$$SiCl_4(g) + 4Na(g) \longrightarrow Si + 4NaCl \tag{3.42}$$

$$2NbCl_5(g) + 5Mg(g) \longrightarrow 2Nb + 5MgCl_2 \tag{3.43}$$

Nitrides are usually prepared using ammonia or ammonia/nitrogen mixtures as the nitrogen source, and carbides using methane (or other hydrocarbons) as the carbon source (see Table 3.3). The presence of hydrogen (not explicitly mentioned in Table 3.3) helps in reducing the quantity of excess carbon in the final product, converting it to methane.

The chemistry of the gas-phase reactions is very complex, and mostly not fully understood. For example, 119 separate reactions have been identified in the gas-phase synthesis of SiC from SiH_4 and propane, and the mechanism probably is still incomplete. The underlying chemistry is often related to what has been discussed in Section 3.2 for CVD processes, but other processes may also occur due to the high temperatures involved.

For example, laser syntheses involving SiH_4 are thought to be initiated by the decomposition of SiH_4 to elemental Si. The concentration of Si in the vapor is supersaturated and, therefore, →nucleation of Si particles occurs. Collision of these nuclei results in the growth of larger particles. The addition of a hydrocarbon or ammonia then results in the carburization or nitridation of the elemental silicon particles.

3.3.4 Film Generation

Although aerosol routes are mainly used for powder synthesis, they also offer a variety of approaches for film generation, which are summarized in Figure 3.32. These processes have been used to deposit many materials at high rates, including

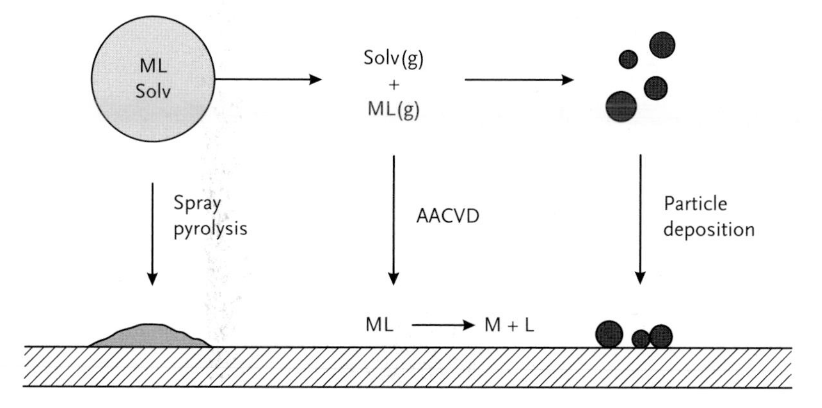

Figure 3.32 Comparison of processes for film formation during aerosol syntheses.

ceramic → superconductors, simple metal oxides, non-oxide ceramics, metals, and → composites.

The biggest application is the manufacture of optical fibers for telecommunication where the fiber core must have a higher refractive index (n_D) to minimize light losses over long distances. This is achieved by depositing (doped) silica with controlled n_D either onto glass rods or inside a glass tube. The coated rod or tube is then drawn into fibers.

In *spray pyrolysis* (*droplet deposition*) for film generation, an → aerosol of a solution containing the reactants is formed and then deposited onto a heated surface where solvent evaporation and chemical reactions take place resulting in a film. This process is usually carried out at atmospheric pressure, often without any enclosure. Some of the advantages of this process include: simplicity and low cost, many choices for the precursors, and high deposition rates (0.1–1 $\mu m\, min^{-1}$). Film thicknesses of 1–10 μm are common.

The primary disadvantages are that porous films are sometimes formed, that purity is low in some cases, and that the method is limited to oxidation-resistant materials. Industrial applications of the technique are for solar cells, oxide → superconductor films, photochemical cell electrodes, gas-sensing elements, antireflection coatings, and thermal coatings.

In *aerosol-assisted chemical vapor deposition* (AACVD), a solution containing volatile precursors is carried in the form of an aerosol close to the substrate that is to be coated (see Section 3.2). Before or near the substrate, the precursors evaporate into the gas phase. This is followed by CVD of gaseous molecular precursors on the heated surface to deposit a thin film. This approach has several advantages for the reproducible generation of multicomponent films (high-temperature superconductors, for example) with thermally sensitive precursors. Deposition rates as high as 5 $\mu m\, min^{-1}$ have been obtained while producing high-quality films. One disadvantage is that a solvent is present in many cases and may be incompatible with the precursors or the chemistry of film formation.

In *particle deposition*, films consisting of particles can be formed by a variety of deposition mechanisms including impaction, diffusion, sedimentation, thermophoresis, and → electrophoresis. This method has been used to form ceramic filters, sensors, and hard coatings. Typically, a carrier gas stream is directed at a surface. Particle deposition takes place by impaction, in the absence of temperature gradients, and porous polycrystalline films are formed. Typically, postprocessing is required to densify the porous films. Under suitable conditions, high deposition rates can be obtained. The primary strength of this method is for generation of thick ceramic films. Deposition rates as high as 1–5 $\mu m\, min^{-1}$ have been obtained.

Further Reading

1 Barron, A.R. (1996) CVD of SiO_2 and related materials: an overview. *Adv. Mater. Opt. Electron.*, **6**, 101–114.

2 Born, M. and Jüstel, T. (2006) Chemie in Lampen: elektrische Lichtquellen. *Chemie i. u. Zeit*, **40**, 294–305.

3 Bhaduri, S.B. and Bhaduri, S. (1999) Combustion synthesis, in *Non-Equilibrium Processing of Materials* (ed. C. Suryanarayana), Pergamon, Amsterdam, pp. 289–309.
4 Chorley, R.W. and Lednor, P.W. (1991) Synthetic routes to high surface area non-oxide materials. *Adv. Mater.*, **3**, 474–485.
5 Colligon, J.S. (1999) Physical vapor deposition, in *Non-Equilibrium Processing of Materials* (ed. C. Suryanarayana), Pergamon, Amsterdam, pp. 225–253.
6 Davies, G.J., Foord, J.S., and Tsang, W.T. (eds) (1996) *Chemical Beam Epitaxy and Related Techniques*, John Wiley & Sons Ltd, Chichester.
7 Gallois, B.M., Lee, W., and Pickering, M. (eds) (1994) Chemical vapor deposition of refractory metals and ceramics III. *Mater. Res. Soc. Symp. Proc.*, **363**.
8 Gruehn, D.M. and Buckley-Golder, I. (eds) (1994) Diamond films: recent developments. *Mater. Res. Soc. Bull.*, **23**, (9).
9 Gruen, R. and Glaum, R. (2000) New results of chemical transport as a method for the preparation and thermochemical investigation of solids. *Angew. Chem. Int. Ed. Engl.*, **39**, 692–716.
10 Gurav, A., Kodas, T., Pluym, T., and Xiong, Y. (1993) Aerosol processing of materials. *Aerosol Sci. Technol.*, **19**, 411–452.
11 Hampden-Smith, M. and Kodas, T.T. (1995) Chemical vapor deposition of metals. *Chem. Vap. Depos.*, **1**, 8–23, 39–48.
12 Hampden-Smith, M., Kodas, T.T., and Ludviksson, A. (1998) Chemical vapor deposition, in *Chemistry of Advanced Materials – an Overview* (eds L.V. Interrante and M. Hampden-Smith), Wiley-VCH, New York, pp. 143–206.
13 Hitchman, M.L. and Jensen, K.F. (eds) (1993) *Chemical Vapour Deposition: Principles and Applications*, Academic Press, New York.
14 Jones, A.C. (1997) Developments in metalorganic precursors for semiconductor growth from the vapor phase. *Chem. Soc. Rev.*, **26**, 101–110.
15 Jones, A.C. and O'Brien, P. (1997) *CVD of Compound Semiconductors*, Wiley-VCH, Weinheim.
16 Kodas, T. and Hampden-Smith, M. (eds) (1994) *The Chemistry of Metal CVD*, Wiley-VCH, Weinheim.
17 Kodas, T. and Hampden-Smith, M. (eds) (1997) *Aerosol Processing of Materials*, Wiley-VCH, Weinheim.
18 Kriechbaum, G.W. and Kleinschmit, P. (1989) Superfine oxide powders – flame hydrolysis and hydrothermal synthesis. *Angew. Chem. Adv. Mater.*, **101**, 1446–1453.
19 Langlais, F. (2005) On the chemical steps involved in the chemical vapour infiltration processing of ceramic matrix composites. *Ann. Chim. Sci. Mater.*, **30**, 579–592.
20 Lenz, M. and Gruehn, R. (1997) Developments in measuring and calculating chemical vapor transport phenomena demonstrated on Cr, Mo, W, and their compounds. *Chem. Rev.*, **97**, 2967–2994.
21 Leskelä, M. and Ritala, M. (2003) Atomic layer deposition chemistry: recent developments and future challenges. *Angew. Chem. Int. Ed. Engl.*, **42**, 5548–5554.
22 Lettington, A.H. and Steeds, J.W. (eds) (1994) *Thin Film Diamond*, Chapman & Hall, London.
23 Lux, B., Haubner, R., and Renard, P. (1992) Diamond for tooling and abrasives. *Diamond Relat. Mater.*, **1**, 1035–1047.
24 Maury, F. (1996) Trends in precursor selection for MOCVD. *Chem. Vap. Depos.*, **2**, 113–116.
25 Morkoc, H. (2001) IIIA nitride semiconductor growth by MBE: recent issues. *J. Mater. Sci.*, **12**, 677–695.
26 Niinistö, L. (1998) Atomic layer epitaxy. *Curr. Opin. Solid State Mater. Sci.*, **3**, 147–158.
27 Pratsinis, S.E. (1998) Flame aerosol synthesis of ceramic powders. *Prog. Energy Combust. Sci.*, **28**, 197–219.
28 Rees, W.S. (ed.) (1996) *CVD of Nonmetals*, Wiley-VCH, Weinheim.
29 Schäfer, H. (1964) *Chemical Transport Reactions*, Academic Press, New York.

30 Selvamanickam, V., Xie, Y., Reeves, J., and Chen, Y. (2004) MOCVD-based YBCO-coated conductors. *Mater. Res. Soc. Bull.*, **29**, 579–582.

31 Spear, K.E. and Cullen, G.W. (eds) (1990) *Chemical Vapor Deposition*, The Electrochemical Society, Pennington, New Jersey.

32 Stringfellow, G.P. (1998) *Organometallic Vapor Phase Epitaxy; Theory and Practice*, Academic Press.

33 Strobel, R. and Pratsinis, S.E. (2007) Flame aerosol synthesis of smart nanostructured materials. *J. Mater. Chem.*, **17**, 4743–4756.

34 Suntola, T. and Simpson, M. (eds) (1990) *Atomic Layer Epitaxy*, Blackie, Glasgow.

35 Teyssandier, F. and Dollet, A. (1999) Chemical vapor deposition, in *Non-Equilibrium Processing of Materials* (ed. C. Suryanarayana), Pergamon, Amsterdam, pp. 257–285.

36 Wagner, E. and Brünner, H. (1960) Aerosil – Herstellung, Eigenschaften und Verhalten in organischen Flüssigkeiten. *Angew. Chem.*, **72**, 744–750.

37 Weimer, A.W. (ed.) (1997) *Carbide, Nitride and Boride Materials – Synthesis and Processing*, Chapman & Hall, London.

38 Zaera, F. (2008) The surface chemistry of thin film atomic layer deposition (ALD) processes for electronic device manufacturing. *J. Mater. Chem.*, **18**, 3521–3526.

4
Formation of Solids from Solutions and Melts

This chapter deals with reactions and processes in which solid products are obtained from a liquid phase. The simplest case, from a chemical point of view, is that the liquid has the same composition as the resulting solid, that is, the solid is formed from its melt, without a chemical transformation, just by change of the physical state. The involved crystallization and glass-forming processes are discussed in Section 4.1. The intention of this book is not to become involved with physical–chemical details, and to keep the number of mathematical formulas as few as possible, without mathematical or physical derivations. Rather, we focus on a broader picture in order to allow an understanding of the chemical processes involved. Crystallization and precipitation processes from solutions are treated in Section 4.2. The physical–chemical fundamentals are very much related to those relevant to melts, but the processes involved are more complex due to the different chemical composition of the solid and liquid phases. In Section 4.3 we will examine how nature controls crystallization processes in biological systems. In solvothermal processes (Section 4.4), the dissolution and recrystallization of compounds is speeded up by increases in temperatures and pressures. In Section 4.5, we will discuss sol–gel processes that allow the obtainment of solid products by gelation rather than crystallization or precipitation.

4.1
Glass

When talking about glass, it is window or bottle glass that first comes to one's mind. However, glasses are also used in "high-tech" applications such as communications technologies or as biomaterials. At the end of this section we will see that even metals are able to form glasses, and we will discuss methods of producing such metallic glasses. We will start with this section by defining what a "glass" is. After touching structural issues, we will find out under which conditions a melt will form a glass upon cooling instead of crystallizing. This is followed by a short introduction into the glass-making process.

Cooling a liquid compound below its melting temperature (T_m) normally results in its crystallization. Crystals are characterized by a long-range, periodic arrangement of atoms or molecules. When a compound crystallizes, its structure

Synthesis of Inorganic Materials, Third Edition. Ulrich Schubert and Nicola Hüsing.

Published 2012 by WILEY-VCH Verlag GmbH & Co. KGaA

rearranges discontinuously from the disordered liquid structure to the ordered crystal structure. Concomitant with that, the enthalpy decreases abruptly from the value for the liquid to that for the crystal (Figure 4.1; the volume dependence on the temperature is very similar). Continued cooling below T_m results in a further enthalpy decrease due to the → heat capacity of the crystal.

A supercooled liquid is obtained if the liquid can be cooled below the melting temperature (T_m) without crystallization. If the supercooled liquid could be cooled indefinitely slowly, the atoms would rearrange into the equilibrium structure of the liquid (which depends on the temperature) without the abrupt decrease in enthalpy observed upon crystallization. However, as the liquid is cooled, its viscosity (η) increases and eventually becomes so great that the atoms can no longer completely rearrange to the equilibrium structure in a finite period of time. The enthalpy thus begins to deviate from the straight equilibrium line until it eventually becomes determined by the → heat capacity of the frozen liquid (Figure 4.1). The viscosity of a frozen liquid is so great that the structure becomes fixed and is no longer temperature dependent. The frozen liquid is now a glass. The temperature region between the limits where the enthalpy is that of the equilibrium liquid and that of the frozen liquid is the *glass-transformation region.*

A glass can thus be defined as an amorphous (non-crystalline) solid without a long-range, periodic structure that exhibits a region of glass-transformation behavior. Any inorganic, organic, or metallic material that exhibits glass-transformation behavior is a glass. Its average behavior in all directions is the same, and therefore the properties of glasses are isotropic.

Figure 4.1 shows that the glass-transformation behavior is a time-dependent phenomenon. When the supercooled liquid is cooled very slowly, the enthalpy

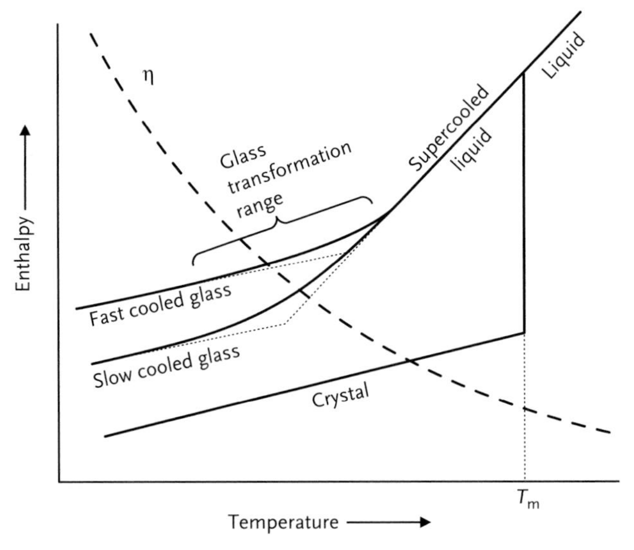

Figure 4.1 Effect of temperature on the enthalpy of a glass-forming melt. η is the viscosity of the melt.

begins to deviate from the equilibrium line at a lower temperature, that is, the glass-transformation region is shifted to lower temperatures. Owing to the higher viscosity at lower temperatures, more time is required to reach the equilibrium structure. A slower cooling rate thus allows the supercooled liquid to adjust its equilibrium structure to a lower temperature. The glass obtained with a lower cooling rate has a lower enthalpy than that obtained with a faster cooling rate.

Although the glass transformation actually occurs over a temperature range, it is convenient to use a single temperature as an indicator for the transition between a melt and a glassy solid. This temperature is called either the glass-transformation temperature or the glass-transition temperature (T_g). Since T_g is a function of both the heating (or cooling) rate and the experimental method used for the measurement, it cannot be considered a true property of the glass. Standardized conditions have to be used to make the T_g of different samples comparable.

4.1.1 The Structural Theory of Glass Formation

Early theories of glass formation were centered on the question why some materials form glasses while others do not. These theories are often called *structural theories of glass formation*. It was assumed that the ability to form *three-dimensional networks* by linking some basic building blocks is the ultimate condition for glass formation. Highly ionic materials do not form network structures.

The basic building blocks of the networks consist of a central electropositive element (called the "cation" in the following discussion, although the networks are not ionic) surrounded by a certain number of electronegative elements (called the "anion" in the following discussion).

Silicates, for example, readily form glasses instead of recrystallizing after melting and cooling (Figure 4.2). They have vitreous ("glass-like") networks rather than close-packed structures. In contrast to the corresponding crystalline compounds, the networks are not periodic and exhibit no symmetry.

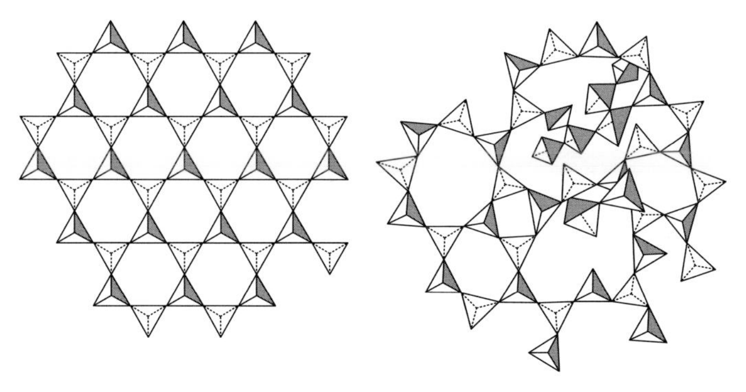

Figure 4.2 Schematic structure of crystalline (left) and amorphous silica (right). In the crystalline forms of silica the $[SiO_4]$ tetrahedra (the silicon atoms are located in the center of the tetrahedra, and the oxygen atoms at the vertices) are regularly arranged. There is no long-range order in the amorphous form.

The most commonly used models for glass structures are based on the original ideas of Zachariasen, and are called "random network theory." The rules for simple oxide, → chalcogenide or halide glasses are:

1. Each anion is linked to no more than two cations (higher coordination numbers for the anion than two prevent variations in bond angles necessary to form a non-periodic, random network).
2. The coordination polyhedra share only corners and not edges or faces (this is a consequence of rule 1).
3. The coordination number of the network-forming cations ("network former") is small (higher coordination polyhedra, such as octahedra, tend to share edges or faces instead of corners).
4. At least three corners of each polyhedron must be shared in order to form three-dimensional networks (only then can the network be three-dimensional. Sharing only two corners results in chain structures instead [as in silicones, see Section 5.2]).

Figure 4.3 provides a schematic drawing of the structure of an alkali silicate glass to illustrate these rules. Silicon (the "cation") in silicates and silicate glasses is always tetrahedrally surrounded by four oxygen atoms (the "anions"), that is, the coordination number of silicon is four (rule 3). The network is formed from connected [SiO_4] tetrahedra. The only way to connect [SiO_4] tetrahedra with each other in a thermodynamically stable way is via shared corners, that is, two adjacent silicon atoms are connected by only one oxygen atom (rule 2). In vitreous silica, each oxygen atom bridges two silicon atoms (rule 1). In silicate glasses, there is a certain portion of non-bridging oxygen atoms. The negative charge of each non-bridging oxygen atom must be compensated by a nearby cation to maintain local charge neutrality. However, rule 4 is obeyed.

The properties of silicate glasses are to a large extent influenced by the portion of non-bridging oxygen atoms. They are formed by cleavage of Si–O–Si bonds

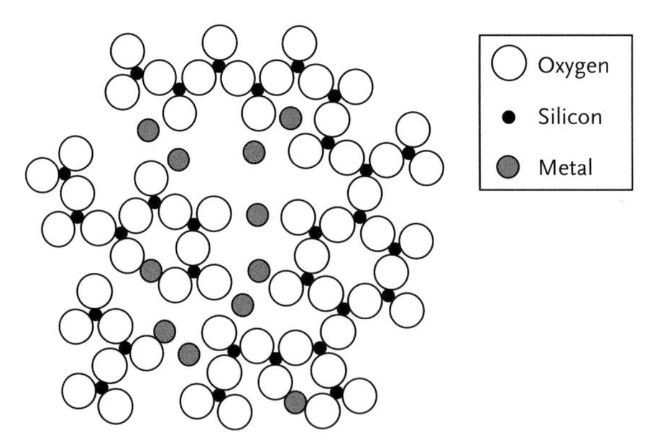

Figure 4.3 Schematic two-dimensional drawing of the structure of a silicate glass. The fourth oxygen atom of each [SiO_4] tetrahedron positioned above or below the silicon atom is not drawn for clarity.

[e.g., Eq. (4.1)]; the portion of non-bridging oxygen atoms thus correlates with the alkali content of silicate glasses.

$$\equiv Si-O-Si\equiv\ + Na_2O \longrightarrow 2\ \equiv Si-O^-Na^+ \qquad (4.1)$$

The comparison of the crystal structures of the SiO_2 polymorphs (Figure 4.2 left) and rutile (TiO_2; Figure 4.4) may additionally illustrate the Zachariasen rules. TiO_2 is much easier to crystallize than SiO_2, or in other words, SiO_2 is a better glass former than TiO_2. This can be traced back to the higher coordination number of the titanium atom. Both silicon and titanium are in the +IV oxidation state and therefore the oxides have the same overall composition (MO_2). However, silicon is always 4-coordinate (tetrahedral [SiO_4] building blocks), while titanium in rutile is 6-coordinate (octahedral [TiO_6] building blocks). Corner sharing of the [SiO_4] tetrahedra results in the overall composition SiO_2 ($=SiO_{4/2}$, because all oxygen atoms bridge two silicon atoms). The situation in rutile is quite different: the MO_2 stoichiometry can only be obtained with the octahedral [TiO_6] building blocks if each oxygen atom coordinates to three titanium atoms ($TiO_{6/3}=TiO_2$, because all oxygen atoms bridge three titanium atoms). In the rutile structure (Figure 4.4), this is achieved by connecting strings of edge-sharing octahedra with the corners of a neighboring string. Because the polyhedra share edges *and* corners, there are fewer degrees of freedom for their long range arrangement compared with crystalline SiO_2 (only corner-sharing polyhedra).

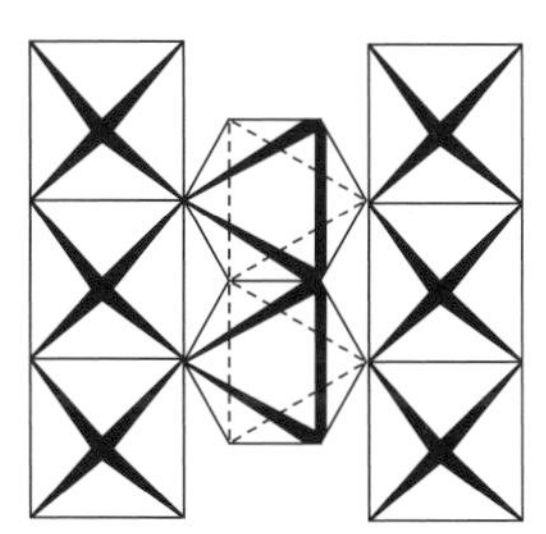

Figure 4.4 Arrangement of the [TiO_6] octahedra in the rutile structure.

Borate glasses exhibit much more complicated structures, owing to a larger number of building blocks. Some structural elements of borate glasses are shown in Figure 4.5. Note that there are both trigonal planar [BO_3] and tetrahedral [BO_4] units.

The structure of both crystalline and glassy boric oxide is composed of planar [BO_3] units only. While the addition of metal oxides to silicate melts results in the formation of non-bridging oxygen atoms [Eq. (4.1)], the addition of metal oxides to borate melts primarily converts planar [BO_3] units into tetrahedral [BO_4] units (2 [BO_3] + O^{2-} → [O_3B–O–BO_3]$^{2-}$) and thus results in a higher degree of cross-linking of the borate network. Non-bridging oxygen atoms are only formed when the alkali content is high. Contrary to silicate glasses, this results in a non-linear dependence of some borate glass properties on the alkali oxide portion, for example,

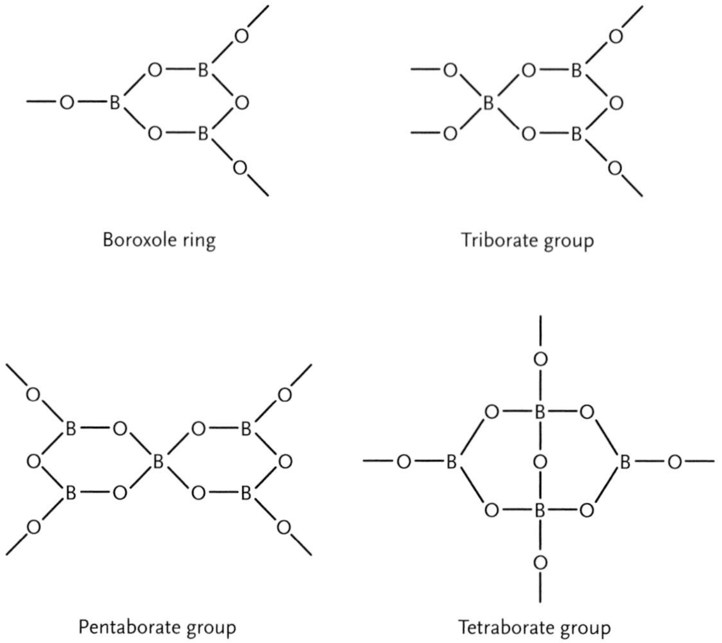

Boroxole ring

Triborate group

Pentaborate group

Tetraborate group

Figure 4.5 Structural elements of borate glasses.

the T_g or the thermal expansion coefficient. This effect is called the "borate anomaly."

The Zachariasen rules were modified for complex glasses that *some* anions are linked only to two network cations, there must be a *high percentage* of network cations that are tetrahedrally or trigonally planar surrounded by the anions, and the tetrahedra or triangles share only corners.

In general, the structure of a glass is determined by:

- **Coordination number of the network-forming cations**. The coordination polyhedra of these cations are the building blocks of the glass structure. There is the possibility that the blocks may be connected to slightly larger units, such as rings or clusters, that have a more ordered arrangement than predicted by a random connection (see Figure 4.5).
- **Network connectivity**; that is, the average number of bridging bonds per network-forming cation.
- **Bond-angle distributions**. Bond angles and dihedral angle distributions introduce randomness into the structure and are therefore inherent to amorphous materials.
- **Dimensionality of the network**. A network does not need to be three-dimensional in order to form a glass. For example, long-chain polymers that have a one-dimensional network may form glasses by three-dimensional entanglement of the polymer chains.

The structural theory of glass formation considers only the relative ease of glass formation. Any compound or mixture that forms a glass during cooling from the melt at a moderate cooling rate is considered to be a good glass former, while materials that require a more rapid cooling rate are considered to be poor glass formers.

It is now recognized that virtually any material will form a glass if cooled so rapidly that insufficient time is provided to allow the reorganization of the structure into a periodic crystal lattice. Therefore, the question is not whether a material will form a glass, but rather how fast it must be cooled to avoid detectable crystallization. This leads to *kinetic theories of glass formation* processes (see below).

4.1.2 Crystallization versus Glass Formation

Crystallization actually includes two processes: the formation of a crystalline nucleus (→ nucleation); and subsequent growth of the crystal. The nucleus may be either homogeneous, that is, formed spontaneously within the melt, or heterogeneous, that is, formed at a pre-existing surface (impurity, crucible wall, etc.). If no nuclei are present, crystal growth cannot occur and the material will form a glass. Thus, melts which exhibit a large barrier to nucleation also exhibit good glass-forming behavior. A melt that is free of potential heterogeneous nuclei can be cooled more easily to form a glass than a melt that contains a large concentration of such nuclei. On the other hand, even if some nuclei are present, but no crystal growth occurs, the solid is still a glass.

Melts composed of many different elements inhibit rearrangement of the melt into the ordered crystalline structure because the redistribution of ions to the appropriate sites on the growing crystals is more difficult. This approach is used routinely in commercial glass technology, and partially explains the complex compositions of many common glasses.

In the process of → homogeneous nucleation, nuclei are formed with equal probability throughout the liquid or melt. In the classical nucleation theory, the nucleation rate I (nuclei per unit volume per second) is given by Eq. (4.2),

$$I \propto \mathrm{e}^{\left[\frac{-(\Delta G_{\mathrm{N}} + \Delta G_{\mathrm{D}})}{kT}\right]} \tag{4.2}$$

where ΔG_N is the free energy change in a system when a crystalline nucleus is formed (*thermodynamic barrier* to nucleation) and ΔG_D is the *kinetic barrier* for diffusion across the liquid/nucleus interface.

For spherical nuclei, the thermodynamic barrier (ΔG_N) is expressed by Eq. (4.3).

$$\Delta G_{\mathrm{N}} = {}^4\!/_3\, \pi r^3 \Delta G_{\mathrm{v}} + 4\pi r^2 \gamma \tag{4.3}$$

where γ is the crystal/melt interfacial energy and ΔG_v is the change in volume free energy per unit volume. Note that ${}^4/_3\, \pi r^3$ is the volume and $4\, \pi r^2$ the surface area of a sphere.

- The first term represents the change of the volume free energy (ΔG_v). ΔG_v is negative, since the crystalline state has a lower free energy than the melt.
- The second term represents the increase in surface energy ($\Delta G_s = 4\pi r^2 \gamma$) due to the formation of a new interface, the interface between the solid phase (nucleus) and the melt.

Since nuclei are small, the surface energy term will dominate at very low radii r. The energy of the system (ΔG_N) will first increase with increasing radius (Figure 4.6), and the nucleus will dissolve or melt. If, however, the nucleus survives (by a statistical event) to grow to a large enough size, the first term of Eq. (4.3) will become larger than the second (surface energy), and the energy of the system will begin to decrease with increasing nucleus size. The nucleus will become stable.

The value of the radius where the nucleus just becomes stable, is known as the critical radius, r^* [Eq. (4.4)],

$$r^* = \frac{-2\gamma}{\Delta G_v} \tag{4.4}$$

(the radius r^* is positive because $\Delta G_v < 0$).

If the temperature is only a little below T_m, the absolute value of the volume free energy (ΔG_v) is very small. It follows that the critical radius, r^* for a stable nucleus [Eq. (4.4)] is very large. Since the probability of a nucleus reaching such a large size is extremely low, the melt will remain effectively free of nuclei, even though the temperature is below T_m. As the temperature decreases further, ΔG_v will increase, thus decreasing the value of r^*. Eventually, r^* will become so small (often only a few tenths of a nanometer) that the probability of formation of a nucleus will become significant. The degree of undercooling may be as little as a small fraction of a degree or as much as a few hundred degrees.

On further lowering the temperature, the thermodynamic barrier to nucleation further decreases, allowing nuclei to form at an ever-increasing rate. However, the viscosity is also highly temperature dependent, so that the kinetic barrier to

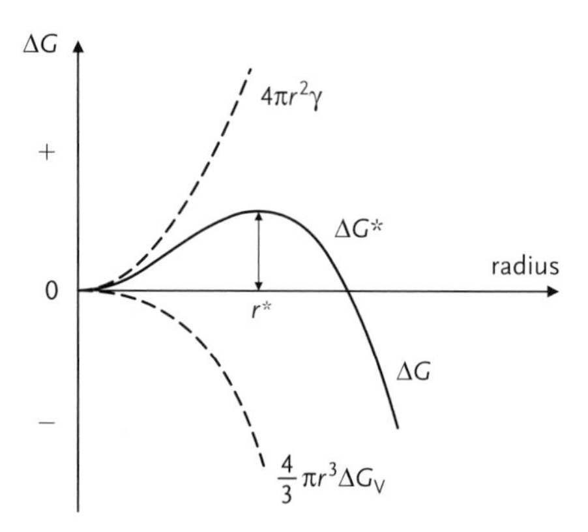

Figure 4.6 Changes in the thermodynamic barrier (ΔG_N) upon formation of a crystalline nucleus.

nucleation will increase rapidly with decreasing temperature. As the kinetic barrier increases, the → nucleation rate begins to decrease and eventually approaches zero. The opposed effects due to changes in the thermodynamic and kinetic barriers thus results in a maximum in the temperature dependence of the nucleation rate.

Heterogeneous nucleation occurs when the energy of the solid/solid interface between the pre-existing surface and the growing crystal is small. Then the energy balance in Eq. (4.3) tips in favor of the reduction of the volume free energy (ΔG_v); the volume of the crystal can form with less energy consumed by the solid/liquid interface.

The temperature dependence of the crystal growth rate is very similar to that for the nucleation rate. The principle difference is that crystals can grow at any temperature below T_m so long as a (homogeneous or heterogeneous) nucleus is available. As for nucleation, if the viscosity is low, the growth rate will be determined by the thermodynamic values, and will tend to be large. As the temperature decreases, the rapid increase in viscosity will eventually stop crystal growth. The resulting curve of the crystal growth rate versus temperature thus also exhibits a maximum.

In reality, → nucleation and crystal growth occur simultaneously during cooling of a melt, with rates that change continuously as the temperature decreases. The dependence of the nucleation rate and crystal growth on the temperature is shown in Figure 4.7 for two cases: crystallizing melts (upper diagram); and glass-forming melts (lower diagram). Upon cooling a melt below T_m, crystals will grow if there are a sufficient number of nuclei. If the maximum of nucleation is at a similar

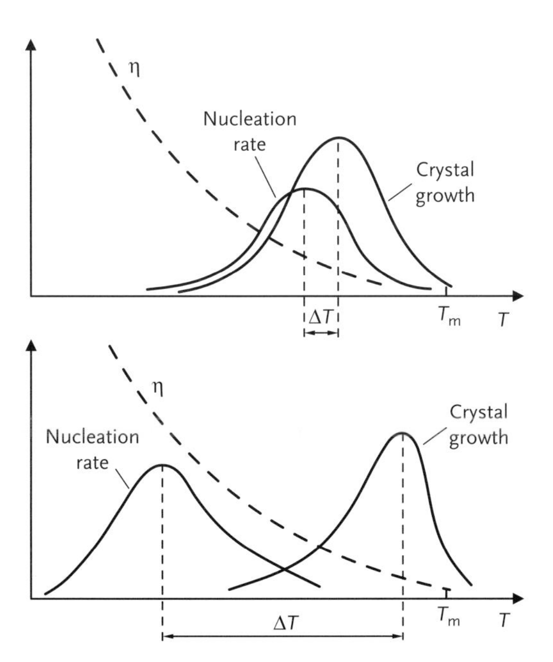

Figure 4.7 The diagrams show the different relation between nucleation rate and crystal growth in a crystallizing system (upper diagram) and glass-forming system (lower diagram). η is the viscosity of the melt.

temperature than the maximum of crystal growth (i.e., if ΔT is small), a large number of nuclei will be produced at a temperature at which crystal growth is optimal. This means that crystallization of the melt will occur easily (upper part of Figure 4.7). On the other hand, if the maximum of nucleation is at a much lower temperature than the maximum of crystal growth (i.e., if ΔT is large), nuclei will be formed, but they cannot grow because the kinetic barrier at this temperature (higher viscosity) inhibits crystal growth. This means that the melt will form a glass (lower part of Figure 4.7).

Formation of a glass requires cooling of the melt in such a manner that significant crystal formation is prevented. If the nucleation rate (I) and the linear crystal growth rate (U) as a function of the temperature are known, then the volume fraction of crystals in a sample, V_x/V, is given by Eq. (4.5) under isothermal conditions.

$$\frac{V_X}{V} = 1 - e^{\frac{-\pi I U^3 t^4}{3}} \tag{4.5}$$

where V_x is the volume of crystals, V is the sample volume, and t is the time the sample has been held at the experimental temperature.

To what volume fraction of crystals a sample is considered a glass is a question of definition; a typical value is 1 ppm. The crystal content acceptable for a window glass will be quite different from that acceptable for an optical fiber or a lens, for example.

From Eq. (4.5) one can calculate the time required to form a particular volume fraction of crystals at a given temperature. At another temperature, the time required for the formation of the same volume fraction of crystals is different, because of the different temperature dependence of the → nucleation and crystal growth rate. One can thus calculate the curve in time–temperature space that corresponds to a particular value V_x/V (Figure 4.8), the so-called time–temperature–transformation (TTT) curve. This curve has the general shape shown in Figure 4.8,

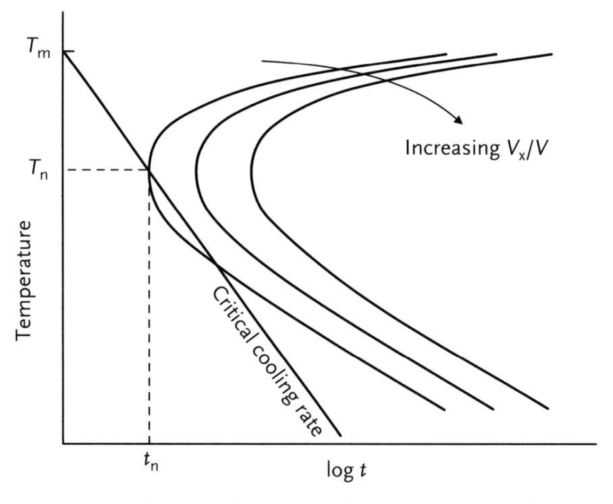

Figure 4.8 Schematic illustration of a TTT curve. The abscissa indicates the time required at each temperature to produce a volume fraction V_x/V of crystals.

and is therefore also called the "nose curve." Since I and U approach zero as the temperature approaches T_m, the time required to form the specified volume fraction of crystals will approach infinity. At very low temperatures, the values of I and U also approach zero due to the very high viscosity of the melt, and the time to reach the specified value of V_x/V also approaches infinity. The least favorable conditions for glass formation occur at the temperature T_n, corresponding to the "nose" of the curve, where the time required for crystal formation, t_n, is minimal.

The critical cooling rate, $(dT/dt)_c$ [Eq. (4.6)], that is, the minimum cooling rate required to yield a glass (with an acceptable crystal content), can be obtained from the slope of the tangent to the curve, with the initial conditions defined as T_m at time zero.

$$\left(\frac{dT}{dt}\right)_c \approx \frac{(T_m - T_n)}{t_n} \quad (4.6)$$

A typical value for the critical cooling rate is 9×10^{-6} K s^{-1} for SiO_2 glass. At the opposite end, the formation of metallic glasses (see Section 4.1.4) requires cooling rates of 10^6–10^{10} K s^{-1}. The melt with the smaller critical cooling rate has the better glass-forming ability.

Unintended crystallization of glass during the production process results in a locally inhomogeneous concentration of crystals of different size. This has to be avoided. However, controlled crystallization by reheating of a glass (Figure 4.9) first to the temperature of maximum nucleation rate (T_1) and then to the temperature of maximum crystal growth (T_2) leads to a class of materials called glass ceramics with many interesting properties and uses. The name reflects the fact that these materials are *not* obtained by sintering (see Section 2.1.5) as the usual ceramic materials. Glass ceramics consist of small (usually some μm) and uniform crystallites irregularly distributed in an amorphous matrix.

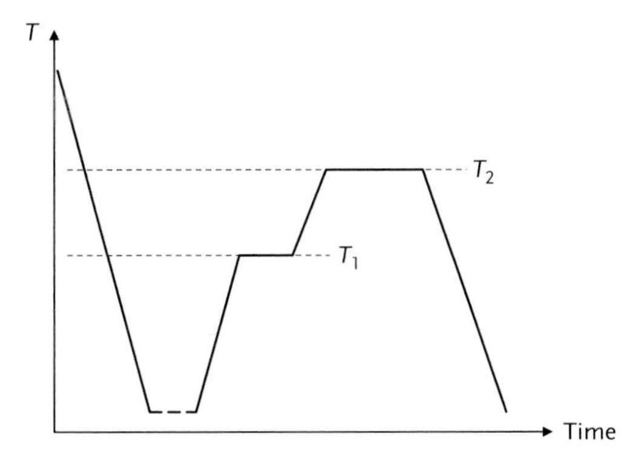

Figure 4.9 Temperature profile during the production of glass ceramics.

Since the thermal expansion coefficient of a sample is a volume-averaged function of the contribution of each of the phases present, formation of crystals with thermal expansion coefficients very different from the initial glass, can radically alter the overall thermal expansion coefficient. The initial glass used for the production of transparent cookware, for example, has a thermal expansion coefficient of about 4 ppm K^{-1}. After processing as glass ceramics, the thermal expansion coefficient is only about one-tenth of this value.

4.1.3
Glass Melting

Although glasses can be made by a wide variety of methods, the vast majority are still produced by melting of batch components at an elevated temperature. The production steps are shown in Figure 4.10.

Examples of the composition of technical and optical glasses are given in Table 4.1.

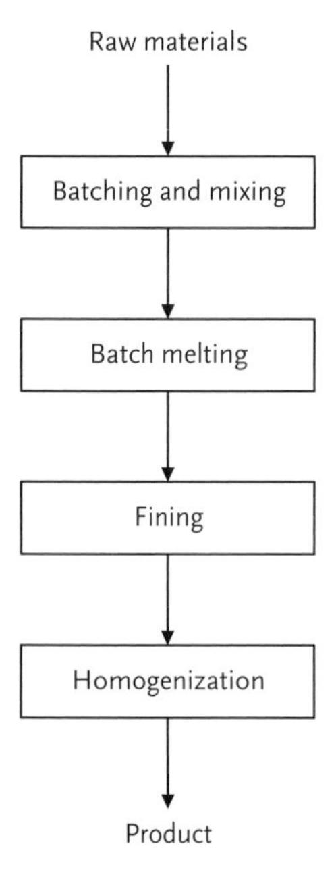

Figure 4.10 Steps in the fabrication of glass.

Table 4.1 Composition of some typical glasses.

Window glass	72% SiO_2, 1.5% Al_2O_3, 3.5% MgO, 8.5% CaO, 14.5% Na_2O
Laboratory glass	80% SiO_2, 10% B_2O_3, 3% Al_2O_3, 1% MgO, 1% CaO, 5% Na_2O
Fluoride fiber glass	53% ZrF_4, 20% BaF_2, 20% NaF, 2% LaF_3, 3% AlF_3, 2% LnF_3

4.1.3.1 Raw Materials of Glass Production

The *raw materials* can be divided into four categories based on their role in the process: glass formers, network modifiers, colorants, and fining agents. The same compound may be used for different purposes. Alumina, for example, serves as a glass former in aluminate glasses, but is a modifier in most silicate glasses. Arsenic oxide may be either a glass former or a fining agent.

Glass formers (network formers) The primary glass formers in commercial oxide glasses are silica (SiO_2), boric oxide (B_2O_3), and phosphoric oxide (P_2O_5), which readily form single-component glasses. A large number of other compounds may act as glass formers, especially in combination with other oxides, such as GeO_2, Bi_2O_3, As_2O_3, Sb_2O_3, TeO_2, Al_2O_3, Ga_2O_3, and V_2O_5. As_2S_3, As_2Se_3, and GeS_2 are important glass formers in → chalcogenide glasses, the three most common halide glass formers are BeF_2, AlF_3, and ZrF_4.

Network modifiers While silica itself forms an excellent glass, with a wide range of applications, the use of pure silica glass for bottles, windows, and other bulk commercial applications would be too expensive due to the high melting temperature (>2000 °C). The processing temperature is lowered by addition of alkali or alkaline-earth oxides that break Si–O–Si bonds [e.g., Eq. (4.1)] and thus lower the melting temperature. The use of PbO is becoming much more limited due to concerns regarding its toxicity. PbO is especially useful in dissolving any → refractory or other impurity particles that might otherwise result in flaws in the final glass. A combination of different network modifiers is often necessary to modify the properties of the glass.

Colorants These are used in small quantities to control the color of the final glass. In most cases, colorants are oxides of either the 3d transition metals or the 4f rare earths. Gold and silver are also used to produce colors by formation of → colloids in glasses. Iron oxides, which are common impurities in the sands used to produce commercial silicate glasses, act as unintentional colorants in many products. When colorants are used to counteract the effect of other colorants to produce a slightly gray glass, they are referred to as *decolorants*.

Fining agents These are added to glass-forming batches to promote the removal of bubbles from the melt. They are usually present in very small quantities (<1 wt%) and have only minor effects on the properties of the final glasses.

4.1.3.2 Batch Melting

Batch melting involves the melting and decomposition of the raw materials to form the initial melt. Initial heating usually results in the release of some moisture, which may have been adsorbed onto the raw materials or may be formed by dehydration of hydroxides. Far more gas is released during the decomposition of carbonates, sulfates, and nitrates. The gases released result in considerable mixing and stirring, which helps to homogenize the melt. However, the gas bubbles must be removed from the melt before processing is completed (see below).

The time required to dissolve the original batch completely is called the *batch-free time* (Figure 4.11). Factors that influence the batch-free time are temperature,

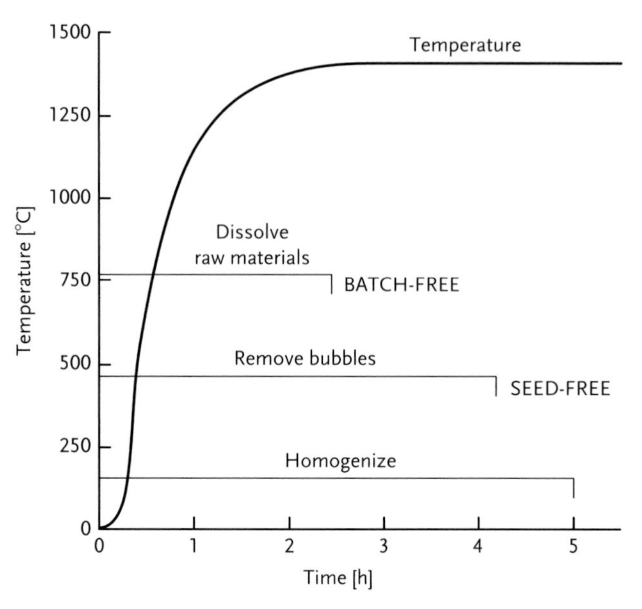

Figure 4.11 Stages of melting of a typical silicate glass.

overall glass composition, the raw materials used, batch homogeneity, → grain size of batch components, and the amount of scrap glass (*cullet*) added to the batch. The use of cullet not only reduces waste, but also reduces the batch-free time both by reducing the amount of → refractory material in the batch and by providing additional liquid throughout the melting process.

A large number of the components of glasses are quite volatile at elevated temperatures. Loss of these components can significantly alter the composition of the glass obtained after prolonged melting. Volatilization losses are particularly significant for alkali, lead, boron and phosphorus oxides, halides, and other components that have high vapor pressures at high temperatures. These losses can usually be reduced dramatically by lowering the melt temperature. Covering the melt increases the partial pressure of the volatile components directly above the melt, establishes equilibrium between the dissolved and vaporized species, and prevents significant loss of those components. However, covering the melt is usually not possible in large commercial melting processes. It may be necessary to allow for losses by providing excess concentrations of components known to vaporize from a given melt.

Many glass-forming melts require special techniques. For example, the components of → chalcogenide glasses are toxic and highly volatile. Furthermore, contamination by very small quantities of oxygen will destroy the infrared transmission of the glass. These melts are usually prepared by heating the mixture of powders in a sealed vitreous silica tube under a vacuum and then quenching the melt to form a glass. Heavy metal halide glasses also require melting in an oxygen-free atmosphere to preserve their optical properties. These glasses are often melted under a reactive atmosphere, such as CCl_4 or SF_6. The atmosphere acts as both a getter for oxygen and as a source of halide to replace volatilization losses and maintain the stoichiometry of the melt.

4.1.3.3 Fining of Melts

The terms fining (or refining) refer to the removal of bubbles from the melt. Bubbles are undesirable in most commercial glasses, and are considered as flaws. Fining of a melt begins during the melting process, but typically extends beyond the batch-free times (Figure 4.11).

Decomposition of batch materials can produce extremely large quantities of gases such as CO_2, SO_3, NO_x, H_2O, and so on. Reactions of materials in contact with the melt can also generate gases. Bubbles can be formed by the physical trapping of gases of the furnace atmosphere during the initial phase of batch melting. Since many gases have a large enthalpy of solution in glass-forming melts, their solubility in these melts is a strong function of temperature and the melt composition. Bubbles can therefore also be formed whenever supersaturation occurs for a specific gas.

Bubbles can be removed from melts either by their rising physically to the surface or by chemical dissolution of the gas into the melt. If bubbles do not rise sufficiently quickly, the melt can be moved by convection or stirring to carry the bubbles to the surface. The velocity of rise of a bubble is inversely proportional to the viscosity of a melt and directly proportional to the density of the melt and to the square of the bubble radius. Procedures that increase bubble sizes will therefore rapidly accelerate fining. Very small bubbles do not rise fast enough, and other processes are needed.

Bubbles can be removed chemically by the addition of fining agents. These release large quantities of gases, which form large bubbles that rise rapidly to the surface of the melt and also carry the smaller bubbles to the surface.

Arsenic and antimony oxides are the most efficient chemical fining agents. They are usually added to the batch in 0.1–1 wt% quantities, and are especially useful when combined with alkali nitrates in the batch. Their fining action results from a series of chemical reactions that occur at different stages of the melting process. One possible reaction might be as given in Eq. (4.7).

$$4KNO_3 + 2As_2O_3 \longrightarrow 2K_2O + 2As_2O_5 + 4NO\uparrow + O_2\uparrow \qquad (4.7)$$

After batch melting is complete, melts are usually heated to higher temperatures and held until completely fined. Since the trioxide is more stable than the pentoxide at these temperatures, the pentoxide produced via reaction with nitrates decomposes by release of oxygen [Eq. (4.8)].

$$\begin{aligned} &E_2O_5 \rightleftharpoons E_2O_3 + O_2 \\ &E = As,\ Sb \end{aligned} \qquad (4.8)$$

The released oxygen can either form new bubbles or diffuse into nearby smaller bubbles, thus increasing their size and rise rate. Any bubbles remaining in the melt at this point will be highly enriched in oxygen relative to the bubbles previously present. The equilibrium expressed by Eq. (4.8) is very temperature dependent. Lowering the temperature will result in a shift toward the pentoxide, which requires absorption of oxygen from the melt. This eventually leads to the complete consumption of the bubbles.

Although other fining agents are usually less efficient, the toxicity of arsenic and antimony oxides often necessitates alternative fining agents. Sodium sulfate, for example, also releases considerable gas volumes during batch melting, and also provides a portion of the sodium for soda-lime-silicate melts [Eq. (4.9)].

$$Na_2SO_4 + n\,SiO_2 \longrightarrow Na_2O \cdot n\,SiO_2 + SO_2 + {}^1\!/_2O_2 \qquad (4.9)$$

Sulfate fining is strongly affected by the reactions with furnace gases or other sources of carbon. Reactions between SO_3 and either C or CO produce SO_2 and CO_2, releasing large quantities of gas.

Halides are most useful as fining agents because they efficiently lower the viscosity of melts. They are particularly effective in fining of high alumina content melts.

4.1.3.4 Homogenization of Melt

The initially formed melt is very heterogeneous. This heterogeneity is gradually reduced by the stirring action of rising bubbles during fining. Production of an acceptably homogeneous glass, however, usually requires additional time for diffusion processes to improve the homogeneity of the melt.

Gross defects such as bubbles and "stones" (particles of undissolved material) are often clearly visible and cannot be tolerated in commercial glasses. The terms "striae" and "cord" are used to describe variation in local composition within a glass. Striae are two-dimensional regions of a composition different from the bulk, while cords are similar regions that are effectively one-dimensional veins in the glass. These regions cause a wavy appearance due to local variations in refractive index. In colored glasses, regions of inhomogeneity can often be detected by variations in color intensity.

Poor homogeneity frequently results from poor mixing of the original batch materials. Striae and cords can be formed by reactions with refractories and at the melt/atmosphere boundary owing to volatilization of batch components, especially alkali, boron, or lead. Decreasing the → grain size of the batch improves homogeneity by reducing the scale of the inhomogeneity of the initial melt. Stirring by mechanical stirrers, creation of convection flow in the melt, or bubbling of a gas through a melt can improve homogeneity.

4.1.4 Metallic Glasses

As discussed above, poor glass formers (such as metals) require a more rapid cooling rate than the good glass formers discussed so far (see TTT curves in Figure 4.8). The first metal → alloy vitrified by rapidly quenching the melt ("rapid solidification") with a cooling rate of 10^6 K s^{-1} had the composition Au_4Si, and was prepared in 1960. Since this time, a number of methods have been developed that allow very high cooling rates. Although solids can be formed from a liquid (melt) or a gas (vapor), the term "rapid solidification" is normally applied to forming solids from a melt. It should be pointed out, however, that metallic glasses can also be obtained by physical vapor deposition (PVD) methods (Section 3.2.1).

Most metals or alloys are polycrystalline when solidified from a melt, with → grains of various size and shape. Atom packing in the grain boundaries is not optimal (see Chapter 7). As a result, metals have a much lower strength than their theoretical maximum and, since energy goes to moving atoms in the grain boundaries, deformation is plastic and permanent. Furthermore, fractures can form and corrosion starts at grain boundaries. In contrast, the atoms in a glass are randomly packed with no long-range order. This results in unique properties for metallic glasses. They are twice as strong as steel (but lighter), have greater wear and corrosion resistance, and are tougher than ceramics. This can be combined with → ductile behavior in bending, shear, and compression. The increased scattering of electrons by the disordered structure gives rise to an increase in electric → resistivity that is two to three times higher than that of the same composition in crystalline form. Some metallic glasses have soft magnetic properties.

The very high cooling rates during rapid solidification of a melt are achieved by generating thin layers, filaments or droplets of the melt in good contact with an efficient heat sink.

Droplet method These are variants of the long-established technology of making lead shots by pouring molten lead through a perforated preheated steel dish. The principle behind is that a pending drop or a free-falling melt stream breaks up into droplets as a result of surface tension. There are several methods to intensify this process. The standard method is *spray atomization (spray forming)*, where high-velocity gas or water jets rapidly fragment a volume of melt into large numbers of small droplets (Figure 4.12). The jets form, propel, and cool the droplets that may

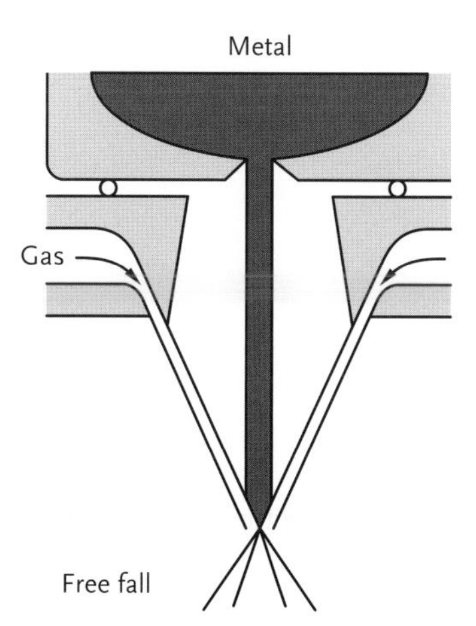

Figure 4.12 Principle of spray atomization by impingement of a high-velocity gas jet on a free-falling melt stream.

freeze completely to form a powder or may form individual splats ("splat quenching") or a thick deposit by impact on a suitable substrate. For example, in the splat-quenching method, the cooling rate can be as high as 10^4 K s^{-1}. The materials need to be consolidated for typical engineering applications.

Spinning methods A single melt stream is stabilized by surface film formation before it can break up into droplets ("melt spinning" or "melt extrusion"). In the standard chill-block melt-spinning method (Figure 4.13), the melt stream is flung onto a chilled roll rotating at a surface speed of some tens of meters per second. A single ribbon is formed that is typically 10–50 μm thick and a few mm wide.

Surface melting methods In its simplest form, a heat source, such as an electron or laser beam, is scanned over the surface of a material to be melted. The molten pool cools very rapidly behind the beam, because the unmelted bulk acts as the heat sink (Figure 4.14). This process can also be used to alloy the surface. An

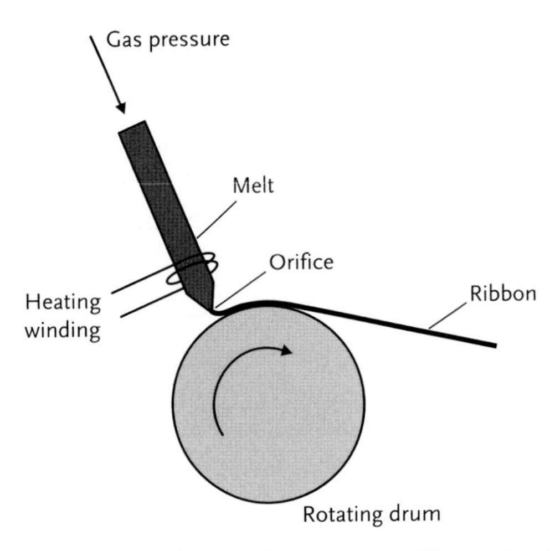

Figure 4.13 Schematic drawing of chill-block melt-spinning for the formation of glass ribbons.

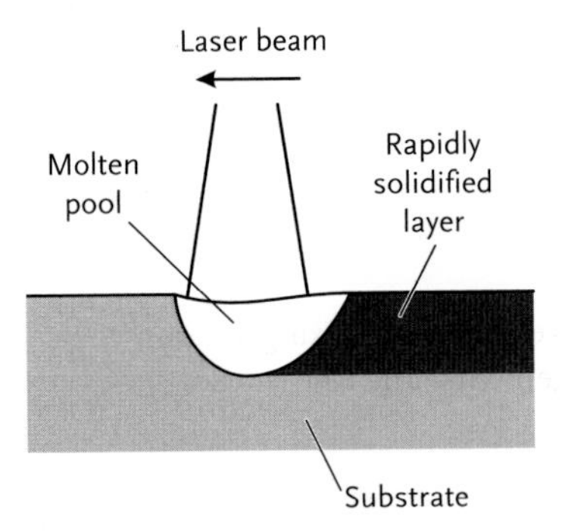

Figure 4.14 Laser surface melting and quenching.

alloying material preplaced on the substrate is fused to and mixed with the surface material. This results in an alloyed layer after solidification.

Ion mixing This is a completely different approach. Multilayer films are first prepared by depositing alternatively the metals A and B on a substrate (Figure 4.15). The as-deposited films are then irradiated by a scanning ion beam (such as Xe ions) to achieve a uniform mixing between metal layers A and B, resulting in forming an A_xB_y → alloy. The effective cooling rates can be as high as 10^{14} K s^{-1}. The supporting substrate can be an inert compound such as silica or NaCl, or it can be metal A as a matrix for forming a surface alloy.

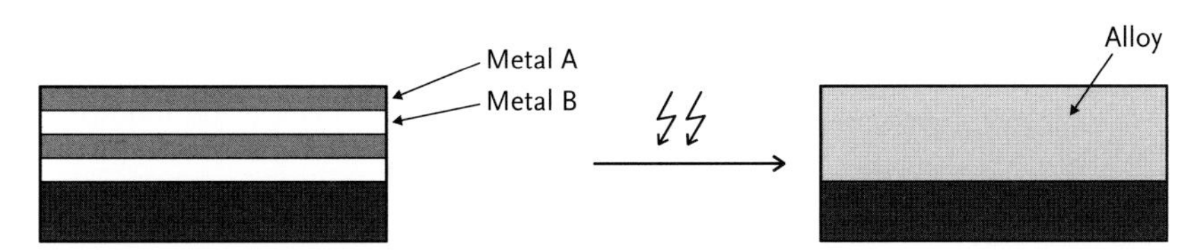

Figure 4.15 Ion-mixing of multilayered films consisting of metal layers A and B.

4.1.4.1 **Bulk Metallic Glasses**

Recently, certain high-order metallic → alloys were synthesized that have exceptional glass-forming abilities, and do not need rapid quenching to form glasses. Such alloy systems include La-Al-Cu-Ni (e.g., $La_{62}Al_{15.7}Cu_{11.15}Ni_{11.15}$), Zr-Ti-Cu-Ni-Be (e.g., $Zr_{41.2}Ti_{13.8}Cu_{12.5}Ni_{10.0}Be_{22.5}$ commercialized as Vitreloy 1®) or $Mg_{54}Cu_{26.5}Ag_{8.5}Gd_{11}$. They have high values of the reduced glass-transition temperature (T_g/T_m) of above 0.6 and form glasses at cooling rates lower than 10^2 K s^{-1}. This allows the casting of glassy rods or bars with maximum sample thickness up to a few centimeters by conventional metallurgical methods. They are called "bulk metallic glasses." The key empirical rules for slow crystallization kinetics (decreased critical cooling rate) include:

- The alloy system should contain more than three elements. This increases the complexity and size of the unit cell; the energetic advantages of forming a periodically ordered structure become progressively more marginal as the systems become more complex.
- There should be a significant difference in atomic size ratios ($>$ about 12%) among the main constituent elements. This results in a higher packing density (smaller free volume) in the liquid state.
- The main elements should exhibit negative heats of mixing among themselves. This retards local atomic rearrangements and the nucleation rate, extending the supercooled liquid temperature range, and leads to a high liquid/solid interfacial energy barrier [see Eq. (4.2)].
- The alloy composition should be close to a deep eutectic. This forms a stable liquid at low temperatures.

Current uses for bulk metallic glass are, for example, for sports goods (golf club heads, tennis rackets, etc.), electronics (casings of cell phones or cameras), medical applications (pacemaker casings, scalpels, etc.), or defense and aerospace applications.

4.2 Precipitation

Every first-year science student knows how to precipitate sparingly soluble salts from solution. However, simply adding a sufficiently concentrated solution of one constituent ion to the solution of the other does not give control over particle size, particle morphology, and so on.

Precipitation is a rather complex phenomenon that can involve the following processes:

- → nucleation,
- crystal growth,
- → Ostwald ripening,
- recrystallization,
- coagulation, and
- agglomeration.

Each of these steps must be controlled in order to create monodisperse particles with a well-defined and reproducible morphology. Together with the chemical composition, the size, size distribution and habit of the crystalline particles contribute to the actual properties and performance of the desired material.

In technical precipitation processes, water is mostly the solvent. The following general considerations can, of course, be applied to any other solvent. A very special "solvent" is a molten salt. For example, the → perovskite $BaTiO_3$ was prepared from $Ba(NO_3)_2$ and partially hydrolyzed $TiCl_4$ in a molten mixture of $NaNO_3$ and KNO_3 (the → eutectic mixture melts at 225 °C) at 500 °C. Note that preparation of $BaTiO_3$ by the ceramic method (Section 2.1.1) would require temperatures >1000 °C.

Some theoretical considerations concerning nucleation and crystal growth were already made in Section 4.1 for the crystallization from melts. Formation of crystalline nuclei in melts occurs by lowering the temperature. For precipitation (crystallization) from solution, the concentration of the precipitating solute species must be increased until nuclei are formed. This can be achieved in different ways, for example:

- direct reaction of ions (e.g., addition of bromide ions to a solution containing silver ions to give AgBr);
- redox reactions (e.g., reduction of $HAuCl_4$ to give → colloidal gold);
- precipitation by poor solvents (e.g., addition of water to ethanolic solutions of sulfur to precipitate sulfur);

- decomposition of compounds (e.g., addition of an acid to an aqueous thiosulfate solution to obtain elemental sulfur); and
- hydrolysis reactions (see Section 4.5.3).

The initial formation of particles from solution proceeds as indicated in Figure 4.16 (LaMer model):

- The concentration of the solute is continuously increased up to the minimum concentration for nucleation, c_0, by one of the processes mentioned before. No precipitation takes place.
- When c_0 is reached, nucleation sets in. The concentration of the solute keeps climbing up to the maximum concentration for nucleation, c_N, and then decreases again due to the consumption of the solute by nucleation and precipitation of particles. The range between c_s and c_N corresponds to a supersaturation. At the critical concentration, c_N, nucleation is very rapid.
- Once the minimum concentration for nucleation, c_0, is reached again, new nuclei are no longer formed. Crystal growth reduces the concentration until the equilibrium solubility, c_s, is reached.

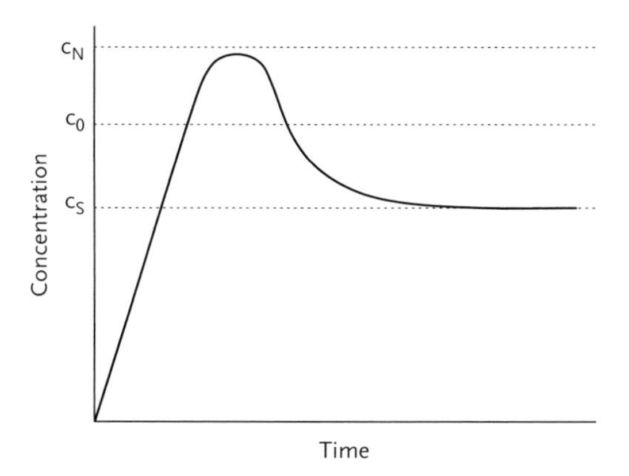

Figure 4.16 LaMer model for the concentration of the solute during crystallization.

If new nuclei are formed during the growth period, then a range of particle sizes results. Thus, separation between the nucleation and the growth stage is the primary requirement for obtaining uniform particles (see below). This means that after nucleation, the particle-forming solute species must continue to be formed at a rate that allows their removal from solution by reaction with the existing particles so that no secondary nucleation can take place. The rate of growth of the particles may be controlled by diffusion of dissolved species to the particle or by the rate of the condensation reactions between the particle and dissolved species.

An alternative supply of particle-forming species is by dissolution of small particles. As will be discussed in Chapter 7, smaller particles are more soluble than larger particles, due to curvature effects. Thus, larger particles grow at the expense of smaller ones. Larger, more stable particles are thus formed – a process called → Ostwald ripening.

Particles formed by these mechanisms should be crystalline, although amorphous, porous particles are frequently obtained. Crystal growth is only possible if the attachment of the molecular species is weak enough to allow the species, if necessary, to redissolve and attach to the surface in a position required for a crystalline structure. This is more likely in dilute solutions. Amorphous particles are obtained if the molecular species attach irreversibly to the particle and are unable to reorient. Note the analogy to CVD processes (Section 3.2), for example, where the probability that a precursor molecule reacts at the first point of contact with the surface should be significantly smaller than unity to obtain smooth films.

The growth mechanism discussed so far does not account for all the available experimental results. For example, monodisperse spheres were obtained under conditions where the concentration of the particle-forming species in solution is above c_0 (Figure 4.16) and new nuclei should be created continuously. This should result in broad particle-size distributions according to the classical theory. Electron micrographs of such materials show that the particles consist of a large number of tiny, aggregated primary particles. A *nucleation–aggregation model* was proposed for such cases that assumes the initial small primary particles to be too unstable owing to their small size. They aggregate to secondary particles, and freshly formed primary particles attach to the aggregates. Monodispersity of the final precipitate is achieved through size-dependent aggregation rates. The aggregate structure apparently causes a collective reduction in surface area. The concomitant reduction in surface energy is the driving force for the ordered aggregation. Rather large particles (up to 250 nm) with this internal structure can be obtained for various compositions. This excludes an alternative explanation of the observed behavior based on Ostwald ripening, because the difference in solubility and hence Ostwald ripening disappears when particles exceed about 5 nm in diameter.

4.2.1 Monodisperse Crystals

The synthesis of → monodisperse crystals, is of great interest for many technical applications, such as the preparation of stable dispersions, uniform ceramic powders, pigments with a reproducible color or catalysts with an optimized activity. Monodispersity is a particularly important topic for nanocrystals (see Chapter 7).

The most important approach involves the separation of nucleation and crystal growth by quickly mixing two precursors together at a high temperature to induce a short burst of nucleation from the supersaturated solution followed by the slow diffusion-controlled growth of particles without significant additional nucleation. This corresponds to passing as quickly as possible through the range between c_0 and c_N in the LaMer model (Figure 4.16). Because all nuclei are generated at almost the same time, their growth history is nearly the same.

A method for the synthesis of large quantities of well-defined dispersions is the *controlled double-jet precipitation* (CDJP) process. The technique was developed for the photographic industry to make precisely defined silver halide crystals, but it is now used to produce various other crystalline materials (see also Chapter 7 for another example). Particle sizes from the nanometer to the micrometer range can be obtained.

The method is based on the simultaneous introduction of the reactant solutions through separate input lines into a stirred reactor under precisely defined conditions (Figure 4.17). In the region where the highly concentrated solutions of the precursors are introduced, the primary mixing zone, a very high supersaturation is reached, typically 10^5–10^8 times the solubility. In this zone a large number of unstable nuclei is formed during addition of the reactants. The unstable nuclei are transported in the well-mixed bulk of the vessel, the secondary mixing zone, where they redissolve if the supersaturation in this zone is low enough (otherwise new nuclei will survive and lead to broad particle-size distributions).

In the early stage, the supersaturation in the secondary mixing zone increases by dissolution of the unstable nuclei. This results in an increasing number of slightly grown stable nuclei. Once a sufficient number of stable nuclei has been built up in the bulk phase, they become able to absorb all species generated by the constant dissolution of the unstable nuclei. From this moment, the number of growing crystals no longer increases, and the unstable nuclei generated in the primary mixing zone act as a monomer source for crystal growth (→ Ostwald ripening). Thus, after a short initiation stage, the only growth of stable crystals occurs in the secondary mixing zone.

For example, mixing aqueous solutions of silver nitrate and sodium bromide results in the immediate precipitation of silver bromide (AgBr), but the obtained precipitate would contain a mixture of crystals of different size and shape. The precipitation process of silver halides can be controlled in such a way that crystals uniform in size and shape can be obtained as needed by the photographic industry. In general, aqueous solutions of silver nitrate ($AgNO_3$) and an alkali

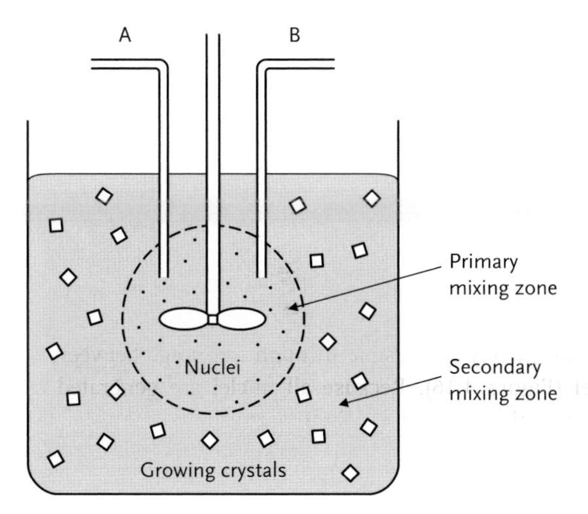

Figure 4.17 Model of a crystallizer for the controlled double-jet precipitation (CDJP) process.

halide are added to an aqueous gelatin solution (about 2–5 wt%) with vigorous stirring at 30–80 °C. The gelatin serves to prevent the coagulation of the microcrystals (see Chapter 7). The soluble alkali salts generated during precipitation and the excess alkali halides must then be removed before the emulsions can be further processed.

4.2.2
Shape Control of Crystals

The history of shape control of crystal goes back to the classical crystallization theories developed in the nineteenth century. Currently, this is a topical issue for nanocrystals (see Chapter 7), because many of their applications require nonspherical morphologies.

Particular crystal shapes originate from either the growth kinetics or from minimizing the surface energy (see Section 4.1.2). If kinetics dominates, then the shape is determined by the rate at which the different crystal faces grow. If the particles grow in thermodynamic equilibrium, their shape results from minimizing their surface energy.

Wulff's rules suggest that the morphology of a crystal in thermodynamic equilibrium is determined by minimization of the surface free energy of individual crystallographic faces, namely Σ(area surface energy) of all exposed faces. As a consequence, faces with the highest energy grow fastest and eventually disappear. An example is shown in Figure 4.18. If solute is preferentially deposited at the {100} face of a cubic crystal, the {100} face grows faster than the {111} face and eventually disappears. Intermediate forms, such as truncated cubes, cuboctahedra, truncated octahedra, are obtained depending on the reaction time period.

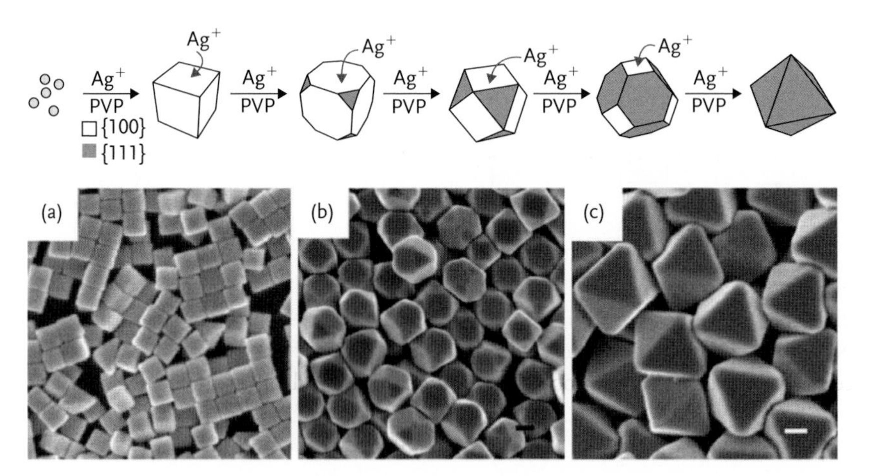

Figure 4.18 Growth of silver crystals by reduction of silver nitrate in the presence of PVP. (a), cubes; (b), cuboctahedra; (c): octahedra (scale bar: 100 nm).

The surface energy of a crystallographic face can be influenced selectively by adsorption of ions, small organic compounds, or surfactants. This opens up the possibility of deliberately influencing the crystal shape by additives.

An example is the growth of monodisperse and uniformly shaped AgBr crystals, developed by the photographic industry. The crystal shape can be controlled by holding the Ag^+ concentration (i.e., the pAg), and thus the bromide concentration at a constant value. AgBr forms cubic crystals only when precipitated at pAg values <7.5. For pAg >8.5, the crystallites become octahedral. The reason for this behavior is the difference in the strength of bromide adsorption to the {100} and {111} crystal faces. At pAg <7.5, more bromide is adsorbed to the {100} than to the {111} faces, so that growth is inhibited at the {100} faces relative to the {111} faces. The octahedral faces thus grow faster and eventually disappear, leaving only the cubic {100} faces. In contrast, at pAg >8.5, the {111} faces are more strongly inhibited by bromide ions.

The shape of AgBr crystals can also be controlled by adding surfactants. In the example in Figure 4.18, PVP [poly(vinylpyrollidone)] is more strongly adsorbed to the {111} face, eventually resulting in the formation of octahedral crystals. Instead of exploiting different adsorption strengths of an additive to different crystal faces, surfactant mixtures can be employed, where the surfactants are selectively adsorbed at different crystallographic faces.

4.3 Biomaterials

Inorganic chemistry and life sciences combined into one discipline appear to be a contradiction at first sight. However, biomineralization processes and bioinspired inorganic materials chemistry are attracting increasing attention, and a vast amount of research is now devoted to them. Two different classes of materials will be discussed in this section: on the one hand, solid biomaterials produced by living organisms such as bones, teeth, spines, and shells (Section 4.3.1) in which nature shows us a huge variety of fascinating morphologies, beauty and complexity with specific mechanical, structural, and optimally adapted functionalities; and on the other hand, substances that are prepared by biomimetic approaches, or which will be used in medical devices designed for contact with the living body (Section 4.3.2). The biomimetic synthesis of soft materials such as chemically responsive hydrogels (artificial muscles, skin) will not be highlighted. Instead, we will focus on biomimetic approaches (Section 4.3.3) to inorganic-based materials, nanophases, and → composites. Some of the more important aspects from a biological point of view cannot be covered within this textbook, for example the activity of the cells that govern all the processes, the hormones and other molecules that are involved in communication between the organism and mineralized cells, as well as all the complicated uptake and transport phenomena of ions to the reaction center. Further information regarding those topics can be found in the literature (see Further Reading).

4.3.1 Biogenic Materials and Biomineralization

Biomineralization refers to the process by which living organisms form inorganic solids. One of the major questions is: why is biomineralization different from other mineralization/crystallization processes (see Section 4.2)? The major component of, for example, seashells (Figure 4.19) that can be found along a beach is crystalline calcium carbonate. However, just try to compare these beautiful biological objects and their complicated architectures and morphologies with their man-made synthetic analogs! It is a fascinating comparison.

First, the "test-tube" calcium carbonate is a mixture of crystals of all kinds of shapes and morphologies, while nature exerts a remarkable control over the crystal's size, shape, and orientation, as well as the properties of the resulting materials such as high strength, resistance to fracture, and aesthetic value. Secondly, abiogenic crystals are formed corresponding to the thermodynamic/kinetic conditions during the synthesis, while the structure of the biogenic material is species-specific and also depends strongly on the local environment in which the crystals are formed.

Biogenic minerals can be found everywhere: oyster shells, corals, ivory, sea urchin spines, magnetic crystals, and so on. They are formed on a huge scale in the biosphere, have a major impact on ocean chemistry, and are important components of marine sediments and ultimately of many sedimentary rocks as well. A major function of biogenic minerals is to provide mechanical strength to skeletal hard parts and teeth. However, biominerals cannot be considered as static systems, but display an active demineralization/regeneration behavior that makes them useful as storage media, for example for iron or calcium. Not all biomaterials are desirable – the biomineral calcium oxalate monohydrate, $CaC_2O_4 \cdot H_2O$, is a major component of urinary stones. An understanding of the processes that lead

Figure 4.19 Different types of seashells.

to calcium oxalate precipitation in the urinary tract could lead to the prevention or treatment of urinary stone disease.

Common constituents of biological minerals are carbonates, phosphates, halides, sulfates, and oxalates of the alkaline-earth metals, particularly calcium, and the oxides of silicon and some transition metals, mainly iron (Table 4.2).

The predominance of calcium-containing minerals over other alkaline-earth metals (group IIA) can be explained by the low solubility products of the carbonate, phosphate, pyrophosphate, sulfate, and oxalate and the relatively high levels of Ca^{2+} in extracellular fluids (10^{-3} M). Magnesium salts, for example, are generally more soluble, and no simple Mg biominerals are known. Most biominerals

Table 4.2 Important biominerals, their chemical composition, and their function.

Chemical composition	Mineral	Function and examples
Calcium carbonate		
$CaCO_3$	Calcite	Exoskeletons (e.g., egg shells, corals, mollusks, sponge spicules)
	Aragonite	
	Vaterite	
	Amorphous	
Calcium phosphates		
$Ca_{10}(OH)_2(PO_4)_6$	Hydroxylapatite	Endoskeletons (bones and teeth)
$Ca_{10-x}(HPO_4)_x(PO_4)_{6-x}(OH)_{2-x}$	Defect apatites	
$Ca_{10}F_2(PO_4)_6$	Fluoroapatite	Calcium storage
$Ca_2(HPO_4)_2 \cdot 2\ H_2O$		
$Ca_2(HPO_4)_2$		
$Ca_8(HPO_4)_2(PO_4)_4 \cdot H_2O$		
$Ca_3(PO_4)_2$		
Calcium oxalate		
$Ca_2C_2O_4 \cdot (1\ \text{or}\ 2)\ H_2O$	Whewellite Wheddelite	Calcium storage and passive deposits in plants, calculi of excretory tracts
Metal sulfates		
$CaSO_4 \cdot 2\ H_2O$	Gypsum	Gravity sensors
$SrSO_4$	Celestite	Exoskeletons
$BaSO_4$	Baryte	Gravity sensors
Amorphous silica		
$SiO_n(OH)_{4-2n}$	Amorphous (opal)	Defense in plants, diatom valves, sponge spicules, and radiolarian tests
Iron oxides		
Fe_3O_4	Magnetite	Chiton teeth, magnetic sensors
α, γ -FeOOH		
$5\ Fe_2O_3 \cdot 9\ H_2O$		

are ionic salts; an important exception is silica, due to the stability of Si–O–Si units in water.

A variety of different structural types of biomaterials is known, such as *amorphous* materials, *ordered mesoscopic crystal aggregates*, for example, in the form of macroscopic functional materials such as bones or teeth, and *nanocrystalline* materials. Typical examples for all three types of biominerals are given in the following paragraphs.

4.3.1.1 **Amorphous: Diatoms**

Diatoms are microscopic unicellular algae that are important components of the phytoplankton. Diatoms possess unique exoskeletons (shells or frustules), which are composed of biogenic amorphous silica (Figure 4.20). The living part is encased inside. When diatoms die, the silica shells collect on the ocean floor. These deposits are used commercially as components in products such as shoe polish and cosmetic articles.

The porous shells or microskeletons found of biogenic silica are based on a covalently linked, but randomly arranged, thus amorphous polymeric network, consisting of tetrahedrally coordinated silicon centers (see also Figure 4.2). The precipitation as an amorphous network – in comparison to other biominerals that are crystalline – is explained by the high stability of the siloxane bonds in water; therefore, no reversibility of the condensation reaction is seen. The high variability

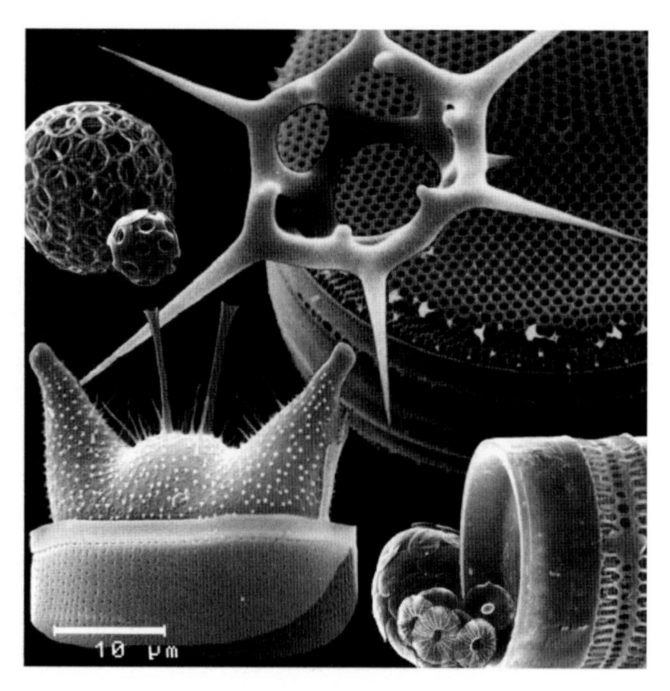

Figure 4.20 Diatom impressions – the fascinating world of diatoms in the micrometer range.

in the Si–O–Si bond angle from 104° to 180° allows for a range of different structures. Molding of amorphous silica into a wide variety of elaborate structures is easier compared to crystalline materials, due to the absence of crystal planes that can fracture during the shaping process.

As mentioned, silica in biological systems is not crystalline, though a microscopic morphological order is often observed. Such order may arise during either the → nucleation or the growth processes. Energy considerations support the formation of a particulate silica aggregate with a densely packed, covalently bonded core and a highly hydrated surface. The surface is likely to interact with organic substrates in a biological environment in a manner analogous to crystal interactions, thereby lowering the free energy of aggregate formation and controlling aggregate morphology on the microscopic scale (Figure 4.21).

Figure 4.22 shows a model of the silicon metabolism of unicellular diatoms. The formation of diatom shells is connected to the vegetative cycle during which two daughter cells with complete exoskeletons are formed by division of a mother cell.

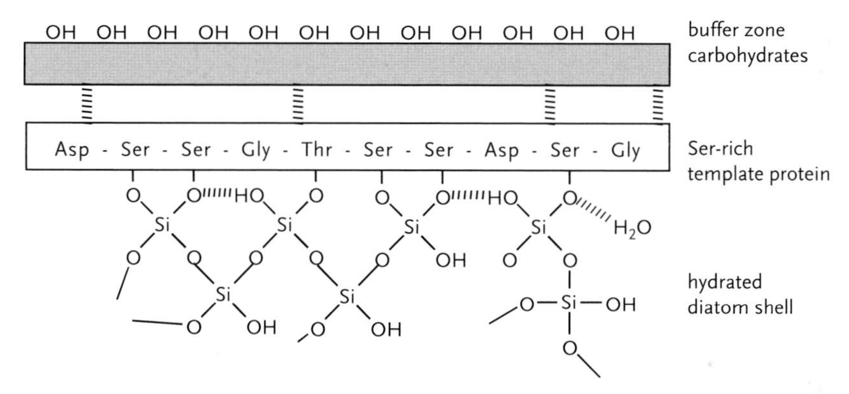

Figure 4.21 Model of a diatom shell.

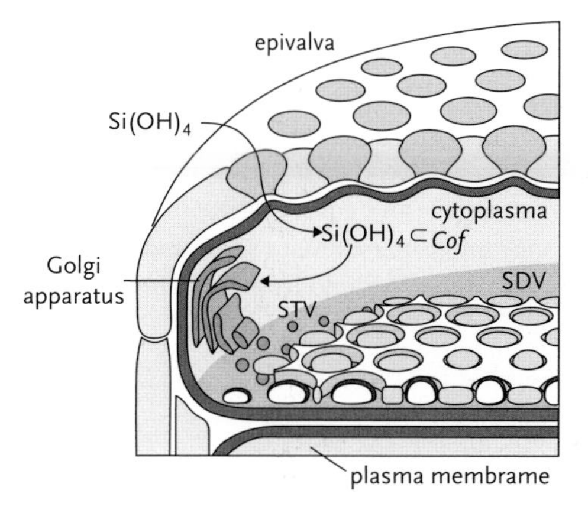

Figure 4.22 Model of the silicon metabolism of diatoms.

Seawater is generally undersaturated with respect to silicic acid ($Si(OH)_4$), the precursor for the deposition of silica. Therefore, mechanisms have evolved to control processes such as uptake and transportation as well as concentration and polycondensation of silicic acid. It is known that the synthesis of the diatom shell occurs in a stepwise manner by transport of monomeric silicic acid from the environment into the interior of the cell by active transport mechanisms. To avoid an uncontrolled → polycondensation of the silicic acid with increasing concentration in the cell it is bonded to a cofactor ($Si(OH)_4 \subset Cof$), the chemical nature of which is still unknown. The Golgi apparatus of the cell probably serves as reservoir for the masked silicic acid. From the silicic acid depots of the Golgi apparatus small → vesicles, silica transport vesicles (STV), are formed that fuse with the silica deposition vesicles (SDV), the mineralization organelles of the cell. Condensation of the silicic acid takes place inside the SDVs, which are located on the bottom part of the new cell wall, and grow rapidly in all directions. Areolae vesicles are also located on this cell wall that are usually one layer of close-packed bubbles serving as a negative pattern for the SDVs. Mineralization of silica occurs around these areolae vesicles, and this often leads to a hexagonal structure of the diatoms shells. Other structures can at least superficially be explained based on this simple model by geometrical deviations in the packing of the areolae vesicles. This mold–prepattern hypothesis describes the formation of diatom shells, but does not explain it. The form and appearance of the silica shell is governed by genetic factors, and is amazingly reproducible.

From a materials science point of view, morphogenesis of the diatom shell is an accomplishment that is unparalleled by all synthetic approaches towards other silica-based porous materials such as zeolites, MCMs and so on (see Section 6.3).

4.3.1.2 **Ordered Mesoscopic Crystalline Aggregates: Bone**

Bone is a fascinating material with very unusual properties, many of which seem to be contradictions, especially in terms of mechanical behavior. Bone serves two essential functions: on the one hand as a structural material that is able to support its own weight, withstand acute forces, bend without shattering, and so on; and on the other hand as an ion reservoir for both cations and anions. Both functions depend to a significant extent on the exact size, shape, chemical composition, and crystal structure of the mineral crystallites, and their arrangement within the organic matrix. Bone mineral can be generally represented by the formula $Ca_{8.3}(PO_4)_{4.3}(CO_3)_x(HPO_4)_y(OH)_{0.3}$, in which y decreases and x increases with increasing age, whereas $x + y$ remains constant and equal to 1.7.

The structure of the bone as a → nanocomposite material can be understood in terms of the different levels of organization which extend from the nanoscopic to the macroscopic scale:

- The lowest level of organization describes the crystals, the organic framework (mostly collagen fibrils), and the relationship between the framework and the crystals.
- The next level of organization (tens of micrometers) describes the longer-range order of collagen and associated crystals.

- The highest level of organization describes the macroscopic build-up of bones. They are usually composed of a relatively dense outer layer (the cortical bone) surrounding a less dense, porous tissue (cancellous bone), which is filled with a gel-like tissue known as bone marrow.

4.3.1.3 Crystalline Materials

Of the many transition metals that display a rich biocoordination chemistry, only iron and, to a smaller extent, manganese, have extensive roles in biomineralization. The bioinorganic solid-state chemistry of these elements is dominated by the redox chemistry as an energy source for biological activities, an affinity for O, S, and OH ligands, and ease of hydrolysis in aqueous solution. Like the calcium-containing biominerals, biological iron oxides are used to strengthen soft tissues and as storage depots (Fe^{3+}, OH^-, and HPO_4^{2-}). Furthermore, the magnetic properties of mixed-valence phases are utilized by bacteria of several types for navigation in the ambient geomagnetic field. Most magnetotactic bacteria synthesize intracellular magnetite (Fe_3O_4 or $Fe_2^{III}Fe^{II}O_4$); species inhabiting sulfide-rich environments deposit the isomorphic mineral greigite (Fe_3S_4). The size and morphology of the crystals is controlled by organic membranes which are species dependent. In both systems, the crystals must be aligned in chains to impart the bacterium with a magnetic dipole moment and must have dimensions compatible with those of a single magnetic domain (about 40–80 nm, ferromagnetism) (Figure 4.23). Smaller particles would show superparamagnetic behavior (see Chapter 7) and larger particles would have several domains that would be unable to function efficiently as a biomagnetic compass.

The growth of the magnetite crystals is a series of complex chemical processes, which is to a large extent governed by the formation of aligned, spherical magnetosome vesicles. These vesicles are essentially the reaction vessel for the formation of well-ordered, single crystals of magnetite. First, Fe^{III} ions are transferred from the environment into the cell, and Fe^{III} is reduced to Fe^{II} during this transport process through the cell membrane. Fe^{II} is transported to the magnetosome vesicles and across this vesicular membrane. Inside the vesicles, iron is

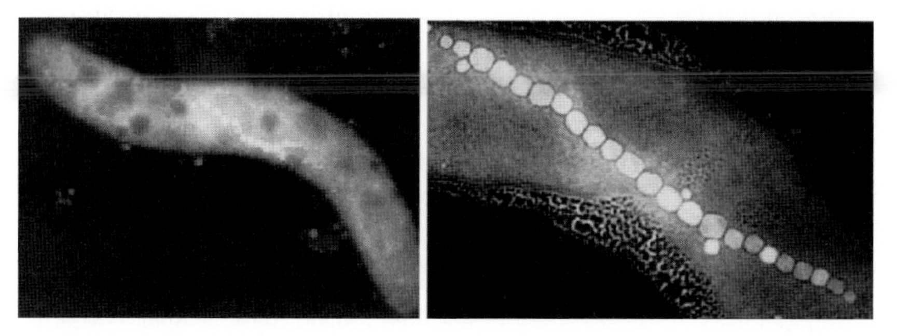

Figure 4.23 Transmission electron micrograph of a magnetospirillum bacterium (left). The chain of magnetite crystals (magnetosomes) can be seen in the picture to the right. Each magnetosome crystal is about 40–60 nm in length.

deposited as amorphous hydrated Fe^{III} oxide and in the last step of this sequence, the amorphous phase is converted to magnetite.

4.3.1.4 **Mineralization Processes**

Much is known about the ultrastructures of biominerals and how they vary in different organisms, but precise details about the molecular interactions governing their formation are still unknown. The precipitation of many of the minerals listed in Table 4.2 from aqueous solutions is a relatively straightforward laboratory procedure, but controlling the size, shape, orientation, and assembly of these crystals, as typical of many biominerals, is a much more complex task. The fundamental physicochemical principles are the same as already discussed in Section 4.2: supersaturation, → nucleation, and crystal growth. In biomineralization processes these steps depend critically not only on the ion concentration of the medium but also on the nature of interfaces (mineral–organic matrix and mineral–environment) present in the system.

Mineralization processes take place in an open system (cell with selective permeable cell membranes) far away from the thermodynamic equilibrium. The cell is in a permanent exchange of energy and material with the environment. In this well-defined spatially delineated site, regulation of the mineralization process is possible. Localized compartments (e.g., → vesicles) which are surrounded by lipid membranes are very common. Furthermore, the exact regulation of the physicochemical processes in these compartments allows the control of structure in biominerals. To achieve supersaturation (Figure 4.16), the compartments in which the mineral is formed must allow passive diffusion of ions and/or the active accumulation of ions against concentration gradients. Ion-specific pumps and channels are therefore necessary components of the machinery required for biomineralization. The site must be activated at specific times in the life of the organism, constrained in size and shape, and highly regulated with respect to the chemistry of the mineralization process.

The process of biomineralization can be divided into four stages (supramolecular preorganization, controlled nucleation, controlled crystal growth, and cellular processing) as follows.

Supramolecular preorganization As already mentioned, one of the prerequisites for the controlled deposition of biogenic inorganic materials in living organisms is the presence of a supramolecularly organized reaction compartment in which the mineralization zone is isolated from the cellular environment. These compartments can be located:

- on or in the membrane wall of bacterial cells (epicellularly);
- outside the cell, such as the extracellular facilitated construction of extended protein–polymer networks as in the collagen matrix for bone formation. Many shells and teeth are constructed within frameworks that may be lamellar, columnar, or reticular;
- intracellularly by self-assembly of enclosed protein cages or lipid → vesicles in which the molecular construction of the compartment is mainly based on the balancing of hydrophobic–hydrophilic interactions that exist for → amphiphilic molecules in aqueous environments.

Controlled nucleation by interfacial molecular recognition Controlled → nucleation of inorganic clusters from aqueous solution into the framework built in the first stage by supramolecular preorganization is one of the major points in biomineralization processes. The underlying concept is that these preorganized organic architectures consist of somehow functionalized surfaces that serve as blueprints for site-directed inorganic nucleation. Electrostatic, structural, and stereochemical recognition processes at the inorganic/organic interface are required (Figure 4.24).

It can be assumed that several of these factors act cooperatively in biological systems. The most fundamental aspect of recognition involves the matching of charge and polarity distributions. Additionally, the curvature of molecular cavities, which can be concave, convex, or planar, provides dimensional control over nucleation and a limit on the size of the nucleation site.

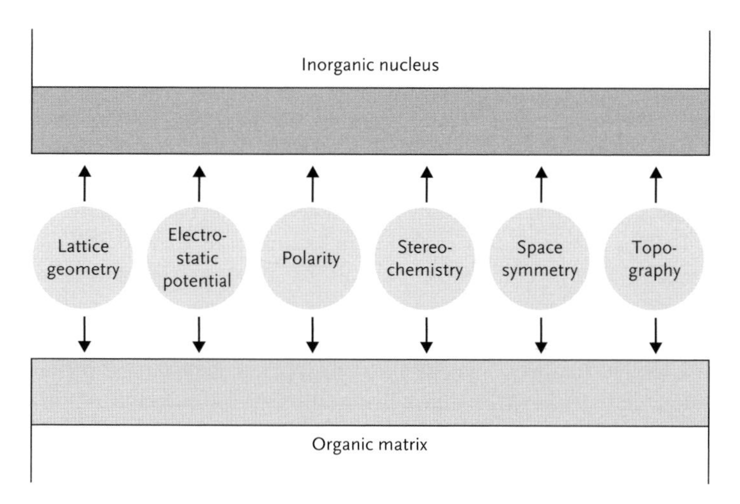

Figure 4.24 Possible modes of molecular complementarity at inorganic/organic interfaces.

Most favorable for a controlled nucleation are probably concave surfaces, since here the local concentration of functional groups is very high. Convex surfaces are less active, since the binding sites are not in close proximity to each other. The same is true for planar surfaces. However, those allow for long-range structural matches, which is called "biological epitaxy" (Figure 4.25). Epitaxy in biomineralization is distinct from inorganic → epitaxy, since the organic substrate does not show the typical smoothness or rigidity but exhibits surface stereochemistry due to exposed functional groups, and so on. The role of the organic surface involved in inorganic crystallization is primarily to lower the activation energy of nucleation (see Section 4.1, heterogeneous nucleation).

It is not only short-range interactions that are important; sometimes larger periodic structures can control inorganic → nucleation by the secondary, tertiary, or quaternary conformation of macromolecules that can act as blueprints for a controlled nucleation along specific crystallographic axes. A good example for such

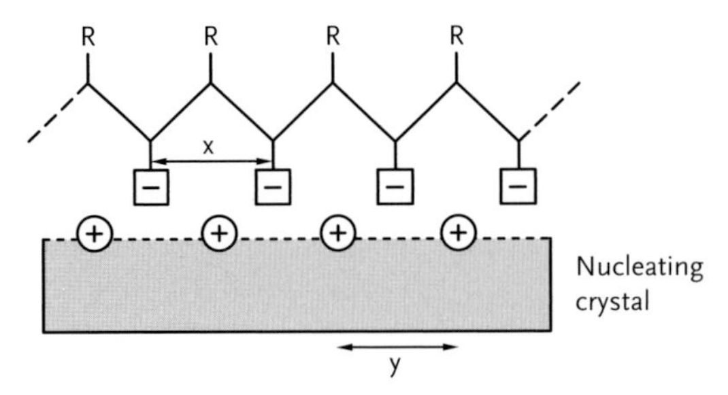

Figure 4.25 The concept of epitaxy as applied to biomineralization. Geometric matching must exist at the interface between a structured organic surface and nuclei of the inorganic crystal. Cation–cation distances in one specific crystal face are commensurate with the spacing of periodic binding sites on an organic surface (i.e., $x \approx y$).

macromolecules is collagen in bone formation. Here, the bone crystals are nucleated in the interstices of a crystalline assembly of collagen fibrils.

Controlled crystal growth By simple nucleation of an inorganic phase in a supramolecular host, followed by crystal growth according to the laws of crystallization, the resulting particles would be constrained in size, but would exhibit normal morphologies. Biological constraints such as genetic mechanisms have to be considered to explain the complexity of shapes in biominerals. They control the cellular environment and therefore the shape of organic compartments for → nucleation and crystal growth. This genetically based morphogenesis causes the mineralized architectures to be unique and species specific.

The chemistry within localized biological environments determines crystal growth, aggregation, and texture. This allows that in the same biomineralized system different compounds or polymorphs can be deposited such as $Fe_2O_3 \cdot n\ H_2O$, γ-FeOOH, and Fe_3O_4 in molluscan teeth, or aragonite and calcite in some shells.

In some systems it is also possible that a spatial location of ion pumps within the enclosed organic compartment allows vectorial shaping of the crystal. If ions flow into the localized compartments only at specific ports of entry, then these sites will be the initial regions of mineral growth. If these sites are now turned off and other pumps along the membrane are switched on, vectorial flow of the ion stream will cause the mineral to develop along a preferred direction.

Figure 4.26 illustrates some generalized processes (a–f) for controlling the supersaturation conditions of solutions encapsulated within preformed supramolecular assemblies.

These mechanisms include:

a. Energy-consuming build-up of a concentration gradient in the membrane by specific ion pumps (A^+ and B^- = extraneous ions).
b. Redox processes at the membrane surface followed by selective transport of the oxidized or reduced species into the cell (e.g., Fe^{3+} is transported through the bacterial cell wall after reduction to Fe^{2+}).

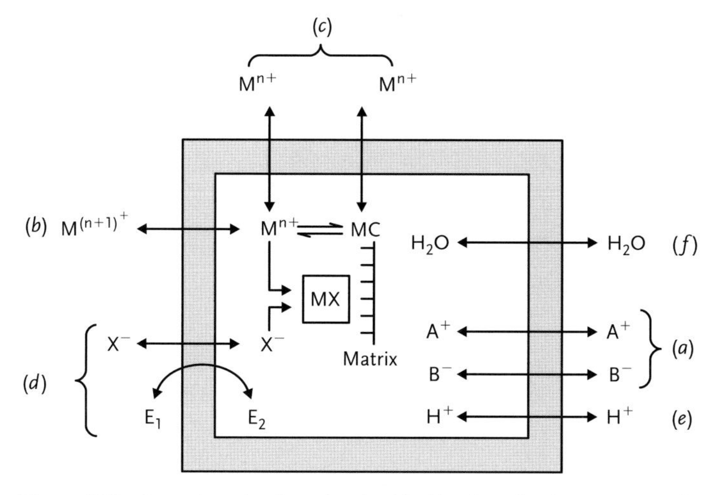

Figure 4.26 Control mechanisms involved in biomineralization processes (MX = biomineral; for an explanation of (a–f) see text).

c. Selective complexation of metal ions (M^{n+}) – usually at the inner membrane surface – followed, in a later stage, by the controlled decomposition of this complex (MC) to release metal cations in aqueous solution.
d. Enzyme (E)-mediated transport processes to increase anion (X^-) concentrations.
e. Variation of the pH.
f. In mineralization reactions in which water is produced by chemical reactions, control over the osmotic pressure governs nucleation, for example, the condensation reactions of Si–OH groups to Si–O–Si units.

In brief, regulation can be achieved by facilitated ion flux, complexation–decomplexation switches, local redox and pH modifications, and changes in local-ion activities.

Cellular processing Biomineralization does not stop with the formation of small particles, but proceeds with the construction of higher-order architectures with elaborate structural properties. An example of such an organized ultrastructure is the relatively rigid and spatially fixed linear assembly of a chain of membrane-bound magnetite crystals in magnetotactic bacteria (see Figure 4.23). Another example is the macroscopic organized architectures found in the nacreous layer of shells with sheet-like organic assemblies (Figure 4.27). These are secreted periodically beyond the mineralization front, leading to progressive infilling, resulting in the construction of a highly organized lamellar architecture.

The details of the recognition and organizational processes involved in the construction of these higher-order biomineralized architectures are currently unknown.

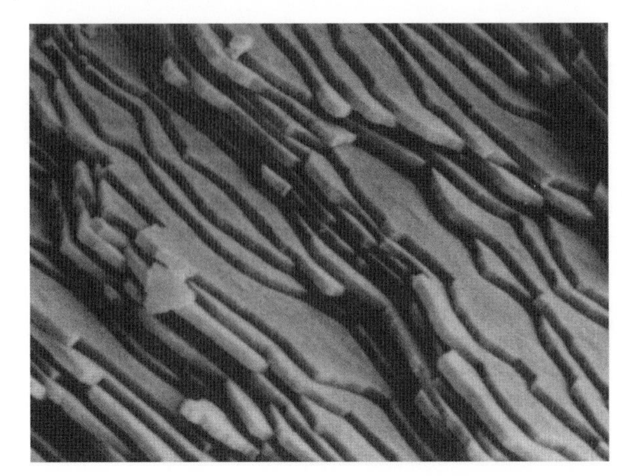

Figure 4.27 Scanning electron microscope image of the fracture of red abalone shell (*Halitois rufescens*) showing multiple tiling of ≈200–500-nm thick aragonite ($CaCO_3$) crystals. Organic layers between the tiles (<10-nm thick) are not discerned at this resolution.

4.3.2
Synthetic Biomaterials

An exact replication of the biological architecture and formation process would be desirable for the development of implants or prostheses. The limit so far is the ability (or better the non-ability) to command and direct living cells in a way to tailor a material deliberately. Therefore, ways must be found to design synthetic materials that can substitute biological materials.

Biomaterials are substances that are used in prostheses or medical devices designed for contact with the living body. Almost all types of materials are represented. Polymers are used in ophthalmology, for skin-wound treatments, and as soft-tissue implants, for example. Metals have a large range of applications, including devices for fracture fixation, partial and total joint replacement, as well as dental amalgams. Pyrolytic carbon is used as coatings, for example, on prosthetic heart valves, and ceramics and glasses as bioactive components for a better adhesion of implants to natural tissue or bone, and as drug-delivery carriers. Bioceramics are specially designed and fabricated ceramics for the repair and reconstruction of diseased, damaged, and "worn-out" parts of the human body. A schematic of some of the clinical uses of biomaterials is shown in Figure 4.28.

Biomaterials for medical applications need optimized mechanical, chemical, and biological properties. In many cases → composite materials and surface-modified materials are in use, because a single material cannot fulfill all the requirements. The form of the biomaterial depends on its intended function in the body. Load-bearing implants are usually made from bulk, non-porous materials, but coatings or composite structures may also be used to achieve improved mechanical and interfacial chemical properties. Implants that serve only to fill space or augment existing bone tissue are used in the form of powders, particulates, or porous materials.

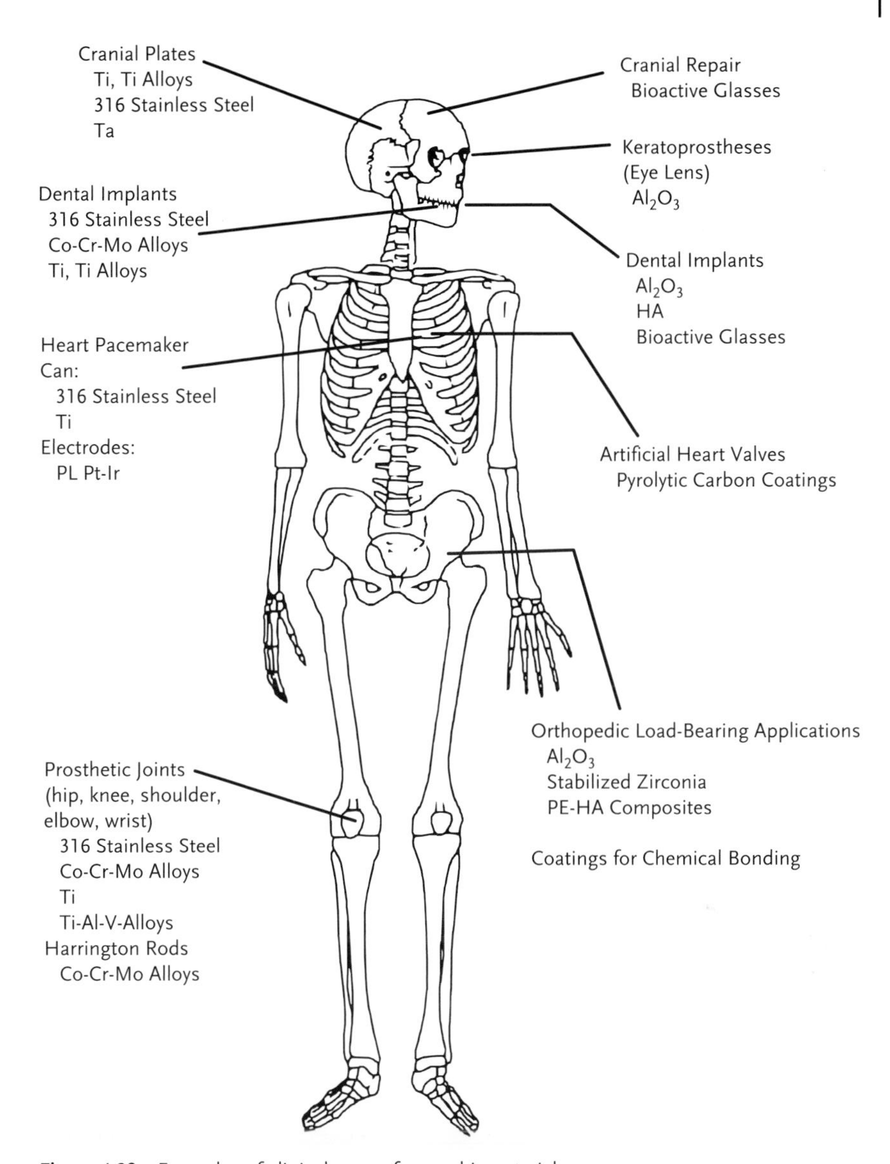

Figure 4.28 Examples of clinical uses of some biomaterials.

Different types of interaction between tissue and biomaterials can be distinguished:

- **Bioinert materials** show minimal interactions with the neighboring tissue. They do not release any compounds to the environment, and there is no damage to the tissue. For example, implants made of metals or non-porous alumina attach by bone growth into surface irregularities, by cementing the device into the tissues, or by press fitting or screwing into a defect (called morphological fixation).

- **Biocompatible materials** do positively interact with the neighboring tissue. Due to this interaction, the mechanical stability of the implant is enhanced. For example, hydroxylapatite implants are mechanically attached by ingrowth (biological fixation).
- **Bioactive materials** increase recovering (healing) and growth of the tissue. This is the ideal case for a biomaterial. Resorbable bioactive materials for implants and prostheses are designed to be slowly replaced by bone. Bioactive, dense, non-porous surface-reactive ceramics, glasses and glass-ceramics attach directly by chemical bonding with the bone (bioactive fixation).

4.3.2.1 **Bioactive Ceramics and Glasses**

Bioactive ceramics and glasses are materials with a high potential for medical applications (Figure 4.28) since they were found to produce a reaction layer on their surface and thereby form a bond with bone tissue. Current clinical applications for these materials are mainly as coatings on metallic prostheses or as bone-graft substitutes. There is also growing interest in the clinical use of calcium phosphate materials and ceramic composites as bone cements and settable bone-grafting materials.

Typical chemical components in bioactive materials are Na_2O, K_2O, MgO, CaO, Al_2O_3, SiO_2, P_2O_5, and CaF_2 in different combinations and ratios. A low silica content and the presence of calcium and phosphate ions in the glass result in rapid ion exchange in physiological solutions and in rapid → nucleation and crystallization of hydroxycarbonate apatite bone mineral on the surface. This growing bone-mineral layer bonds to collagen, which is grown by the bone cells, and a strong interfacial bond is formed between the inorganic implant and the living tissues. Depending on the bulk composition the behavior with respect to interactions to the neighboring tissue varies. The sequence of reactions that occur on the surface of bioactive glass as a bond with tissue is formed is summarized in Figure 4.29.

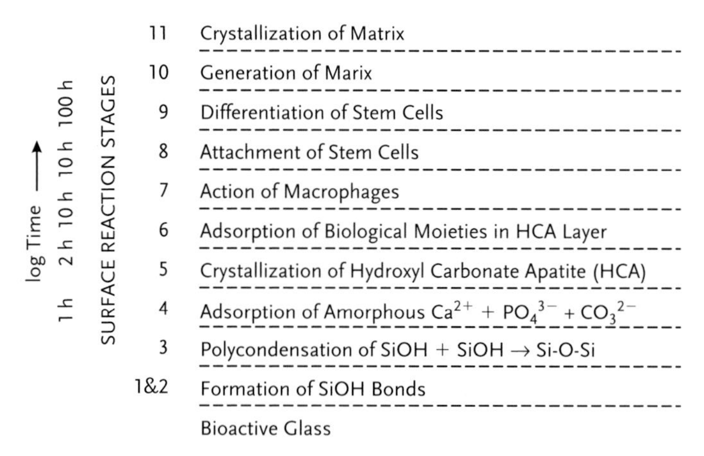

Figure 4.29 Proposed sequence of interfacial reactions involved in forming a bond between tissue and bioactive glass.

Stages 1–5 are well understood, while the overall understanding of stages 6–11 is sparse.

4.3.2.2 **Bone Substitutes**

Materials that can be used as bone substitutes are also very important. The complicated hierarchical structure of bone is not easily mimicked by materials scientists. Different approaches have been taken to substitute bone material. Modified biological samples such as sterilized and calcined bones from animals can be used. Alternatively, corals and algae can be chemically treated by hydrothermal treatment (see Section 4.4) to transform calcium carbonate to calcium phosphate [Eq. (4.10)].

$$5CaCO_3 + 3(NH_4)_2HPO_4 + H_2O \longrightarrow Ca_5(PO_4)_3OH + 3(NH_4)_2CO_3 + 2H_2CO_3 \quad (4.10)$$

Using this approach, it is possible to retain the porous structure in the calcium phosphate material, which is important in allowing the newly built bone to grow into the pores.

Other methods use chemically modified collagen on biologically degradable polymers, on metallic implants, on “bioglasses” and on combinations of those materials. However, despite much research an optimal bone substitute with good mechanical stability and biocompatibility is not yet available. New approaches are based on calcium phosphate [Eq. (4.11)].

$$Ca(H_2PO_4)_2 \cdot H_2O(s) + \alpha\text{-}Ca_3(PO_4)_2(s) + CaCO_3(s) + Na_2HPO_4(aq) \longrightarrow Ca_{8.8}(HPO_4)_{0.7}(PO_4)_{4.5}(CO_3)_{0.7}(OH)_{1.3} \quad (4.11)$$

The solid components are mixed with sodium phosphate solution to give an injectable paste that cures *in situ* after only 5 min. The carbonate-containing hydroxylapatite formed during curing is very similar to the bone mineral with its very small crystals (about 20 nm).

4.3.3
Biomimetic Materials Chemistry

Bioinspired materials processing (which means exploiting the basic principles of biomineralization) is also an emerging discipline in materials science. Materials such as bones, teeth, and shells are complex → composites and the organization and interfacial chemistry of the components are optimized for functional use. Mimicking such structures would be a significant step towards so-called “smart” materials. For materials scientists, biomineralization provides a unique opportunity to study solutions to key problems in materials design. Despite some success, no system has yet been devised that approaches the integrated molecular controls in natural biomineralization, and this rudimentary development is reflected in the relatively simple materials architectures so far formed. Tools that have been used for bioinspired routes towards controlling crystal morphologies are the application of templates and confined reaction environments (see below), spatially controlled mineral deposition, the presence of soluble additives as crystallization modifiers (see Section 4.2.2), inhibitors, or nucleation agents, and the vectorial alignment of smaller crystalline subunits.

An example of a synthetic processing technique inspired by the compartmental strategy of, for example, magnetite crystal (Fe_3O_4) formation (see above) is the use of → microemulsions, → phospholipid vesicles, proteins, and → reversed micelles formed by surfactant–water mixtures to produce inorganic nanoparticles (see Chapter 7) with precisely controlled sizes and shapes. The organic supramolecular cages formed by lipids or surfactants (micelles or vesicles) contain a microenvironment in which controlled precipitation occurs, thus mimicking the biomineralization processes. They are very versatile systems because the reaction environment can have variable diameters (from 1 to 500 nm), and the surface functional groups can be modified. Many different nanoparticles have been synthesized in this way. As each particle is surrounded by an organic membrane, particle–particle interactions are negligible, and reaction rates can be diffusion controlled (Figure 4.30; see also Chapter 7).

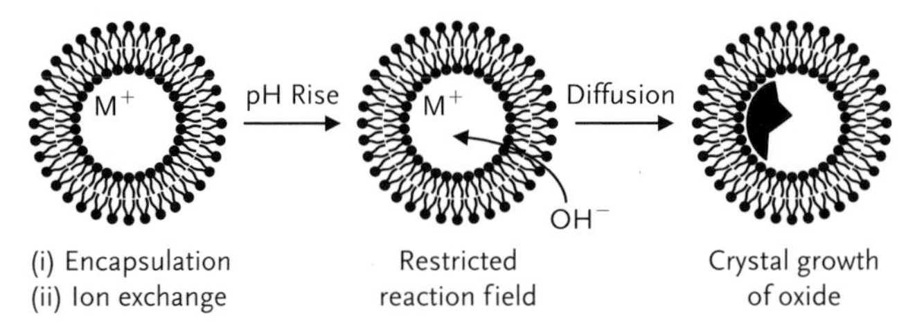

Figure 4.30 Membrane-mediated precipitation of metal oxides in phospholipid vesicles. The chemical reaction will be discussed later [Eq. (4.22)].

As is seen, surfaces play an important role in biomineralization processes. A second example for biomimetic approaches in materials science is the use of synthetic surfaces to initiate crystal → nucleation and growth. Other techniques, for example, chemical vapor deposition methods (see Section 3.2) also use oriented inorganic substrates such as gold and silicon for → epitaxial growth. The use of solutions means a significant step forward, since they can be applied to complex shapes and a wide variety of surfaces. Biomimetic approaches towards a controlled nucleation and growth of inorganic materials involve the use of surfactant monolayers or surfaces that have the advantage that their functionality and packing can be modified in a way that they exactly serve as a molecular blueprint.

Surfactant monolayers on surfaces of aqueous solutions can be considered to model the surface of biological membranes. The potential of monolayers formed at gas/liquid interfaces for promoting crystallization was identified for the first time, when → amphiphilic films of chiral amino acids were found to induce → enantioselective crystallization of organic crystals (α-glycine).

Size-quantized microcrystals and ultrathin, particulate *films* of → semiconductor sulfides were also synthesized by using surfactant monolayers (Figure 4.31).

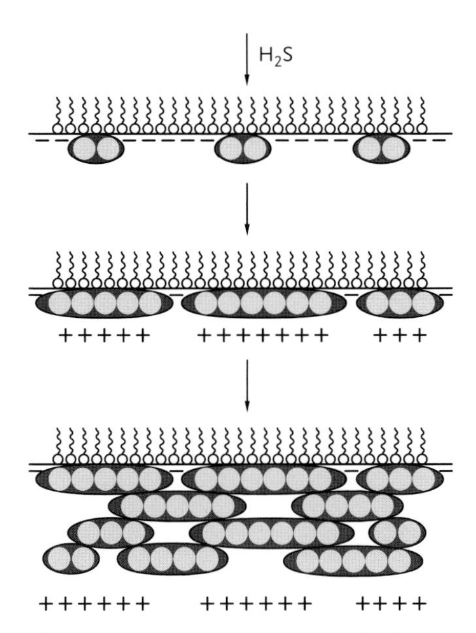

Figure 4.31 Schematics of the growth of nanoparticulate metal sulfide films under monolayers. The time of hydrogen sulfide treatment increases from top to bottom.

A surfactant monolayer is spread on an aqueous solution of the metal salt precursor. Hydrogen sulfide gas infuses through the monolayer, nucleation of the metal sulfide takes place at different sites at the monolayer/water interface (upper part of Figure 4.31), and well-separated nanocrystalline metal sulfide particles grow.

The particles coalesce, forming a "first layer" of a porous sulfide semiconductor particulate film. Fresh metal species diffuse to the monolayer head-group area and form a second layer of the sulfide film. These steps are repeated subsequently to build up layer-by-layer the sulfide semiconductor film up to a plateau thickness, which depends on the chemical composition (CdS $\sim$30 nm and ZnS $\sim$350 nm). The presence of the surfactant monolayer is absolutely essential to the formation of the semiconductor film or the nanoparticles. This can be seen in an experiment in which hydrogen sulfide gas is infused over an aqueous metal-ion solution without surfactant. This experiment results in the formation of large, irregular, and polydispersed metal sulfide particles.

Functionality and packing of the supramolecular surface can be modified to provide complementarity between the surface chemistry and structure of a film and the crystal faces of the nucleus. An example is the nucleation and growth of crystals on → template surfaces. For example, barium sulfate (baryte, $BaSO_4$) was precipitated from supersaturated solutions in the presence of a monolayer of *n*-eicosyl sulfate, $C_{20}H_{41}OSO_3^-$, an → amphiphilic aliphatic long-chain sulfate. Crystallization of barium sulfate in the absence of monolayers results in the precipitation of rectangular tablets. Under monolayers of *n*-eicosyl sulfate, barium sulfate crystals nucleate with the (100) face parallel to the plane of the monolayer (Figure 4.32). The arrangement of the three oxygen atoms of the sulfate head groups at the interface mirrors a similar arrangement of sulfate anions on the (100) face of barium sulfate. Binding of Ba^{2+} cations to the monolayer may simulate the (100) face and, thus, initiate oriented nucleation from the monolayer. When a monolayer of eicosanoic acid was used instead, in which only minimal structural recognition appeared to take place, because the carboxylate end group was only → bidentate, the oriented growth of $BaSO_4$ was not observed.

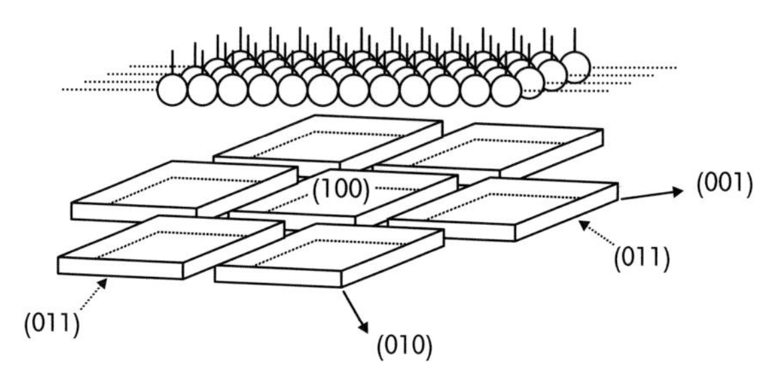

Figure 4.32 Scheme of barium sulfate precipitation in the presence of a monolayer of *n*-eicosyl sulfate.

Besides classical crystallization relying on the attachment of ions/molecules to a primary particle as discussed in Chapter 4.2, strong indications were found that biomimetic mineralization can also take place via non-classical particle-mediated reaction channels. Here, crystallization can also proceed via self-assembly or transformation of metastable or amorphous precursor particles. Such crystallization pathways especially apply to systems far from equilibrium, for which classical thermodynamic considerations are no longer valid to predict the morphology or size of the crystals. This "non-classical crystallization" involves multiple nucleation events of nanoparticles, which form a nanoparticle superstructure in contrast to a single nucleation event to form a single crystal. Figure 4.33 gives a schematic presentation of the different routes for classical and non-classical crystallization.

Path a represents classical crystallization (see Section 4.1.2) where nuclei form and grow until they reach the critical size, followed by crystal growth. The primary nanocrystals can also arrange to form an iso-oriented crystal, where the nanocrystalline building units can crystallographically lock in and fuse to form a single crystal (oriented attachment, path b). If the primary nanoparticles get covered by a polymer or another additive before they undergo a mesoscale assembly, they can form a mesocrystal (path c). There is also the possibility that amorphous particles are formed, which can transform before or after their assembly to complicated morphologies (path d).

One of the main differences between chemical and biological crystallization is the rate of precipitation. While in many chemical processes, precipitation occurs

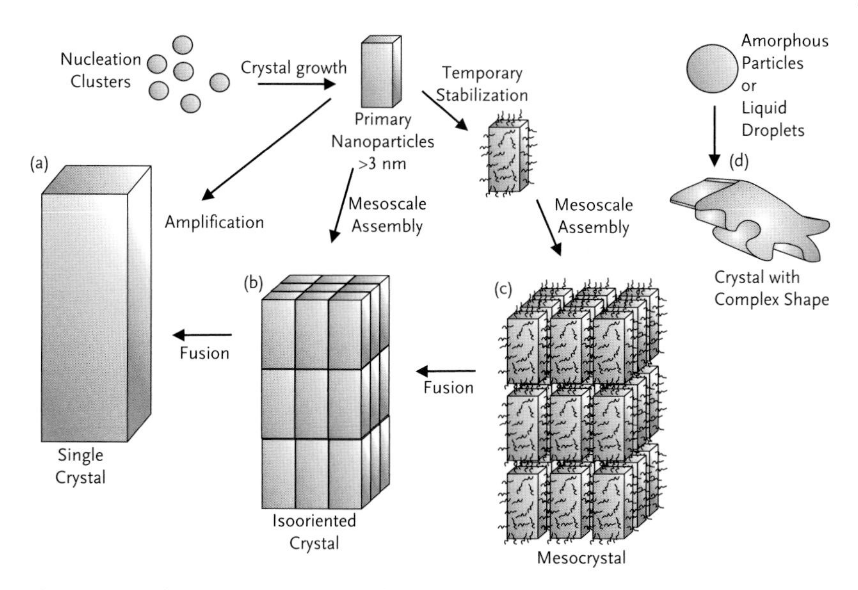

Figure 4.33 Schematic representation of classical and non-classical crystallization.

very fast, in biology the crystals grow during days, weeks, or months. One way to mimic the biological processes is thus to slow down crystallization by separation of the different components with a membrane or medium acting as diffusion barrier, as shown in Figure 4.33c, such as collagen matrices, matrices of polyglycolide, and so on, that in an optimal case carries functional groups able to template certain crystal morphologies.

One recent example is the formation of fluoroapatite mesocrystals in dumbbell and spherical shape in collagen or gelatin (Figure 4.34). The remarkable structural similarity of these composite aggregates, grown *in vitro* for days and weeks, to teeth enamel without the controlling influence of cells shows that it is in principle possible to reduce many of the sophisticated controlling factors in living organism to some relevant parameters.

In this work, the extracellular matrix for the enamel formation is simulated by a gelatin gel with a high content of intact R-helices, which are also present in the original collagen and with nucleation sites showing a high resemblance to collagen. However, the amino acid composition of gelatine is quite different from the original biological organic matrix responsible for the enamel crystallization. To avoid a fast precipitation of calcium phosphates, the buffered calcium and phosphate solutions are diffused from opposite sites into the acidified gel by means of a double-diffusion chamber at 25 °C. Fluoride, which is necessary for fluoroapatite formation, is added to the phosphate solution. The fluoride content and the regular hexagonal habitus in the resulting fluoroapatite aggregates closely resemble that of the enameloid of shark teeth. Further on, a transformation of hexagonal prisms into completely different shapes, for example, spheres by fractal growth is

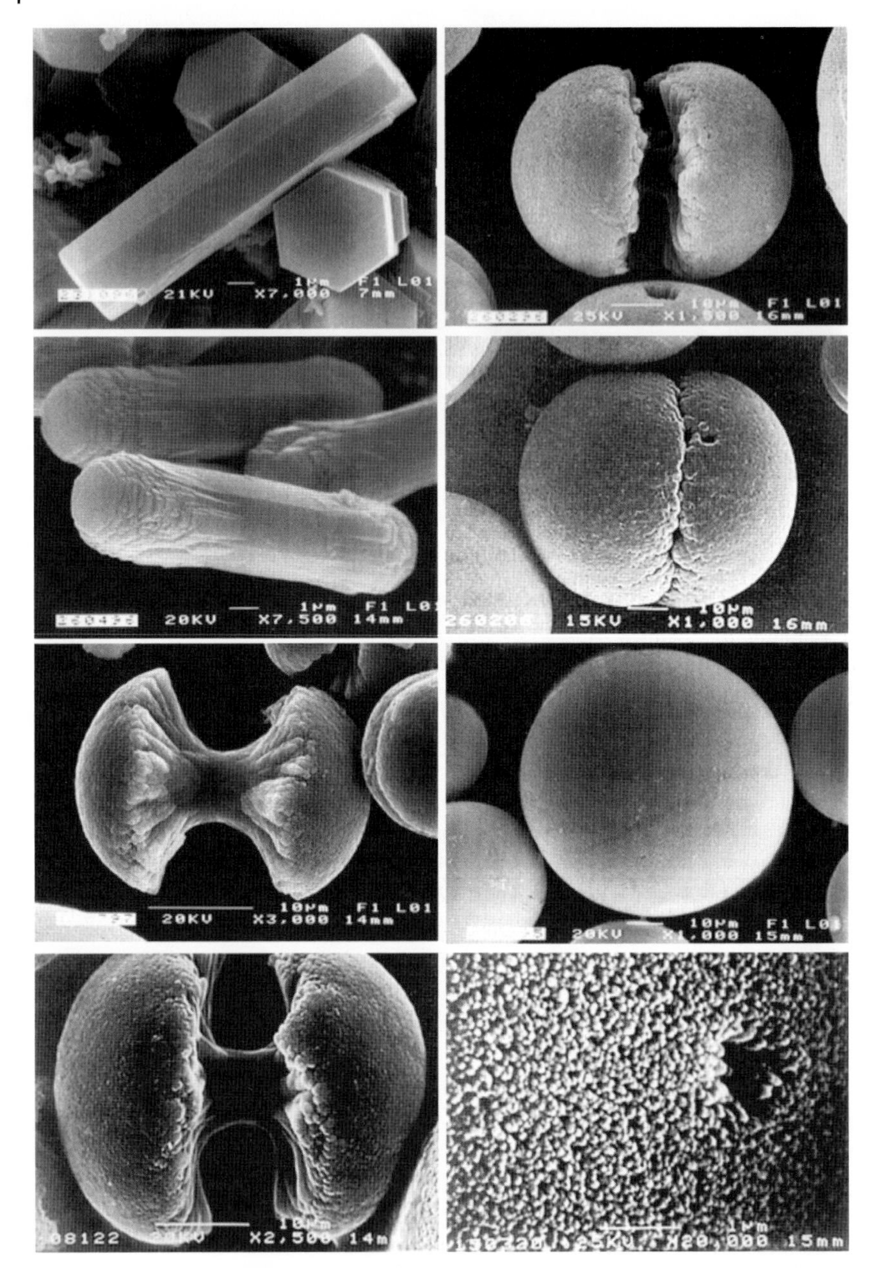

Figure 4.34 Sequence of SEM images of progressive stages of self-assembled (hierarchical) growth of fluoroapatite aggregates in a gelatin gel: from an elongated hexagonal-prismatic seed (top left) through dumbbell shapes to spheres. The surface of a just closed sphere also consists of needle-like units (bottom right), following the general principles of self-similarity.

observed, obviously controlled by means of diffusion and self-organization only. It has to be noted that the final structure is not a result of a simple mineralization in a preformed gel matrix, but is due to the continuous interplay of the molecules of the organic matrix with the inorganic ions during diffusion, nucleation, and mineralization.

4.4 Solvothermal Processes

Solvothermal processes can be defined as any homogeneous or heterogeneous reaction in the presence of a liquid medium and/or mineralizer above room temperature at pressures >1 bar in closed systems. Besides water (hydrothermal synthesis), other solvents such as ammonia (ammonothermal synthesis), glycols (glycothermal), alcohols (alcothermal) are the most important solvothermal reaction media. Nature serves as a role model for solvothermal methods, since many geologically important minerals are grown under hydrothermal conditions.

From a materials point of view, solvothermal processing provides the excellent possibility for synthesizing advanced materials either as bulk single crystals, as fine powders or even as nanoparticles that are not accessible by other methods discussed in this book. From the many examples in which hydrothermal processing plays a decisive role, this section will focus on the following three topics:

1. preparation of large crystals and gemstones;
2. hydrothermal syntheses for the production of fine-particle oxides such as zeolites;
3. Hydrothermal leaching, for example for the treatment of ores to recover metals.

A detailed discussion of the fundamental aspects of hydrothermal crystallization is given in Section 4.4.1, followed by a short description of the hydrothermal synthesis (Section 4.4.2) and the hydrothermal leaching process (Section 4.4.3).

4.4.1 Hydrothermal Synthesis of Single Crystals

As early as 1845 Schafhäutl mentioned the aqueous synthesis of small quartz crystals in a "Papinscher Topf" (Papin's pot). Based on the work of Spezia, who synthesized crystals up to 1 cm^3 under isothermal conditions, suitable ways for an industrial application were developed, such as the temperature gradient method after World War II. In 1985, the world production of quartz crystals was about 1500 tons (Figure 4.35), and today the industrial growth of quartz crystals is performed in large autoclaves of $\geq$100 l capacity.

Since the discovery of the → piezoelectric properties of quartz in 1880 by Pierre Curie, quartz has become a significant factor in the growth of the electronics industry. The pressure resulting from a voltage being applied is displayed in the

Figure 4.35 Synthetic quartz crystal (left) and rock crystal (right).

form of oscillations at a particular resonant frequency. This frequency is a function of the thickness of the crystal. By carefully preparing a crystal, it can be made to oscillate at any frequency. Piezoelectric quartz is widely used, from electronic watches and electronic instruments to processors in automobiles to satellite communications, with frequencies ranging from 10 to 200 MHz, as clock oscillators, and as crystal wafers or blanks. For electronic applications the crystals must be free of electrical and optical twinning, voids, and liquid or solid inclusions. Other applications for quartz crystals are in the area of optical materials such as prisms, eyeglasses, or windows of high-power lasers.

Syntheses of gemstones are also an interesting feature of hydrothermal crystallization. Quartz variants such as amethyst, citrine, smoky quartz, ruby, sapphire, and emerald can be synthesized by incorporation of ions into the quartz during hydrothermal treatment.

4.4.1.1 **General Aspects of Solvothermal Processes**

Crystallization by solvothermal synthesis is, in the simplest case, a straightforward (isothermal) equilibrium reaction. Substances that are insoluble under ambient conditions are dissolved, and the crystalline phase, which is stable under the hydrothermal conditions, is formed. For the growth of single crystals, however, temperature gradients in the reaction vessel often have to be applied. Therefore, hydrothermal synthesis can often be treated as a special case of a transport reaction (Section 3.1).

Most of the solvothermal processes are carried out in water, and thus are termed hydrothermal. Water serves two different functions: it is the solvent, and also the pressure-transmitting medium. Therefore, the pVT-data (p = pressure, V = volume, T = temperature) and the physicochemical properties of water at high temperatures and pressures are of significant importance. For example, the ion product of water increases with increasing pressure and temperature, the viscosity decreases with higher temperatures (facilitating diffusion processes), and the polarity, that is, the → dielectric constant, decreases with higher temperatures, but increases with higher pressures.

Since hydrothermal reactions must be carried out in closed vessels, the pressure–temperature relation of water at constant volume is very important. Figure 4.36 shows the *V–T* diagram of water in which the volume usually is replaced by the density (a similar diagram can be drawn for the *V–P* relation).

Typically phase diagrams are shown as *P–T* diagrams. A *T–V* diagram is different from a typical *P–T* diagram, because the coexistence of the two phases, liquid and vapor (steam), is visible by the closed, two-phase region in which the liquid and the gaseous phase are in equilibrium. The dotted lines in Figure 4.36 are lines of constant pressure (isobars, with $p_1 < p_2 < p_{cr}$ = critical pressure $< p_3$). At any point in the two phase region, the density of the liquid and vapor phase at a fixed temperature is given by the two intersections of a horizontal line through the two-phase region with the *equilibrium curve (saturated liquid line and saturated vapor line)*. For example, at 300 °C, the density of the liquid water phase is approximately 0.75 g cm^{-3} (A in Figure 4.36), and that of the gas phase 0.05 g cm^{-3} (B in Figure 4.36). With increasing temperature, the density of the liquid phase decreases, while the density of the corresponding gas phase increases since more molecules are transferred to the gas phase. At the critical point C (with a *critical temperature* of 374.15 °C and *a critical density* of 0.321 g cm ^{3}, in addition to *a critical pressure* of 22.12 MPa) the densities of both phases are the same, as are all other properties. That means, the difference between liquid and gaseous phases has disappeared. Above the critical temperature only one phase exists, which is called *supercritical fluid*.

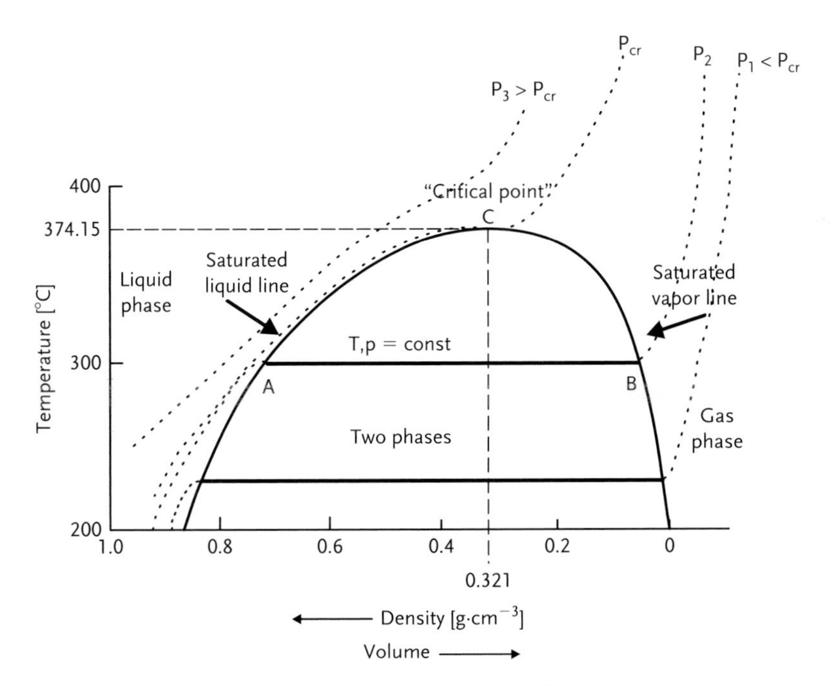

Figure 4.36 Volume (density)/temperature dependence of water.

What is the significance of Figure 4.36 for practical application in hydrothermal syntheses? As learned in the paragraph before, heating of water in a closed vessel results in a decrease in the density and in an increase in volume of the liquid phase. Depending on the volume of the closed container or, more precisely, on the degree of filling of the container, this phenomenon has consequences for the synthesis, as can be seen in the corresponding *P–T* diagram (Figure 4.37). The curve between the triple point (Tr) (the point at which gas, liquid, and solid phase coexist) and the critical point (C) represents the saturated steam curve at which the liquid and gaseous phases coexist.

At pressures below Tr and temperatures above Tr, that is, above the sublimation curve, liquid water is absent and the vapor phase is not saturated. Above this curve liquid water is under compression and the vapor phase is absent. The solid lines in Figure 4.37 (isochores) may be used to calculate the pressure that develops inside a vessel after it has been partially filled with water, after closing and heating to a certain temperature.

When an autoclave is filled to more than 32%, the meniscus between the liquid and the gas phase rises when the temperature is increased due to the fact that the density of the liquid phase is decreasing, while the density of the gas phase is increasing (see above and Figure 4.38). At some temperature below the critical temperature (373 °C) the saturated liquid line is crossed and the whole autoclave is filled with liquid. For example, if the percentage of filling of the autoclave is greater than 80% (Figure 4.38, example C and the corresponding *T–V* diagram), the autoclave is totally filled by the liquid phase at 245 °C (this can be seen by following a vertical line from this density until the saturated liquid line is crossed), because the corresponding isochore separates from the saturated steam curve at this temperature (Figure 4.37). The higher the degree of filling of the autoclave in the beginning, the lower is the temperature at which it is filled completely with

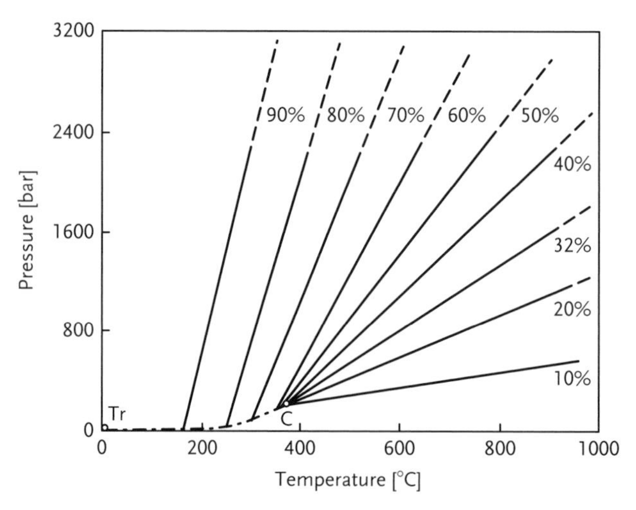

Figure 4.37 Pressure/temperature dependence of water for different degrees of filling of the reaction vessel

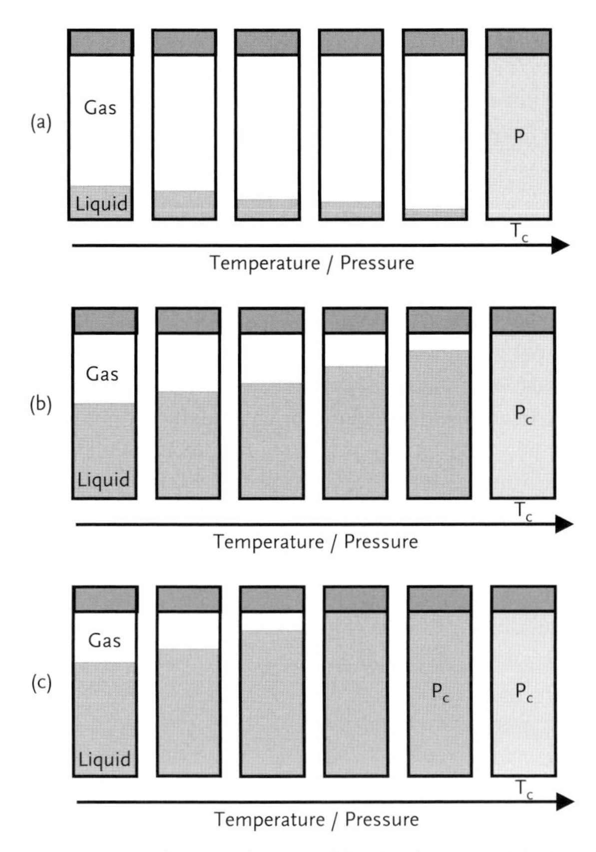

Figure 4.38 Schematic drawing of the development of the liquid/gas meniscus for different degrees of filling; (a) = degree of filling below 32%; (b) = degree of filling about 60%; (c) = degree of filling above 80%; the supercritical state is reached if the critical pressure and temperature is passed.

liquid. However, with a high degree of filling, the pressure is dramatically increasing with increasing temperature, as can be seen from the isobars in Figure 4.36.

If the degree of filling at room temperature is below 32%, the liquid–gas meniscus drops, and at some temperature below 373 °C the autoclave is filled only with gas (the saturation vapor line has been crossed). The lower the starting percentage of filling, the lower is the temperature at which this "dry boiling" occurs and a superheated vapor is obtained. For an autoclave filled to exactly 32%, the meniscus does not move with increasing temperature, and it disappears at the critical temperature of 373 °C. Above this temperature the medium exists only as a supercritical fluid. The density and solvent properties of this fluid, however, are very similar to those of a conventional solvent. Usually hydrothermal crystallization is performed with degrees of filling of 32%, and, with even more success, above 65% and pressures of 200–3000 bar. These data refer to pure water. Assuming that the solubility of the substances is small and that only a very small portion is dissolved, these data may also be used for the solutions.

The solvent properties of pure water are often – even at high temperatures – not sufficient to dissolve substances for crystallization. Therefore, it is often necessary to add a *mineralizer*. A mineralizer is any compound that when added to the aqueous solution speeds up crystallization, and it usually operates by increasing the solubility of the solute through the formation of soluble species that would not normally be present in the water.

Typical mineralizers are the hydroxides of the alkali metals, especially for the growth of amphoteric and acidic oxides, of silicates, germanates or elemental metals. In addition, alkali salts of weak acids, for example, Na_2CO_3, Na_3BO_3, Na_2S, or the chlorides of alkali metals as well as acids are applied.

For instance, the solubility of quartz in water at 400 °C and 2 kbar is too small to permit the recrystallization of quartz, in a temperature gradient, within a reasonable time. Quartz crystals are grown in an alkaline medium in a temperature gradient of 400–380 °C and 1 kbar. Usually polycrystalline quartz serves as a precursor material, which is termed "Lasca". Typical mineralizers for SiO_2 are NaOH, KOH, Na_2CO_3, or NaF. Eq. (4.12) gives an example of how bases act as mineralizers for silica.

$$SiO_{2(\text{solid } \alpha-\text{quartz})} + 2OH^- \rightleftharpoons SiO_3^{2-} + H_2O \tag{4.12}$$

Figure 4.39 shows the solubilities of some substances under hydrothermal conditions as a function of the OH^- concentration.

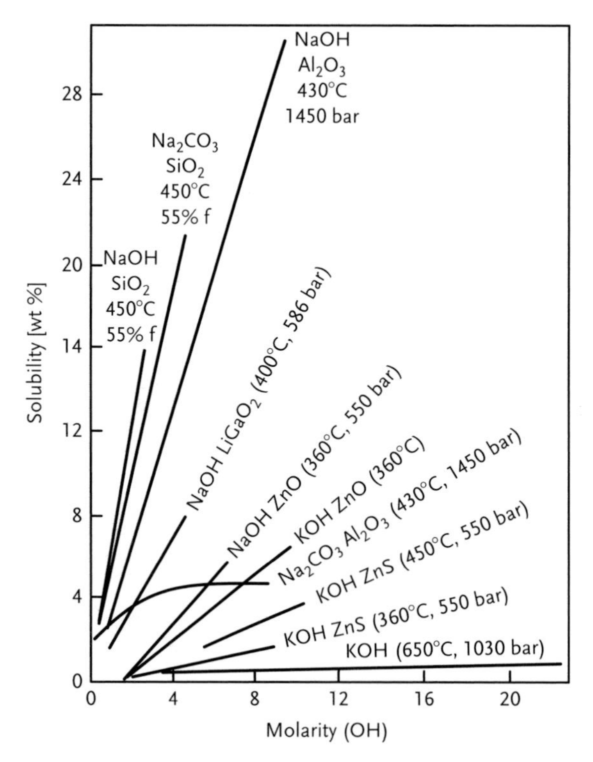

Figure 4.39 Solubility as a function of the hydroxide concentration for some hydrothermal systems.

Typical saturation concentrations of the substances for crystallization are in the region of 2–5 wt%. With this concentration, growth rates of around 1 mm per day can be achieved. As can be seen in Figure 4.39, this saturation concentration is reached, for example, for SiO_2 (quartz) in solutions of 0.5 M NaOH or for ZnO in solutions of 6 M NaOH or 6 M KOH.

A typical hydrothermal procedure is as follows: the precursor substance (nutrient) is put at the bottom of the autoclave that is filled to the desired degree with the solvent (water and mineralizer). The autoclave contains a baffle – a perforated metal disc (with defined openings) in the middle – to separate the growth zone of the single crystals from the zone in which the precursor dissolves. Seed crystals, cut with defined crystallographic orientation, are hung into the growth zone to facilitate → nucleation (Figure 4.40).

High pressures and temperatures during the hydrothermal treatment require a special design for the reaction vessel. Up to 300 °C and 10 bar, thick-walled glass tubes can be used. These have the advantage that the crystallization reactions can be followed easily. A steel tube that is closed at one end is usually used.

The autoclave is put into an oven that heats the nutrient section more than the growth zone ($T_1 < T_2$). Therefore, a temperature gradient exists in the reaction vessel. On increasing temperature, the substance at the bottom part of the autoclave dissolves and is transported into the cooler growth zone by convection. Here,

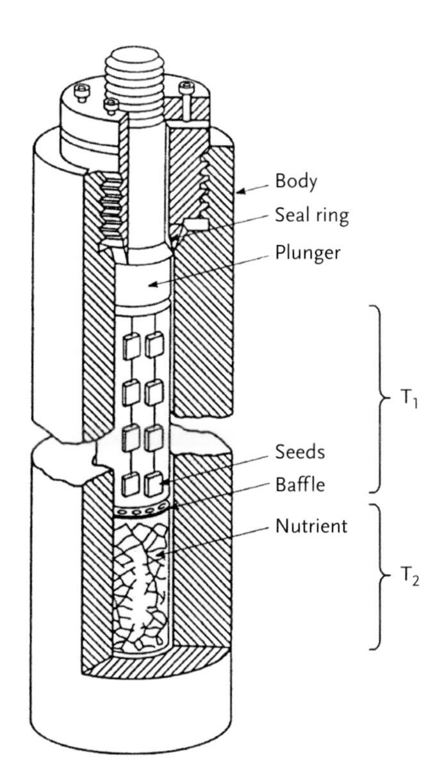

Figure 4.40 Scheme of an autoclave for hydrothermal single-crystal growth.

the solution is supersaturated due to the lower temperature, and crystal growth occurs at the single-crystal nuclei. For crystallizations by the temperature gradient method, solubilities (s) of some weight per cent are required. Additionally, the temperature coefficient of the solubility, ds/dT, should be around 0.01–0.1 wt% per 10 °C. Because of the convection processes, supersaturation conditions are maintained during the synthesis. The baffle hinders total mixing of the two zones and localizes the temperature gradient, mainly at the metal disc. This creates very uniform conditions within the different zones, which facilitates uniform crystal growth. The duration of growth runs varies between days and months, depending on the growth conditions.

More rarely, the case of *retrograde solubility* can be found, which means transport from cold to hot regions. For example, the solubility of SiO_2 in pure water shows up to 600–700 bar (=60–70 MPa) a retrograde behavior (Figure 4.41); thus the temperature coefficient is negative. Only at higher pressures does the temperature coefficient become positive, which allows for dissolution at high temperatures and crystallization in the colder zone of the autoclave.

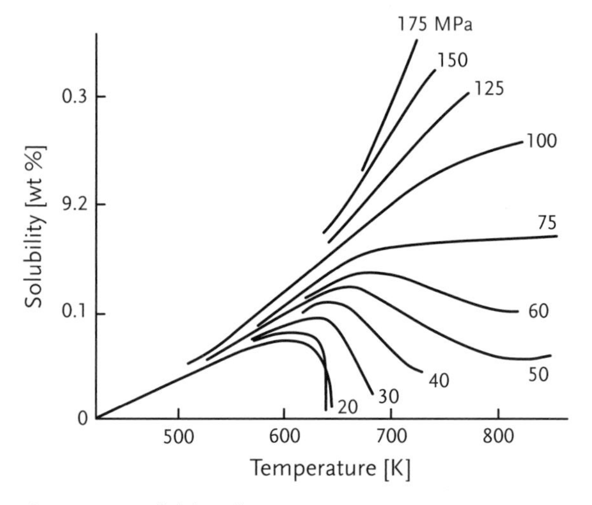

Figure 4.41 Solubility of SiO_2 in water.

4.4.1.2 **Non-aqueous Solvents**

Although water is considered to be the most important solvothermal reaction medium, other solvents may also be used (Table 4.3).

Non-aqueous solvents (with the exception of ammonia) are not very common for solvothermal processes. However, in cases in which the solubilities are not high enough in water, in which the solvent itself is part of the reaction, or in which the product reacts with water, other solvents have to be used. Table 4.4 lists the critical data for some non-aqueous solvents. Formic acid is different because it decomposes at high temperatures to CO and H_2O or CO_2 and H_2. Many oxides and carbonates have been synthesized in this reducing and CO_2-rich atmosphere.

Table 4.3 Examples for solvothermal processes in non-aqueous environments.

Solvent	Examples of synthesized compounds
NH_3	nitrides, imides, amides, CsOH, Cs_2Se_2
HF	$MO_{3-x}F_x$ (M = Mo, W)
HCl, HBr	$AuTe_2Cl$, AuSeCl, AuSeBr, $Mo_3S_7Cl_4$
Br_2	SbSBr, SbSeBr, BiSBr, BiSeBr, $MoOBr_3$
S_2Cl_2	MoS_2Cl_3, $Mo_3S_7Cl_4$
S_2Br_2	$Mo_3S_7Br_4$
$SeBr_2$	SbSeBr, BiSeBr
$H_2S + (C_2H_5)_3NHCl$	β-Ag_2S
C_2H_5OH	SbI_3, BiI_3
CS_2	monoclinic Se
CCl_4	$SeCl_4$, $TeCl_4$
C_6H_6	selenium
CH_3NH_2	CH_3NHLi

Table 4.4 Critical constants of some solvents.

Solvent	Critical temperature [°C]	Critical pressure [bar]
H_2O	374.1	221.2
NH_3	132.3	111.0
Cl_2	144	77.1
HCl	51.4	83.2
CO_2	31.3	73
SO_2	157.8	78.7
H_2S	100.4	90.1
CS_2	279	79
C_2H_5OH	243	63.8
CH_3NH_2	156.9	40.7
CH_3OH	240	81
HCOOH	308	–

4.4.2 Hydrothermal Synthesis

Commercial processes that utilize hydrothermal syntheses include those for the manufacture of a wide range of zeolites (see Section 6.3). In addition, a number of processes have been developed for a variety of electronic, magnetic, and ceramic materials. For example, in magnetic tapes chromium(IV) oxide is used due to its excellent magnetic properties. In this case, Cr(III) or Cr(VI) precursors are transformed by hydrothermal treatment at 300–400 °C and 50–800 bar to needle-like CrO_2 crystals.

For most hydrothermal syntheses only moderate temperatures (100–300 °C) at the corresponding low solution vapor pressures (subcritical region) are applied. The feed slurry or nutrient solution, usually consisting of oxides, hydroxides, and salts of the corresponding metal is fed into the autoclave. This slurry is heated, and at the desired reaction temperature the nutrient materials react and/or transform, primarily through dissolution and precipitation, to the stable compound. An example is the formation of barium titanate (see Section 2.1.1) from barium hydroxide and titanium oxide at temperatures between 150 and 250 °C [Eq. (4.13)].

$$Ba(OH)_2(aq) + TiO_2 \longrightarrow BaTiO_3(s) + H_2O \qquad (4.13)$$

After cooling the autoclave, the product can be isolated by filtration, and after a series of washing steps the pure product is obtained.

In recent years, solvothermal synthesis approaches become increasingly important in the synthesis of nanoscale materials (see also Chapter 7). A great variety of nanomaterials, including nanoparticles and nanocomposites from metals, metal oxides, semiconductors, silicates, carbon, ceramics, and so on, have been processed using hydrothermal techniques in combination with other techniques, such as microwave processing, or by utilization of modifying agents to control the crystal shape and size (see Section 4.2.2).

4.4.2.1 **Synthesis of Zeolites**

Zeolites are important materials for catalysis, adsorbents and molecular sieve applications (see Section 6.3 for details). Zeolites are crystalline silicate materials with a well-defined and highly ordered structure and porosity, and have the general composition $M_{x/n}(Al_xSi_yO_{2(x+y)}) \cdot z\ H_2O$ (M = metal cation with the charge n). Most (but not all) zeolites are prepared via hydrothermal synthesis. The synthesis conditions depend on the desired composition of the material, the desired particle size, morphology, and so on.

For the synthesis of the commercial zeolite ZSM-5, for example, a mixture of polysilicic acid, sodium hydroxide, aluminum sulfate, water, *n*-propylamine, and tetrapropylammonium bromide is heated in an autoclave to 160 °C for several days. The synthesis process is sensitive to numerous variables such as temperature, pH, origin of silica and alumina, the type of alkaline cation, reaction time, the templating agent, and the use of seeding to control rate of → nucleation and particle size.

The hydrothermal approach to the synthesis of zeolites is considered to be "chimie douce." Compare the typical technical preparation conditions, however, with the biological synthesis of structured SiO_2 (Table 4.5) discussed in Section 4.3.1.

Table 4.5 Technical and biological synthesis of SiO_2 materials.

	Zeolites	**Unicellular algae**
Reaction time	days	shell formation within hours
Concentration of the inorganic precursor	>1 M	<0.001 M
pH	6–14	6–8
Temperature	125–200 °C	4–25 °C
Pressure	1–100 bar	1 bar
Structural description	Compact Translatorial repetitive Microporous with uniform pores (typical: 0.3–2 nm)	Shell Hierarchical Micro- and macroporous (pore diameters 5 nm to μm)

4.4.3
Hydrothermal Leaching

The Bayer process is an important and large-scale industrial process that uses hydrothermal leaching to extract and precipitate high-grade aluminum hydroxide from bauxite ore. Bauxite is a mixture of $Al(OH)_3$ and AlOOH, and can contain large quantities of silica and Fe_2O_3. Extraction is carried out under hydrothermal conditions to remove these components by reacting bauxite with a concentrated sodium hydroxide solution. The soluble sodium aluminate complex $Na[Al(OH)_4]$ is formed [Eq. (4.14)], while Fe_2O_3 is insoluble under these conditions.

$$\begin{aligned} Al(OH)_3 + NaOH &\longrightarrow NaAl(OH)_4(aq) \\ AlOOH + NaOH + H_2O &\longrightarrow NaAl(OH)_4(aq) \end{aligned} \tag{4.14}$$

After filtration, $Al(OH)_3$ is precipitated from the $Na[Al(OH)_4]$ solution by cooling, diluting, and seeding with $Al(OH)_3$. The pure $Al(OH)_3$ obtained is calcined to yield corundum, Al_2O_3, which is mainly used for the production of aluminum metal.

4.5
Sol–Gel Processes

The goal of this section is to present the general principles of sol–gel processing as well as its potential for material synthesis. This process allows solid products to be obtained by gelation rather than by crystallization or precipitation.

The structure of oxide glasses consists of dense amorphous networks of connected polyhedra, as has been discussed in Section 4.1.1. Glasses are obtained by cooling melts fast enough to avoid crystallization. An alternative way to obtain amorphous oxide networks is a "bottom-up" approach, that is, by connecting molecular building blocks, such as [SiO_4] tetrahedra, with each other in a stepwise manner. During sol–gel processing, oxide materials are obtained through progressive → polycondensation reactions of molecular precursors in a liquid medium via a sol and a gel stage. This method is considered as a soft chemical approach for the synthesis of amorphous oxide materials because of the mild reaction conditions.

Sol–gel materials are formed in kinetically controlled reactions. Contrary to glasses from melts, where the energetically favorable crystalline structure is avoided by fast cooling, fast and (nearly) irreversible reactions are the reason for the amorphous network structures in sol–gel materials. An immediate consequence is that all reaction parameters have a decisive influence on the structure and thus properties of sol–gel materials (as the cooling rate for glass structures – see Figure 4.1).

Before turning to the mechanisms by which sol–gel materials are formed, the processing steps and applications of the final materials, the terms "sol" and "gel" must be defined:

- A *sol* is a stable suspension of → colloidal solid particles or polymers in a liquid. The particles can be amorphous or crystalline. Note the difference from → aerosols discussed in Section 3.3, which are dispersions of particles or droplets in a *gas* phase.
- A *gel* consists of a porous, three-dimensionally continuous solid network surrounding and supporting a continuous liquid phase ("wet gel"). In "colloidal" ("particulate") gels, the network is made by agglomeration of dense colloidal particles, whereas in "polymeric" gels the particles have a polymeric substructure made by aggregation of subcolloidal chemical units. In general, the sol particles can be connected by covalent bonds, van der Waals forces, or hydrogen bonds. Gels can also be formed by entanglement of polymer chains. In most sol–gel systems used for materials syntheses, gelation (= formation of the gels) is due to the formation of covalent bonds and irreversible. Gel formation can be reversible when other bonds are involved in gelation.

Figure 4.42 presents a scheme of the different processing routes leading from the sol to a variety of materials. Powders can be obtained by spray drying of a sol. Gel fibers can be drawn directly from the sol, or thin films can be prepared by standard coating technologies such as dip or spin coating, spraying, and so on. Here, gelation occurs during the preparation of the film or fiber due to rapid evaporation of the solvent.

Gelation can also occur after a sol is cast into a mold, in which case it is possible to make monolithic objects of a desired shape. Drying by evaporation of the pore liquid gives rise to capillary forces that causes shrinkage of the gel network. The resulting dried gel is called a xerogel ("xero" means dry). Compared to the original wet gel its volume is often reduced by a factor of 5–10. Due to the drying stress, the

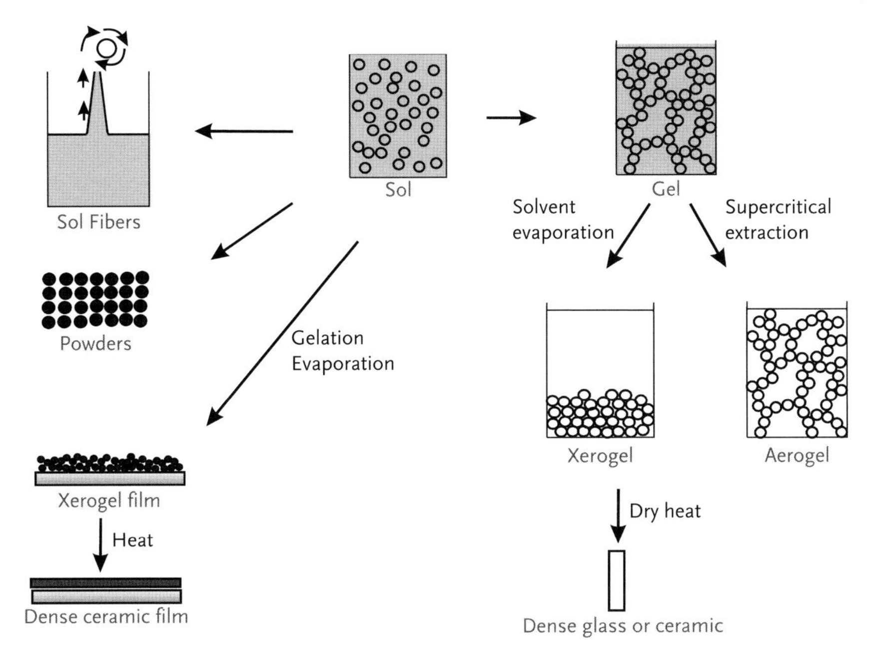

Figure 4.42 Sol–gel processing options.

monolithic gel body is often destroyed and powders are obtained. When the wet gel is dried in a way that the pore and network structure of the gel is maintained even after drying, the resulting dried gel is called an "aerogel" (see also Section 4.5.6). Dense ceramics or glasses can be obtained after heat treatment of xerogels or aerogels to a temperature high enough to cause sintering.

Typical applications of sol–gel materials include all kinds of coatings (see below), catalysts and catalyst supports, ceramic fibers, ceramic powders, insulating materials, highly porous materials, and so on. The many processing options allow a unique access to multicomponent oxide systems, as demonstrated in the following example.

The → perovskite lead zirconate → titanate, $PbZr_{1-x}Ti_xO_3$, (PZT) is an important → ferroelectric and → piezoelectric ceramic material. The properties of PZT depend very critically on the exact stoichiometry; optimal values are obtained for $x = 0.47$.

PZT powders can be obtained by the ceramic method (Section 2.1.1), for example, by heating an intimate mixture of PbO, TiO_2, and ZrO_2 to 750–1100 °C for several hours. Owing to the volatility of lead oxide at the required high temperatures, the precise stoichiometry is difficult to obtain. PZT thin films are difficult to make by the ceramic method, because the film thickness cannot be controlled and the required temperatures for the formation of the PZT phase are too high for many substrates.

PZT films can be prepared by the CVD method. However, the problem to control exactly the stoichiometry is similar to that of the ceramic method. A typical flow chart for the preparation of PZT films by sol–gel processing is shown in Figure 4.43.

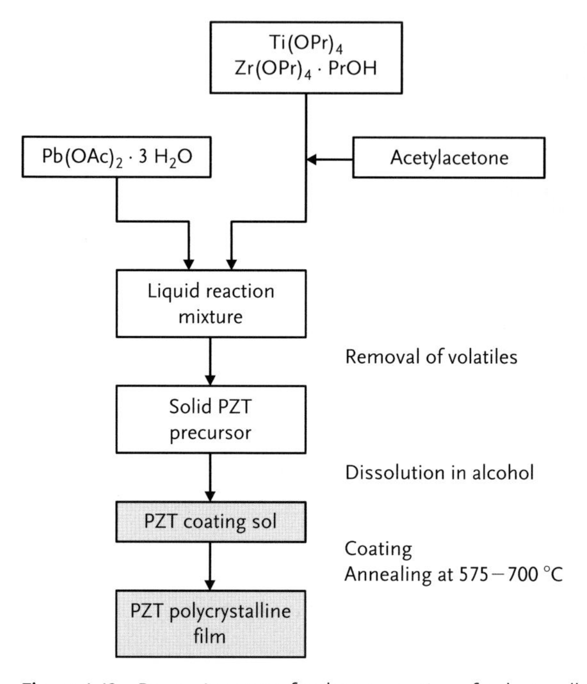

Figure 4.43 Processing steps for the preparation of polycrystalline PZT films.

The precursors for the preparation of PZT by the sol–gel method according to Figure 4.43 are $M(OPr)_4$ (M = Ti, Zr) and lead acetate. The role of acetylacetone to modify the alkoxide precursors will be discussed later [see Eq. (4.28)]. The water necessary for the hydrolysis reactions is brought in by the hydrated lead acetate. In the resulting coating sol, with particle sizes between 6 and 10 nm, the metals are homogeneously distributed. The sol is applied to a substrate by conventional coating techniques (such as dip or spin coating, doctor blading, and so on). Subsequent heat treatment at relatively mild conditions (low temperatures and short heating periods) results both in the removal of the organic groups by pyrolysis and thermolysis reactions and the formation of a crystalline PZT film up to 1.5 μm thick. Thicker films are obtained by repeating the coating and annealing cycle. Figure 4.44 shows a metallic printing drum coated with PZT as an example of a technical application of such films.

PZT powders or aerogels can be made by essentially the same method (without the coating step). The gels can also be used for making ceramic fibers, similar to the preceramic polymers that will be discussed in Chapter 5. Due to the high

Figure 4.44 Metallic printing drum coated with a 8-μm thin film of ferroelectric $PbZr_{0.53}Ti_{0.47}O_3$ (PZT). Patterning is achieved by application of a directed electric field. This creates a remnant charge in the PZT layer that is used to adsorb the printing ink. After use, the printing pattern can be electrically erased, and the drum can be used again.

melting points of ceramic materials, ceramic fibers cannot be made by processing the ceramic material itself. The thin PZT fiber shown in Figure 4.45 was prepared by first preparing a PZT gel from alkoxide and carboxylate precursors similar to the process shown in Figure 4.43. The properties of gels allow their extrusion through a spinneret. The → green fiber is then sintered at 925 °C. The process of fiber spinning is described in more detail in Section 5.5.1.

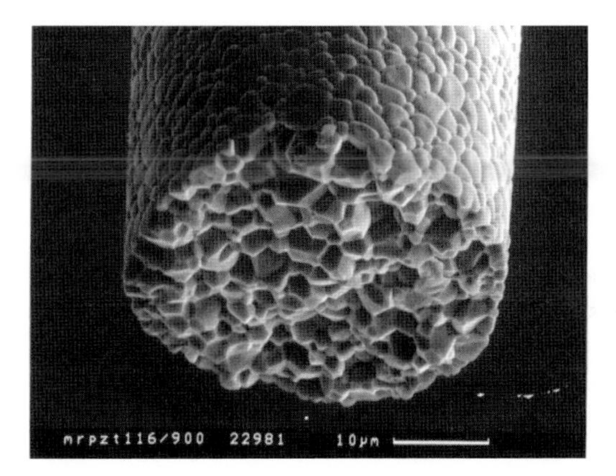

Figure 4.45 Cross section of a dense, polycrystalline PZT fiber with the composition $PbZr_{0.53}Ti_{0.47}O_3$. The grain size is 2–4 μm.

The example of PZT shows that sol–gel processing even allows the preparation of ceramic materials with complex compositions. Another example of such a compound is the high-temperature superconductor yttrium barium copper → oxide (YBCO, see also Section 2.1.1).

In the following, we will first discuss some of the fundamental physical principles that govern sol and gel formation (Section 4.5.1), followed by a detailed treatment of the individual steps in the sol–gel processing of silicate materials (Section 4.5.2). In silicate systems, these steps were particularly well investigated; the fundamental issues are common to sol–gel processing of other materials that will be treated in Section 4.5.3. Inorganic–organic hybrid materials, which are composed of inorganic and organic building blocks and are easily prepared by sol–gel processing, are the topic of Section 4.5.4. Section 4.5.5 deals with "non-hydrolytic sol–gel processes," where oxo groups are formed by alkyl chloride or ester elimination instead of water or alcohol elimination as in classical sol–gel processes. Preparation and properties of aerogels will be discussed in the final section (Section 4.5.6).

4.5.1
The Physics of Sols

The stability and coagulation of sols (colloids) is of great importance to sol–gel chemistry since the structure of the final oxide network (gel) depends mainly on the size and shape of the aggregating species: the sol particles. Therefore, some elementary aspects will be outlined in the following paragraphs.

The easy agglomeration of fine particles is caused by attractive van der Waals forces and/or forces (V_A) that tend to minimize the total surface or interfacial energy of the system. Van der Waals forces result from dipole–dipole interactions, and are rather weak; thus the attractive potential extends over distances of only a few nanometers.

In order to prevent aggregation, repulsive forces of comparable dimensions are required. This can be done by

- Adsorbing an organic layer ("*steric barrier*"). Steric stabilization can occur in the absence of an electric barrier and is particularly efficient in dispersing high concentrations of particles. Surfactant molecules can adsorb at the surface of the particles (see also Section 4.2). Their hydrophobic chains will then extend into the solvent and interact with each other (Figure 4.46). The solvent–chain interaction increases the free energy of the system and creates an energy barrier to the closer approach of the particles. The thickness of the layer must therefore be in a range (usually <3 nm) that the closest possible distance between the particles is outside the range for a sufficient van der Waals attraction. Furthermore, the motion of the chains extending into the solvent becomes restricted and produces an entropic effect when the particles come into closer contact with each other.

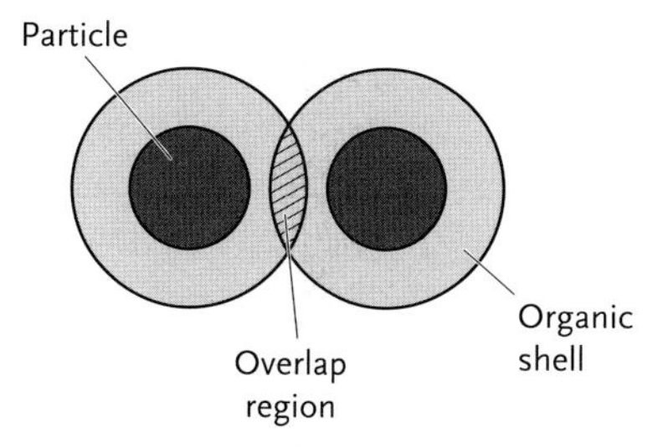

Figure 4.46 Representation of the steric barrier by surfactant molecules adsorbed to the particle surface.

- Creating *electrostatic repulsion* between particles. This repulsion results from charges adsorbed on the particles. Electrostatic stabilization occurs when the electrostatic repulsive forces overcome the attractive van der Waals forces between the particles. The electrostatic approach is mostly used to stabilize → colloidal dispersions in aqueous systems.

The repulsive electrostatic barrier results from an electric *double layer*: the surface of the particle is covered with ionic groups that control the potential of the surface. Counterions in solution in the vicinity of the particle screen the surface charges.

For hydrous oxides, the charge-determining ions are H^+ and OH^-, which create the charge at the particle surface by protonating or deprotonating M–OH groups [Eqs. (4.15) and (4.16)].

$$\text{M–OH} + \text{H}^+ \longrightarrow \text{M–OH}_2^+ \tag{4.15}$$

$$\text{M–OH} + \text{OH}^- \longrightarrow \text{M–O}^- + \text{H}_2\text{O} \tag{4.16}$$

The pH at which the particle is electrically neutral is called the *point of zero charge* (PZC). At pH >PZC, Eq. (4.16) dominates, and the particle is negatively charged (surrounded by positively charged counterions in solution), whereas at pH <PZC the particle is positively charged [Eq. (4.15)]. Typical values for the PZC are: MgO 12, Al_2O_3 9.0, TiO_2 6.0, SnO_2 4.5, SiO_2 2.5. Note that the acidity of the surface M–OH groups depends somewhat on the particle size and the degree of condensation (see below). The magnitude of the surface potential ϕ_o depends on the difference between pH and PZC. This potential attracts the counterions present in solution.

The double layer is schematically shown in Figure 4.47 for a positively charged surface. The repulsive potential V_R drops linearly through the tightly bound layer of water and counterions (called the Stern layer). Beyond the so-called Helmholtz plane, the linear dependence changes into a region in which the electrostatic potential varies exponentially with the distance from the particle (h).

When an electric field is applied to a sol (colloid), the charged particles move towards the electrode with the opposite charge (→ "electrophoresis"). When the particle moves, it carries along the adsorbed layer and part of the cloud of

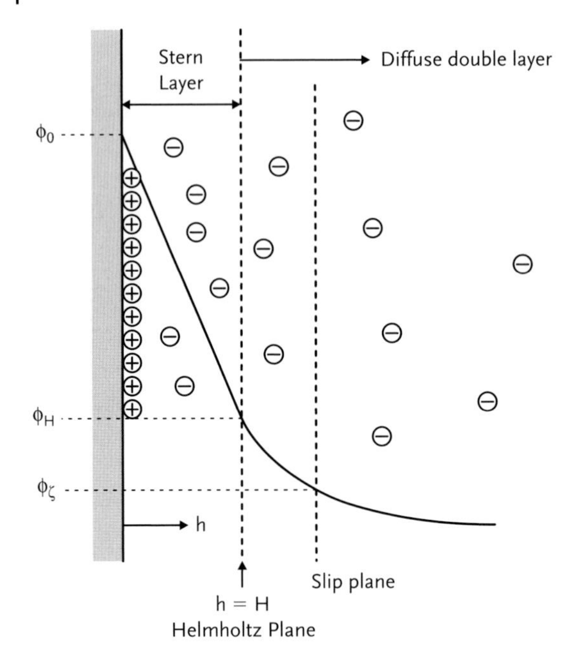

Figure 4.47 Electric double layer over a positively charged surface. h is the distance to the particle surface and V_R the repulsive potential.

counterions, while the more distant portion of the double layer is drawn to the opposite electrode. The "slip plane" or "shear plane" separates the region of fluid that moves with the particle from the region that flows freely. The potential at the slip plane is the zeta (ζ) potential (ϕ_ζ). The pH at which ϕ_ζ is zero is called the isoelectric point (IEP), which in general is not equal to the PZC. The stability of the → colloid correlates with the magnitude of the ζ potential. The higher the ζ potential the more stable are the sols; stability requires a repulsive potential $\geq$ 30–50 mV.

For the same surface potential, the repulsive barrier is greater for larger particles. This effect results from the different dependencies of the attractive potential V_A and the repulsive potential V_R on the particle size. It explains why stable sols can be formed. Once the initial nuclei have grown to sufficient size, the repulsive barrier becomes large enough to prevent coagulation.

Coagulation of sol particles results from

- Decreasing the surface potential ϕ_o (by changing the pH).
- Increasing the electrolyte concentration in the solution. As the concentration of counterions increases, the double layer is compressed, because the number of charges required to balance the surface charge is now available in a smaller volume surrounding the particle. The colloid will eventually coagulate because the attractive force is unchanged, while the repulsive barrier is reduced.

In some cases, a coagulated colloid can be redispersed. This process is called peptization. This can be done by removing the counterions that caused coagulation (washing) or by adsorbing charge-determining ions that re-establish the double layer.

4.5.2
Sol–Gel Processing of Silicate Materials

The coordination number of silicon is generally four, although coordination expansion can occur in transition states (see below). Compared to the transition metals, silicon is less electropositive and therefore it is not very susceptible to a nucleophilic attack. This makes silicon compounds quite stable and easy to handle.

Sol–gel processing proceeds in several steps that will subsequently be discussed in sequence:

1. hydrolysis and condensation of the molecular precursors and formation of sols;
2. gelation (sol–gel transition);
3. aging;
4. drying.

4.5.2.1 Hydrolysis and Condensation Reactions

Silica gels certainly belong to the best-investigated materials in inorganic chemistry. The basic chemical principle behind sol–gel processing is the transformation of Si–OR and Si–OH-containing species (Figure 4.48) to siloxane compounds by condensation reactions. To obtain a particle or even a gel, the number of siloxane bonds has to be maximized and consequently the number of silanol and alkoxide groups has to be minimized.

Various types of precursors can be used to prepare silica gels by sol–gel processing. The most common precursors are aqueous solutions of sodium silicates ("water glass") and silicon alkoxides, $Si(OR)_4$. The formula of water-glass is often given as "Na_2SiO_3". In reality, however, water glass solutions contain mixtures of different silicate species. They are only stable under strongly alkaline conditions.

The chemical reactions during sol–gel processing can be formally described by three equations [Eq. (4.17)]. In alkoxide-based systems, hydrolysis reactions are required to generate Si–OH groups from Si–OR groups before condensation can take place. Condensation can happen either via an alcohol-producing mechanism, or via a water-producing mechanism.

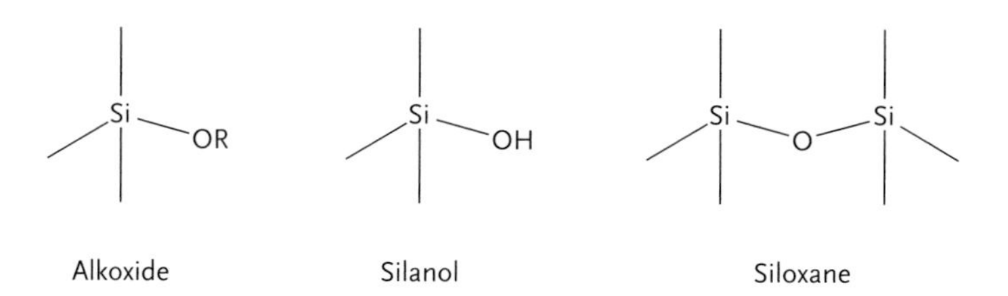

Figure 4.48 Silicon–oxygen groups relevant to sol–gel processing.

$$\equiv\text{Si}-\text{OR} + \text{H}_2\text{O} \longrightarrow \equiv\text{Si}-\text{OH} + \text{ROH} \quad \text{Hydrolysis}$$

$$\left.\begin{array}{l}\equiv\text{Si}-\text{OH} + \equiv\text{Si}-\text{OR} \longrightarrow \equiv\text{Si}-\text{O}-\text{Si}\equiv + \text{ROH} \\ \equiv\text{Si}-\text{OH} + \equiv\text{Si}-\text{OH} \longrightarrow \equiv\text{Si}-\text{O}-\text{Si}\equiv + \text{H}_2\text{O}\end{array}\right\} \text{Condensation} \tag{4.17}$$

The most important differences between these two types of precursors are:

- Gelation is initiated in aqueous silicate systems by pH changes (which generates $\equiv$Si$-$OH groups from $\equiv$Si$-$O$^-$ groups), or by addition of water to alkoxide precursor systems (which generates $\equiv$Si$-$OH groups from $\equiv$Si$-$OR groups).
- The solvent in water-glass-based reactions is always water, while the alkoxides are employed either neat or dissolved in an organic solvent (mostly an alcohol).
- Alkoxide-based systems are more complex because more parameters are involved that influence the sol–gel reactions (see below). This gives more possibilities to control the morphology and properties of the materials obtained.

The reaction mechanisms are the same for both types of silicon-containing precursors, and for monomeric, oligomeric, or polymeric species. However, two chemical different situations have to be considered: reactions under acidic or basic conditions, respectively. To understand the difference it is necessary to remember that the point of zero charge (PZC) of Si–OH-containing species is between pH 1.5 and 4.5 (the higher the degree of condensation of the silica species the lower the PZC). Acidifying the solution to a pH below the PZC means that the siliceous species are positively charged, and increasing the pH above the PZC (more basic) means that the species are negatively charged.

Under acidic conditions, the oxygen atom of a Si–OH or a Si–OR group is protonated in a rapid first step [Eq. (4.18)]. A good leaving group (water or alcohol) is created by the protonation. Additionally, electron density is withdrawn from the central silicon atom, rendering it more electrophilic and thus more susceptible to attack by water (hydrolysis reaction) or silanol groups (condensation reaction).

$$\equiv\text{Si}-\text{OX} + \text{H}^+ \rightleftharpoons \equiv\text{Si}-\text{O}^+(\text{H})(\text{X}) \qquad \text{X} = \text{R, H}$$

$$\text{Y}-\text{OH} + \equiv\text{Si}-\text{O}^+(\text{H})(\text{X}) \longrightarrow \text{Y}-\text{O}-\text{Si}\equiv + \text{HOX} \qquad \text{Y} = \text{H}, \equiv\text{Si} \tag{4.18}$$

hydrolysis reaction: $\text{X} = \text{R}, \quad \text{Y} = \text{H}$

condensation reaction: $\text{X} = \text{R or H}, \quad \text{Y} = \text{Si}\equiv$

Under basic conditions, to which we generally refer when the reactions occur at a pH >3, the reaction proceeds by nucleophilic attack of either an OH$^-$ or a Si–O$^-$

ion to the silicon atom [Eq. (4.19)]. The entering OH^- or $\equiv SiO^-$ are formed by dissociation of H^+ from a water molecule or a Si–OH group.

$$\equiv \text{Si-OX} + \text{YO}^- \rightleftharpoons \left[\begin{array}{c} \text{OY} \\ | \\ >\text{Si}- \\ | \\ \text{OX} \end{array} \right]^- \rightleftharpoons \text{YO-Si}\equiv + \text{XO}^- \qquad \text{Y} = \text{H}, \equiv \text{Si} \tag{4.19}$$

hydrolysis reaction: $X = R$, $Y = H$

condensation reaction: $X = R$ or H, $Y = Si\equiv$

In hydrolysis reactions, the hydroxide anion attacks the silicon atom by a S_N2-type mechanism in which OH^- displaces OR^-. In the condensation reaction, a nucleophilic silanolate ion attacks the neutral silicate species and displaces OH^- or OR^-.

One has to keep in mind that all mechanisms discussed are, in principle, reversible. This means that Si–O–Si bonds can be cleaved by OH^-, and silanol groups can react with alcohol to alkoxy groups. The degree to which the back-reaction may occur mainly depends on the reaction conditions.

With these general mechanisms in mind we will now continue to discuss how various parameters influence the reaction rates of the hydrolysis and condensation reactions.

Silicon alkoxides The overall reaction for sol–gel processing of silicon tetraalkoxides is given in Eq. (4.20),

$$\text{Si(OR)}_4 + 2\text{H}_2\text{O} \longrightarrow \text{SiO}_2 + 4\text{ROH} \tag{4.20}$$

The most common tetraalkoxysilanes used for sol–gel processing are tetramethoxysilane (TMOS) and tetraethoxysilane (TEOS). Since many alkoxysilanes are immiscible with water, alcohols are used as solvents to homogenize the reaction mixture. However, alcohols not only serve as solvent but are also reactants due to the reversibility of the reactions. A mixture of a tetraalkoxysilane in water and alcohol would react very slowly. Therefore, acid or base catalysis is necessary to start the hydrolysis and condensation reactions.

The sol–gel chemistry of silicon alkoxides as described by the overall reaction [Eq. (4.20)] or the hydrolysis and condensation reactions [Eq. (4.17)] seems to be straightforward and easy to understand. However, the whole picture is much more complex since hydrolysis and condensation reactions compete with each other during all steps of the sol–gel process and are additionally influenced to a different degree by the parameters discussed in the following paragraphs. Figure 4.49 shows some of the species that can be obtained in the initial stages of the reaction of $Si(OR)_4$. Each species can further react by either hydrolysis or condensation. Furthermore, these reactions can occur on chemically different silicon atoms once the trimer stage has been reached (the silicon atom in the center is chemically different from the terminal silicon atoms).

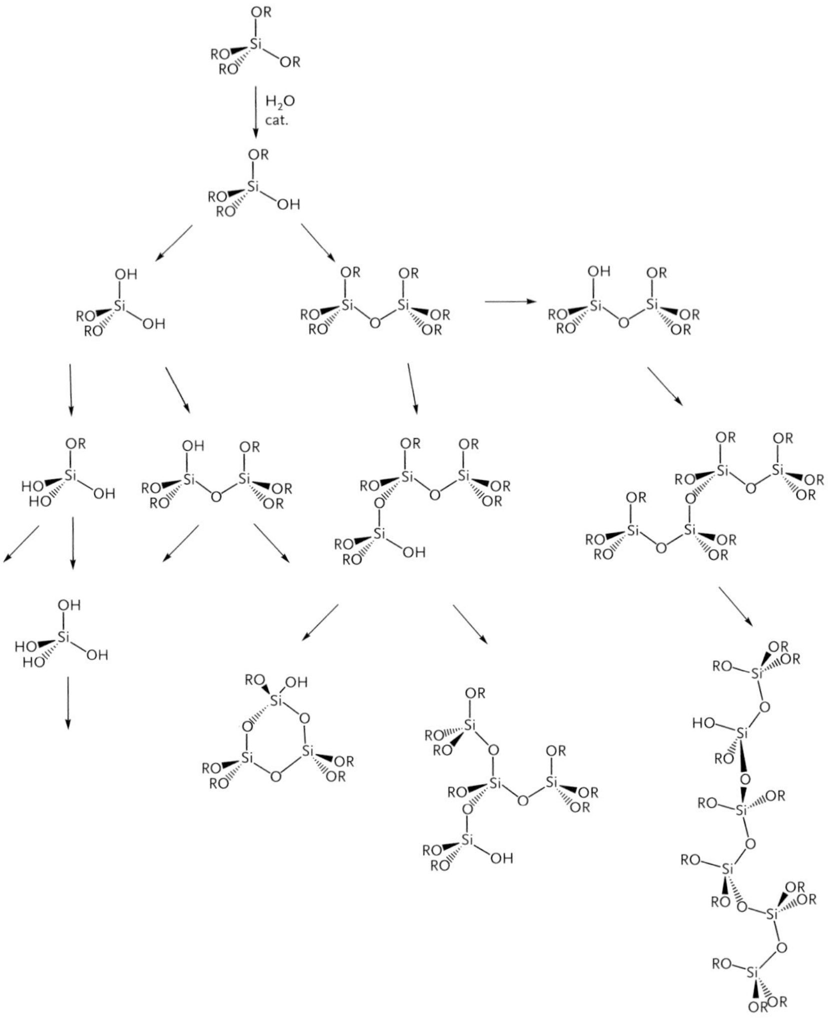

Figure 4.49 Some initial intermediates during sol–gel processing of $Si(OR)_4$. Note that each intermediate can, in principle, undergo either hydrolysis or condensation reactions.

From Figure 4.49 it is obvious that the systems are very complex and many different routes from the $Si(OR)_4$ precursor to the final silica gel are possible. The chemical parameters discussed below determine which route is taken. The final properties of the obtained gel very much depend on the structural evolution. The most important factor for the later material properties is the relative rate of the hydrolysis and condensation reactions. The influence of the different parameters on the network formation is very complex, since many parameters change progressively as → polycondensation proceeds. Therefore, it must be well understood that we are discussing not a static, but a continuously changing system.

Because of this complexity, a detailed understanding of the parameters influencing the reaction rate and the course of the reaction is necessary to tailor the properties of the final material. Parameters that influence hydrolysis and condensation and whose deliberate variation are used for materials design are, *inter alia*

- the kind of precursor(s),
- the alkoxy group to water ratio (R_w),
- the kind of catalyst,
- the kind of solvent,
- the temperature,
- the pH, and
- the relative and absolute concentration of the components in the precursor mixtures.

a) Steric *and inductive effects of the precursor(s)*

The hydrolytic stability of alkoxysilanes is influenced by steric factors. Any branching of the alkoxy group or increasing of the chain length lowers the hydrolysis rate of the alkoxysilanes. That means, that the reaction rate decreases in the order

$$\mathrm{Si(OMe)_4 > Si(OEt)_4 > Si(O^nPr)_4 > Si(O^iPr)_4}$$

Inductive effects of the substituents attached to a silicon atom are very important, because they stabilize or destabilize the transition states during hydrolysis and condensation. The electron density at the silicon atom decreases in the following order:

$$\equiv\mathrm{Si{-}R} > \equiv\mathrm{Si{-}OR} > \equiv\mathrm{Si{-}OH} > \equiv\mathrm{Si{-}O{-}Si}\equiv$$

For acid catalysis, the electron density at the silicon atom should be high because the positive charge of the transition state is then stabilized best. Therefore, the reaction rates for hydrolysis and condensation under acidic conditions increase in the same order as the electron density. For base catalysis a negatively charged transition state has to be stabilized. Therefore, the reaction rates for hydrolysis and condensation increase in the reverse order of the electron density.

This has several consequences, for example:

- As hydrolysis and condensation proceed (increasing number of OH and OSi substituents), the silicon atom becomes more electrophilic. This means, for example, that in acidic media, monomeric $Si(OR)_4$ hydrolyzes faster than partially hydrolyzed $Si(OR)_{4-x}(OH)_x$ or oligomeric species (which have more Si–O–Si bonds), and vice versa in basic media.
- More branched (i.e., more highly condensed) networks are obtained under basic conditions and chain-like networks under acidic conditions, because reactions at central silicon atoms (i.e., atoms with two or three Si–O–Si bonds) are favored at high pH, and reactions at terminal silicon atoms (i.e., atoms with only one Si–O–Si bond) are favored at low pH.

- Organically substituted alkoxysilanes $RSi(OR)_3$ are more reactive than the corresponding $Si(OR)_4$ under acidic conditions and less reactive under basic conditions.
- The acidity of a silanol group increases with the number of Si–O–Si bonds at the silicon atom. Note that this is one of the reasons why the PZC changes with the degree of condensation.

b) *The catalyst (pH)*

Sol–gel reactions of silicon alkoxides are typically catalyzed by an acid or a base. As discussed above, the reaction mechanisms for acid or base catalysis are very different. Additionally, the reaction rates for hydrolysis and condensation have a different dependence on the pH (Figure 4.50). The minimal reaction rate for hydrolysis is at pH = 7, and for condensation around pH 4.5. For the condensation reaction, the minimum pH corresponds to the IEP of silica, where the sol particles can approach each other without a large electrostatic barrier. The pH value is the decisive parameter for the relative rates of hydrolysis and condensation of tetra-alkoxysilanes ($Si(OR)_4$). At pH <5 hydrolysis is favored, and condensation is the rate-determining step. A great number of monomers or small oligomers with reactive Si–OH groups is simultaneously formed. In contrast, hydrolysis is the rate-determining step at pH >5 and hydrolyzed species are immediately consumed because of the faster condensation.

c) *The alkoxy group/H_2O ratio (R_w)*

The overall reaction for sol–gel processing of silicon tetraalkoxides [Eq. (4.20)] implies that two equivalents of water (R_w=2) are needed to convert $Si(OR)_4$ to SiO_2. Four equivalents of water (R_w=1) are needed for the complete hydrolysis of $Si(OR)_4$ if no condensation takes place. Increasing the water content (i.e., lowering R_w) generally favors the formation of silanol groups over Si–O–Si groups, especially since the condensation reaction is, in principle, reversible. Even with an $R_w > 2$, only mixtures of hydroxylated species $[SiO_x(OH)_y(OR)_z]_n$ $(2x+y+z=4)$ are created. As a general rule, an $R_w >> 2$ favors the condensation reaction, and an

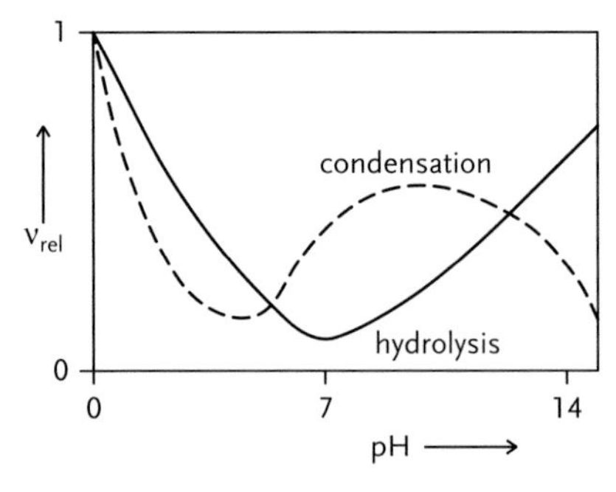

Figure 4.50 Dependence of the relative rates of hydrolysis and condensation reactions of $Si(OR)_4$ on the pH.

$R_w \leq 2$ favors the hydrolysis reaction. The R_w, together with the kind of catalyst, influences the properties of the silicate material very strongly.

d) *The solvent*

The solvent is important to homogenize the reaction mixture, especially at the beginning of the reaction. Polarity, dipole moment, viscosity, protic or non-protic behavior of the solvent influence the reaction rate and thus the structure of the final sol–gel material.

Polar and particularly protic solvents (H_2O, alcohols, formamide) stabilize polar siliceous species such as $[Si(OR)_x(OH)_y]_n$ by hydrogen bridges; non-polar solvents (tetrahydrofuran, dioxane) are usually used for organotrialkoxysilanes $[R'Si(OR)_3]$ or incompletely hydrolyzed systems.

Aqueous silicic acid solutions A generalized overall reaction for sol–gel processing of silicic acid is given in Eq. (4.21).

$$Si(OH)_4 \longrightarrow SiO_2 + 2H_2O \tag{4.21}$$

Silicic acid, $Si(OH)_4$, which is formed when water glass is reacted with an acid or by ion exchange of the sodium ions for H^+, is only stable in solution at low pH. At pH >7 deprotonated species, mainly $H_3SiO_4^-$, are found in diluted solutions ($<10^{-4}$ mol l^{-1}), and mainly condensed anionic species are present in concentrated solutions (>0.1 mol l^{-1}).

Contrary to sol–gel processing of alkoxide precursors, there are mainly three possibilities to control the morphology and properties of the products: the concentration of the precursor, the presence of salts and the pH of the solution.

Three important pH regimes have to be distinguished: pH<2, pH 2–7, and pH >7. At pHs lower than about 2, the silicic acid species are positively charged, and according to the mechanism given in Eq. (4.18), the reaction rate of the condensation is proportional to the concentration of the H^+ ions. Between pH 2 and 7 the reaction rate of condensation is proportional to the concentration of the OH^- ions [see Eq. (4.19)]. This means, that the condensation rates are increasing when the pH is increased. At pHs higher than 7, the rates of solubility and for redissolution of particles are maximal. The solution now contains mainly anionic species that reject each other. Therefore, the presence of electrolytes has a strong influence on the gelation behavior (see also Section 4.5.1).

4.5.2.2 The Sol–Gel Transition (Gelation)

We will now discuss how the sol particles aggregate to give gel networks. In the initial stage of the sol–gel reactions, small three-dimensional oligomeric particles are formed, with Si–OH groups on their outer surface. The oligomers serve as nuclei. They may either grow, or agglomerate at a certain size, depending on the experimental conditions (Figure 4.51). Agglomeration of the oligomeric particles does not necessarily result in gelation; instead, larger particles with a polymeric substructure can be formed.

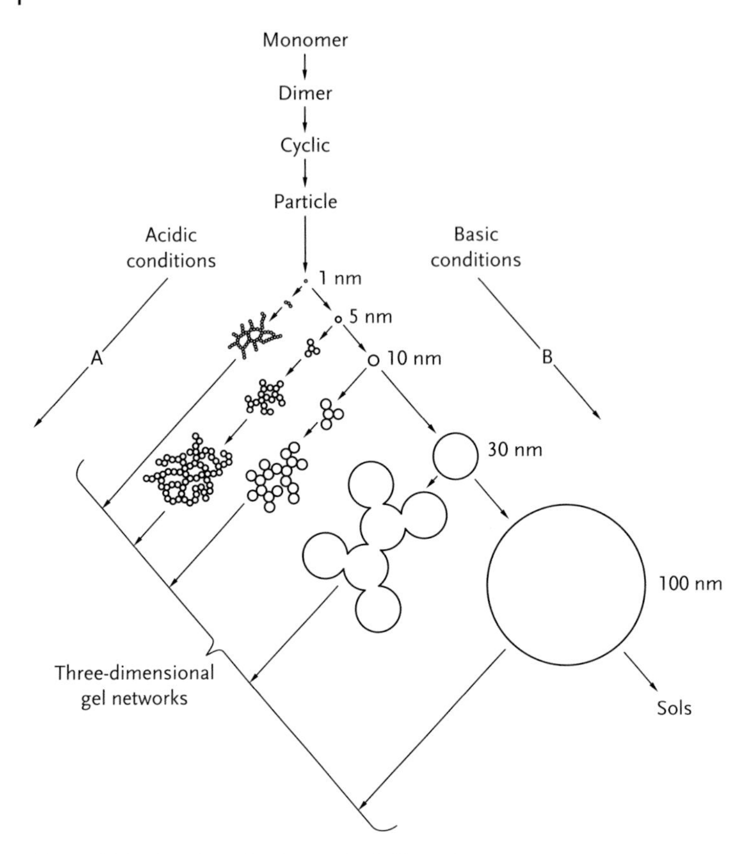

Figure 4.51 Structural development of silica gels.

Whether the larger particles (with a polymeric or dense substructure) remain suspended in solution (i.e., whether a stable sol is obtained) or agglomerate themselves to form a three-dimensional network (i.e., whether gelation occurs) again depends on the system and the experimental conditions that influence the stability of the sol.

As the sol particles aggregate and condense, the viscosity of the sol gradually increases and a gel is eventually formed (provided that the sol is sufficiently concentrated). The sol–gel transition (gel point) is reached when a continuous network is formed.

From a practical point of view, the gel time (t_{gel} = time at which the gel point is reached after starting hydrolysis and condensation reactions) is determined by turning the reaction vessel upside down. Before the gel point has been reached, the → colloidal dispersion behaves like a more or less viscous liquid, that is, the liquid will flow out of the vessel. At the gel point, the viscosity increases sharply, and a form-stable, elastic gel body is obtained. Since all liquid is retained in the gel body, no liquid can flow out of the vessel if it is turned upside down. For the same reason, the volume of the gel in this stage is the same as that of the original precursor solution.

The t_{gel} is generally lowered by all parameters that increase the rate of condensation reactions, as discussed above. These parameters thus allow to deliberately influencing the gel times. Typical t_{gel} values for $Si(OMe)_4$ in the presence of 0.05 mol of a catalyst are 92 h with HCl as the catalyst, H_2SO_4 106 h, NH_4OH 107 h, and HI 400 h. Without a catalyst, t_{gel} would be about 1000 h.

The simplest picture of gelation is that the particles grow by aggregation or condensation until they collide to give clusters of particles. The clusters become bigger and bigger by repeated collisions. In this picture, which is described mathematically by percolation theory, the gel is formed when the last link between two giant clusters of particles is formed. Percolation theory offers a simple description of gelation. Figure 4.52 illustrates percolation on a two-dimensional lattice.

Starting with an empty grid, spheres (representing the sol particles) are placed randomly on the intersections of the grid lines. The fraction of filled sites is designated p. If two neighboring sites are filled, they are joined by a bond. This process, called *site* percolation, produces clusters of certain sizes (s). As more and more sites are filled, the average cluster size (s_{av}) increases. In Figure 4.52a, there is a broad range of cluster sizes, but none of them is a "spanning cluster," that is, a cluster that reaches across the vessel that contains it. The percolation threshold p_c is defined as the value of p at which the spanning cluster first appears. This is equivalent to gelation. In Figure 4.52b, p is clearly larger than p_c, the percolation threshold has been crossed. Note, however, that:

- at the gel point a certain number of unbounded clusters is still present. This is important for the aging of gels (see below); and
- the last bond resulting in the formation of the spanning cluster is not different from the previously formed bonds, that is, gelation is not a special thermodynamic event.

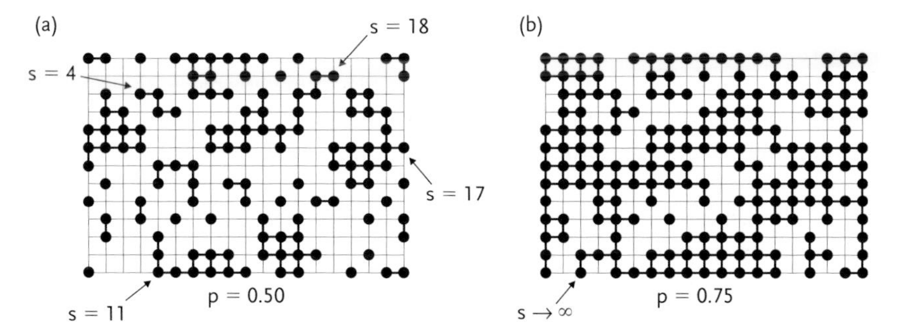

Figure 4.52 Site percolation on a square lattice: (a) for $p=0.50$, and (b) for $p=0.75$ (p is the fraction of filled spheres).

A different version of percolation, which is more appropriate as a model for gelation, is *bond* percolation. All sites are initially filled with spheres, and the bonds are filled in randomly.

An alternative description of gelation is given by kinetic growth models. These also explain the different microstructures upon changing the reaction conditions. Depending on the conditions, growth in silicate systems may occur predominantly by condensation of clusters with monomers or with other clusters. The rate of the condensation reactions may be diffusion- or reaction-limited. Note that the term "cluster" is used with different meanings. In the kinetic growth models, "cluster" is used equivalent to "particles" or "oligomeric species" in the preceding discussion.

As has been discussed before, hydrolysis of silicon alkoxides is much faster than condensation under acidic conditions. Since all species are hydrolyzed at an early stage of the reaction, they can condense to form small oligomeric species (clusters) with reactive Si–OH groups. Under these conditions, reactions at terminal silicon atoms are favored for electronic reasons. This results in polymer-like gels, that is, small clusters undergo condensation reactions with each other to give a polymer-like network with small pores. This process is called reaction-limited cluster aggregation (RLCA).

Monomer-cluster growth requires a continuous source of monomers. Hydrolysis is the rate-determining step under basic conditions. The hydrolyzed species are immediately consumed by reaction with existing clusters because of the faster condensation reactions. Furthermore, the rate of hydrolytic cleavage of (terminal) Si–O–Si bonds, is much higher than under acidic conditions. This additionally insures that a source of monomers is available. This model is called reaction-limited monomer cluster growth (RLMC), or Eden growth. Condensation of clusters among each other is relatively unfavorable because this process requires inversion of one of the silicon atoms involved in the reaction. Due to a different mechanism, reaction at central silicon atoms of an oligomer unit is favored (see above). The resulting network has a particulate character with big particles and large pores (colloidal gels).

The formation of larger particles, mainly in aqueous systems, is also favored by → Ostwald ripening by which small particles dissolve and larger particles grow by condensation of the dissolved species. The solubility of a particle is inversely proportional to its radius. The solubility of small particles (<5 nm) therefore is rather high. Growth stops when the difference in solubility between the smallest and the largest particles in the system becomes only a few ppm. Solubility depends on the given conditions (temperature, pH of the solution, etc.). For example, silica particles grow to a size of about 5–10 nm at pH >7, but growth stops at a size of 2–4 nm at lower pH. At higher temperatures, larger particles are obtained because the solubility of silica is higher.

An example of how the growth mechanisms discussed above can be used to obtain special morphologies of silica materials, is the production of smooth monodispersed silica spheres by the so-called Stöber method. TEOS is reacted at high pH (with ammonia as catalyst) and an R_w between 0.5 and 0.05 (large excess of water). The particle diameter can be accurately controlled through the process parameters (temperature, concentrations, etc.). Silica spheres with diameters between tenths of a micrometer and a micrometer can be produced (Figure 4.53).

Growth of the silica spheres represents a good example for RLMC growth. Small primary particles form by → nucleation and growth; repeated dissolution and reprecipitation insures that a source of monomers is available, and that reaction-limited conditions exist. Larger monosized spherical aggregates are formed with the primary particles as "monomers" (also by RLMC).

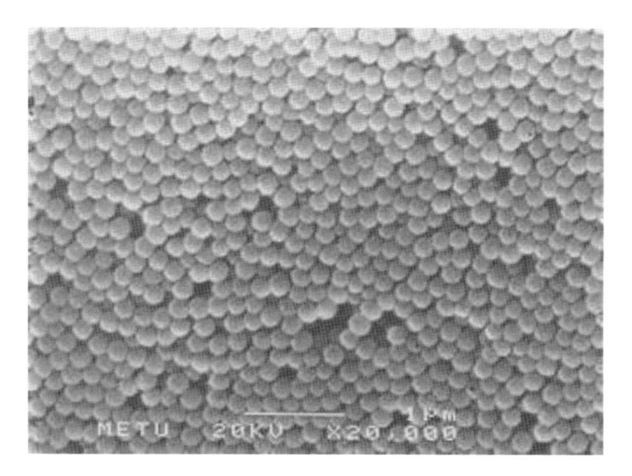

Figure 4.53 Monodisperse silica spheres with a diameter of about 0.2 μm made by the Stöber process.

4.5.2.3 **Aging**

The sharp increase in viscosity at the gel point freezes in a particular network structure. Thus, gelation is related to glass-forming processes (see Section 4.1). However, this structure may change appreciably with time, depending on the temperature, solvent or pH conditions, or upon removal of the pore liquid.

It is very important to realize that the chemical reactions are not finished with gelation, and structural rearrangements take place in the wet gels (i.e., in the gels still containing their pore liquid). This phenomenon increases the stiffness of the gels and is called "aging." Aging is due to several processes:

- The gel network still contains a continuous liquid phase. The pore liquid initially is a sol, that is, it contains condensable particles or even monomers (see Figure 4.52), which eventually condense to the existing network. This causes gradual changes in the structure and properties of the gels.
- The gel network originally is still very flexible. This allows neighboring M–OH or M–OR groups to approach each other and to undergo condensation reactions. This causes contraction of the network and expulsion of pore liquid.

This spontaneous shrinkage of some gels is called syneresis and continues as long as the gel network exhibits sufficient flexibility. The driving force is the reduction of the large solid/liquid interface in the gels.
- Hydrolysis and condensation reactions are, in principle, reversible. Therefore, mass is dissolved from thermodynamically unfavorable regions, mostly regions with a high positive curvature or small particles. The solutes condense to thermodynamically more favorable regions, particular in pores, crevices, particle necks, and so on. This process ("ripening" or "coarsening") results in the reduction of the net curvature, disappearance of small particles, and filling up of small pores (the processes are similar to those discussed in Section 2.1.5 for sintering processes).

4.5.2.4 **Drying**

The evaporation of the liquid from a wet gel by conventional methods, that is, by temperature increase or pressure decrease, proceeds in a very complex way in which three different stages can be distinguished:

1. The gel shrinks by the volume that was previously occupied by the liquid. The liquid flows from the interior of the gel body to its surface. If the network is compliant, as it is in alkoxide-derived gels, the gel deforms. Upon shrinkage, OH groups at the inner surface approach each other and can react with each other. For example, new siloxane bridges are formed in SiO_2 gels. As drying proceeds, the network becomes increasingly stiffer and the surface tension in the liquid rises correspondingly because the pore radii become smaller.
2. This stage of the drying process begins when the surface tension is no longer capable of deforming the network and the gel body becomes too stiff for further shrinkage. The tension in the gel becomes so large that the probability of cracking is highest. In this stage of drying, the liquid/gas interface retreats into the gel body. Nevertheless, a contiguous funicular liquid film remains at the pore walls, that is, most of the liquid still evaporates from the exterior surface of the gel body.
3. Here, the liquid film is ruptured. Eventually, liquid is only in isolated pockets and can leave the network only by diffusion via the gas phase.

Two processes are important for the collapse of the network. First, the slower shrinkage of the network in the interior of the gel body results in a pressure gradient that causes cracks. Secondly, larger pores will empty faster than smaller pores during drying, that is, if pores with different radii are present, the meniscus of the liquid drops faster in larger pores. The walls between pores of different size are therefore subjected to uneven stress, and crack (Figure 4.54).

For these reasons, xerogel powders are usually obtained when wet gel bodies are conventionally dried. Although strategies were developed to obtain crack-free xerogel bodies, the large shrinkage cannot be avoided. Due to the shrinkage problem, one of the most important applications of sol–gel materials is for films and coatings, where shrinkage is easier to control.

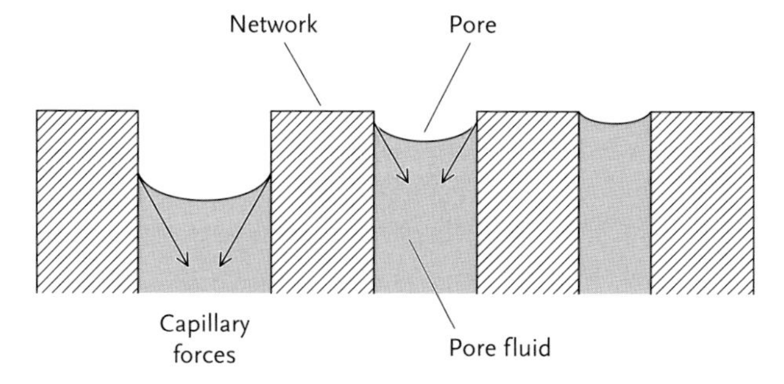

Figure 4.54 Contracting surface forces in pores of different size during drying.

4.5.3
Sol–Gel Chemistry of Metal Oxides

The fundamental issues of sol–gel processing were mainly investigated for silica-based systems. One of the reasons for this is that network structures with no long-range order form easily (see comparison of SiO_2 and TiO_2 in Section 4.1.1). Nevertheless, any other metal oxide can, in principle, be prepared by sol–gel processing. Although the principles of sol and gel formation, aging, and so on are the same, there are some differences in the chemistry of the precursors.

As in silicate sol–gel processes, inorganic or metal–organic (alkoxide) precursors can be used. The chemistry of these precursors and ways to influence their reactivity is discussed in the following paragraphs.

4.5.3.1 **Inorganic Precursors (Metal Salts)**

Many metal salts are hydrolytically unstable, that is, they form oxide/hydroxide precipitates from aqueous solutions when the pH is increased. Hydrolysis reactions occur because water molecules coordinated to metal ions are more acidic than in the non-coordinated state due to charge transfer from the oxygen to the metal atom. When the pH is increased (i.e., if a base is added), the series of equilibria shown in Eq. (4.22) is shifted to the right. Water (aqua) ligands (M–OH_2) are thus converted to hydroxo (M–OH) and oxo (M=O or M–O–M) ligands. This process is also called *forced hydrolysis*.

$$\left[\mathrm{M{-}O}\begin{matrix}\diagup\mathrm{H}\\ \diagdown\mathrm{H}\end{matrix}\right]^{n+} \underset{}{\overset{-\mathrm{H}^+}{\rightleftharpoons}} \left[\mathrm{M{-}OH}\right]^{(n-1)+} \overset{-\mathrm{H}^+}{\rightleftharpoons} \left[\mathrm{M{=}O}\right]^{(n-2)+} \quad (4.22)$$

For example, when aluminum salts are dissolved in water, the hydrated cation $[Al(H_2O)_6]^{3+}$ exists only below pH 3. As the pH is increased, the water ligands are deprotonated, and the ions $[Al(OH)_x(H_2O)_{6-x}]^{(3-x)+}$ are formed. Mononuclear species with $x = 0-4$ are only stable in very dilute solutions; at higher concentrations, → polynuclear species are formed by condensation reactions, that is, by formation of Al–O–Al links.

Figure 4.55 shows that the formation of oxides is favored for highly charged metal ions and/or high pH. This diagram also explains qualitatively why the hydrolysis of low-valent cations ($Z<4$) yields aquo, hydroxo, or aquo-hydroxo complexes over the complete pH scale, while high-valent cations ($Z>5$) form oxo or oxo-hydroxo compounds.

There are three possibilities to shift the equilibria in Eq. (4.22) to the right, that is, to obtain oxides from hydrated metal salts. Raising the pH, as in the case of $[Al(H_2O)_6]^{3+}$, corresponds to moving from left to right in Figure 4.55 at a given charge (Z). Alternatively, the solution can simply be aged at elevated temperatures. A higher temperature promotes dissociation of protons from the hydrated metal ions. Solutions of metallate ions, such as alkaline metal or ammonium titanates, vanadates, niobates, tantalates, or tungstates form gels when acidified (right to left in Figure 4.55). Finally, solutions of oxide species with the metal in high oxidation states can be reduced to give gels (top-down in Figure 4.55 at a given pH); one of the best known examples is the formation of MnO_2 gels from MnO_4^-.

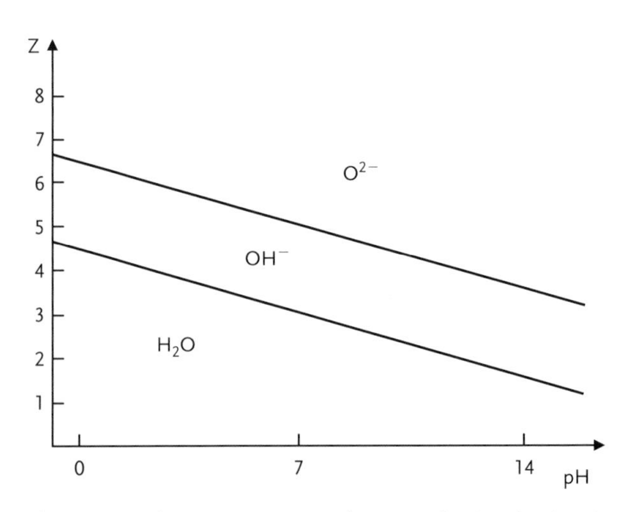

Figure 4.55 Charge (Z) versus pH diagram indicating the domains of aquo, hydroxo, and oxo species.

Depending on the kind of bridging group formed between the metal centers, two condensation processes are distinguished:

1. **Olation** is a condensation process in which a *hydroxy* bridge is formed by nucleophilic substitution. Since water is the leaving group, the kinetics of olation is related to the lability of the aqua ligand. The ability to dissociate from the metal center depends on the size, electronic configuration, and Lewis acidity of the metal ion. In general, the smaller the charge and the larger the size of the metal ion, the greater the rate of olation. Several types of OH bridges can be formed by olation, as shown in Eq. (4.23).

$$\mathrm{M{-}OH} + \mathrm{M{-}OH_2} \longrightarrow \mathrm{M{-}\overset{H}{O}{-}M} + \mathrm{H_2O}$$

$$\mathrm{(M)_2OH} + \mathrm{M{-}OH_2} \longrightarrow \mathrm{(M)_2O(H){-}M} + \mathrm{H_2O} \tag{4.23}$$

$$\mathrm{H_2O{-}M(OH)} + \mathrm{(HO)M{-}OH_2} \longrightarrow \mathrm{M(\mu{-}OH)_2M} + 2\,\mathrm{H_2O}$$

2. **Oxolation** is a condensation process in which an *oxo* bridge is formed. Oxolation is a two-step addition/elimination process [Eq. (4.24)]. Under basic conditions, the first step is catalyzed, because a M–OH group is deprotonated and the resulting M–O^- is a stronger nucleophile. In the presence of an acid, the second step is catalyzed, because protonation of a terminal M–OH group creates M–OH_2^+, and water is a good leaving group.

$$\mathrm{M{-}OH} + \mathrm{M{-}OH} \longrightarrow \mathrm{M{-}\overset{H}{O}{-}M{-}OH}$$

$$\mathrm{M{-}\overset{H}{O}{-}M{-}OH} \longrightarrow \mathrm{M{-}O{-}M} + \mathrm{H_2O} \tag{4.24}$$

Thus, the rate of oxolation reactions is smallest at the isoelectric point. Owing to catalysis by both H^+ and OH^-, oxolation occurs over a wider pH range than olation. The mechanisms of base and acid catalysis are similar to what has been discussed for silicate systems [Eqs. (4.18) and (4.19)].

There is an important stereochemical difference between silicate systems and most transition-metal systems. As has been discussed in Section 4.1.1, all silicate materials are composed of corner-sharing tetrahedra. When a silica network grows, the question that decides of the morphology of the gels obtained is whether condensation occurs preferentially at the end of a chain or at a central atom. For transition metals this issue is more complicated and hardly understood in detail in most cases. First, the charges of transition metals are higher than their coordination numbers. As a result, the polyhedra (octahedra, square antiprisms, and so on) must also share edges and faces to satisfy the coordination requirements of the metal (see Section 4.1.1). This results in more compact structures and a higher tendency for crystallization. Second, hydroxo or aquo → ligands coordinated to the same metal may be chemically non-equivalent due to the effect of different *trans* ligands. For example, an aquo ligand *trans* to a M=O unit is a better leaving group than *trans* to a M–OH group. The nucleophilicity of OH groups or the acidity of hydoxo and aquo ligands are also influenced by the *trans* ligands.

The whole situation may even be more complicated, because the counterion (X) of the metal salt precursor (MX_n) can compete with the aquo ligands for coordination to the metal center. An anionic ligand (X) may remain coordinated to the metal through all stages of the overall process and even turn up in the final product. The stability of a M–X bond in a metal complex depends on several electronic and steric factors (see textbooks on coordination chemistry for details). For example, in basic zirconium salts of the overall composition $Zr(OH)_2X_2(H_2O)_n$ (all of which have oligomeric or polymeric structures) NO_3^-, SO_4^{2-}, or HPO_4^{2-} are coordinated to zirconium, while ClO_4^- or Cl^- are not able to displace the coordinated water molecules at the zirconium atom and are not involved in the formation of condensed species. A strong metal–anion interaction (coordination) can influence sol–gel processing in several ways:

- The hydrolysis and condensation reactions may proceed differently when different salts of the same metal are employed.
- Strong coordination of a counterion blocks coordination sites and leads to a smaller degree of condensation (fewer M–O–M links per metal atom).
- Coordinated counterions can direct the site of nucleophilic attack (*cis* or *trans* to the coordinated X, for example) during hydrolysis and condensation reactions and thus influence the microstructure and morphology of the gels or precipitates.
- The counterions may influence the electric double layer of the sol particles and the ionic strength of the solution, and hence the aggregation of the sol particles.
- The counterions may (partially) stay in the material. Their (complete) removal may be difficult and may require special postsynthesis procedures.

4.5.3.2 Alkoxide Precursors

There are some important differences between metal alkoxides and $Si(OR)_4$:

- Owing to their lower electronegativity, metal alkoxides are stronger Lewis acids than silicon alkoxides. Nucleophilic attack at the metal is thus facilitated, and the hydrolysis rates are strongly increased. For example, the hydrolysis rate of $Ti(OR)_4$ is about 10^5 times faster than that of $Si(OR)_4$ with the same alkoxide substituents.
- Most metals have several stable coordination numbers, or the expansion of the coordination sphere in transition states is easier.

Owing to both effects, the reactivity of some tetravalent iso-propoxides in hydrolysis reactions increases in the order

$$Si(O^iPr)_4 <<< Sn(O^iPr)_4, Ti(O^iPr)_4 < Zr(O^iPr)_4 < Ce(O^iPr)_4$$

The reactivity of many metal alkoxides towards water is so high that precipitates are formed spontaneously. While the reactivity of alkoxysilanes has to be promoted by catalysts, the reaction rates of metal alkoxides must be moderated to obtain gels instead of precipitates.

Hydrolysis of metal alkoxides occurs by an addition/elimination mechanism [Eq. (4.25)]. The mechanisms of the condensation reactions are very similar to what was discussed before for inorganic precursors. Both oxolation [Eq. (4.23)] and olation [Eq. (4.22)] are possible; a ROH molecule may be the leaving group instead of H_2O. Oxolation with elimination of an alcohol is also called alcoxolation [Eq. (4.26)].

$$\mathrm{H_2O} + \mathrm{M{-}OR} \longrightarrow \mathrm{H_2O} \rightarrow \mathrm{M{-}OR} \longrightarrow \mathrm{HO{-}M} \leftarrow \mathrm{O(H)R} \longrightarrow \mathrm{M{-}OH} + \mathrm{ROH} \tag{4.25}$$

$$\mathrm{M{-}OH} + \mathrm{M{-}OR} \longrightarrow \mathrm{M{-}O(H){-}M{-}OR} \longrightarrow \mathrm{M{-}O{-}M} + \mathrm{ROH} \tag{4.26}$$

In principle, the same parameters influence the rates of the hydrolysis and condensation reactions of metal alkoxides for a given metal as those already discussed for silicon alkoxides. An additional factor is the degree of oligomerization of the alkoxide precursors. Silicon alkoxides are always monomeric, while metal alkoxides may be associated via μ_2- or μ_3-OR bridges (the subscript denotes the number of metal atoms coordinated to the bridging ligand). The reason for the association is that the metals do not reach their full coordination number by the terminal alkoxo groups. For example, the usual coordination number of titanium is six, but there are only four alkoxo ligands in $Ti(OR)_4$. When neat or dissolved in non-polar solvents, coordination expansion occurs by association via OR bridges.

The degree of association depends on:

- The size of the metal: the tendency to oligomerize increases with the size of the metal. For example, the average degree of oligomerization in $M(OEt)_4$ (the oligomerization processes are equilibria) is 2.9 for M = Ti, 3.6 for Zr, 3.6 for Hf, and 6.0 for Th.
- The size of the groups R: larger groups R favor smaller units because of steric hindrance. For example, $Ti(OEt)_4$ in ethanol has a trimeric structure, while $Ti(O^iPr)_4$ is monomeric in propanol solution.

The degree of oligomerization influences not only the solubility of the metal alkoxides (highly associated oligomers may even be insoluble) but also the reaction kinetics. Coordinatively unsaturated species have a higher reactivity in hydrolysis and condensation reactions. For example, the monomeric species $Ti(O^iPr)_4$ is hydrolyzed more rapidly than trimeric $Ti(OEt)_4$ despite the larger alkoxo ligands. When the degree of oligomerization of two metal alkoxides $M(OR)_n$ with different groups R is the same, then the size of R has the same influence on the reaction kinetics, as was discussed for silicon alkoxides.

Association via alkoxide bridges is not the only way for coordination expansion. When the alkoxides are dissolved in polar solvents such as alcohols, addition of solvent molecules may occur and compete with association. For example, hydrolysis of $Zr(O^nPr)_4$ in PrOH is very fast and results in a precipitate, while hydrolysis of the same alkoxide in cyclohexane gives a homogeneous gel. The explanation of this difference is that OR-bridged oligomeric species are formed in cyclohexane solution, while an alcohol molecule is coordinated to $Zr(O^nPr)_4$ in propanol solution. Upon hydrolysis, the coordinated ROH is more easily cleaved than the OR bridge.

Neither association to oligomeric species nor solvate formation is observed for silicon alkoxides.

The high reactivity of metal alkoxides can be moderated by their chemical modification. The general approach is to replace one or more alkoxo → ligands by groups that are less easily hydrolyzed (i.e., which are more strongly bonded to the metal) and additionally block coordination sites at the metal. The most common ligands are carboxylate [Eq. (4.27)] or β-diketonate groups [Eq. (4.28)], but any other anionic → bidentate ligand can, in principle, be used. It should be pointed out, that this explains the use of acetylacetone, the simplest β-diketone, for the synthesis of PZT in Figure 4.43.

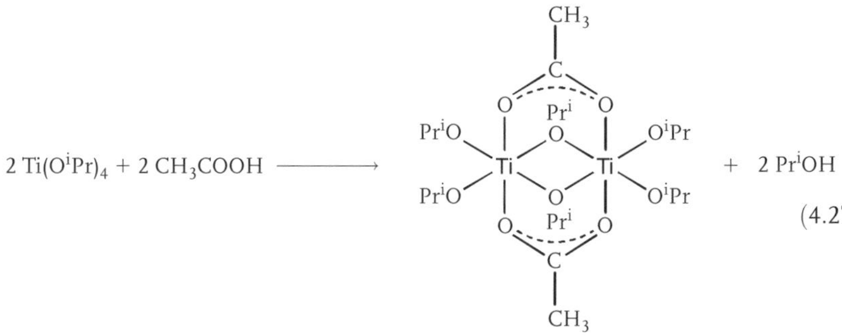

$$2\,Ti(O^iPr)_4 + 2\,CH_3COOH \longrightarrow \quad + \; 2\,Pr^iOH \qquad (4.27)$$

CH3, C, O, O, Pr^iO, Pr^i, O^iPr, Ti, Ti, Pr^iO, O, O^iPr, Pr^i, O, O, C, CH3

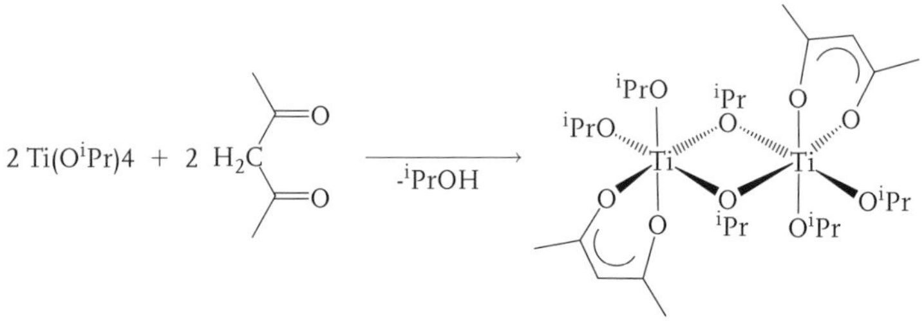

$$2\,Ti(O^iPr)4 \;+\; 2\;H_2C(C(=O)CH_3)_2 \xrightarrow{-^iPrOH} \qquad (4.28)$$

^iPrO, ^iPr, O, ^iPrO, O, O, Ti, Ti, O, O, O^iPr, O, ^iPr, O^iPr

The bidentate ligands can be → chelating (i.e., bonded to the same metal [Eq. (4.28)] or bridging [Eq. (4.27)]. Bidentate ligands are inherently more strongly bonded than comparable monodentate ligands and therefore are less readily hydrolyzed than the remaining OR groups. The → bidentate ligands are introduced by substitution reactions, as shown in Eqs. (4.27) and (4.28).

The substitution of OR ligands by anionic bidentate ligands has several consequences:

- A new precursor is created in which the remaining alkoxo groups have a different reactivity in hydrolysis and condensation reactions than the parent alkoxide. The situation is comparable to alkoxysilanes: $Si(OMe)_4$ has a different reactivity from $MeSi(OMe)_3$, for example.
- The substitution of a monoanionic monodentate ligand (OR^-) by a monoanionic → bidentate ligand maintains the charge balance within the precursor.

However, an extra coordination site is blocked. This additionally reduces the reactivity in hydrolysis and condensation reactions but also introduces building blocks with a different connectivity. The bidentate ligand may stereochemically direct the hydrolysis or condensation site as discussed before for coordinated counterions. The formation of gels instead of precipitates is facilitated by the lower degree of crosslinking of the network. Note that the acetate or acetylacetonate derivatives in Eq. (4.28) and (4.29) are dimeric because in monomeric $Ti(O^iPr)_3$(acetate) or $Ti(O^iPr)_3$(acetylacetonate) the titanium atom would still be five-coordinate.

- Functional or non-functional organic groups are introduced by the bidentate ligands (see Section 4.5.4).

4.5.4 Inorganic–Organic Hybrid Materials

Sol–gel materials are formed by kinetically controlled reactions, as discussed above. One of the advantages of such amorphous materials is that their chemical composition is not bound to given stoichiometries. Thus, different molecular precursors, that is, *building blocks* of different composition, can be coprocessed, resulting in mixed-oxide materials. In principle, any chemical composition is feasible unless phase separation occurs because of the different reactivities of the precursors.

This is not restricted to inorganic precursors. One of the major advances of sol–gel processing is undoubtedly the possibility of synthesizing hybrid inorganic–organic materials, where organic and inorganic building blocks are combined. These are called *hybrid* materials because they combine, to some extent, the properties of inorganic and organic compounds in *one* material. The idea for developing such materials is similar to that of → composite materials, where two or more materials are combined that differ in form and (mostly) chemical composition. While macroscopic constituents with defined phase boundaries are used for composite materials, the building blocks of inorganic–organic hybrid materials are mixed on a molecular level.

Sol–gel processing is a very suitable way to make such materials because of the mild processing conditions. The high-temperature synthesis route to ceramic materials, for example, does not allow the incorporation of thermally labile organic moieties. Inorganic–organic hybrid materials of another type and structure have been referred to previously in Section 2.3 (intercalation of inorganic layered structures by organic molecules or polymers) or Section 5.8.2 (MOF).

The inorganic–organic hybrid materials made by sol–gel processing have been called ORMOSILs (*or*ganically *mo*dified *sil*icates), ORMOCERs (*or*ganically *mo*dified *cer*amics), CERAMERs (*cer*amic poly*mers*), or POLYCERAMs (*poly*meric *ceram*ics). As these names already imply, these hybrid materials are composed of building blocks found in inorganic oxidic glasses or ceramics, silicones, or organic polymers. The concept is to combine properties of organic groups and polymers (functionalization, processing at low temperatures, toughness, etc.) with properties of glass- or ceramic-like materials (→ hardness, chemical and thermal stability, etc.) in order to generate new and synergetic properties not accessible otherwise.

There is a wide range of possibilities to vary the composition and structure, and thus the properties of the materials:

- chemical composition of the organic and inorganic precursors,
- ratio of the inorganic to organic components,
- structure of the building blocks, and
- distribution of the building blocks (statistical, block-like, etc.).

Two different approaches can be used for the incorporation of organic groups into an inorganic network by sol–gel processing, namely embedding of organic molecules into gels without chemical bonding, and incorporation of organic molecules via covalent bonding.

4.5.4.1 **Embedding of Organic Molecules**

This is achieved by dissolving the molecules in the precursor solution. The gel matrix is formed around them and traps them. In such so-called class I hybrid materials, the organic and inorganic entities thus interact only weakly with each other. A variety of organic or organometallic molecules can be employed, such as dyes, catalytically active metal complexes, sensor compounds, or biomolecules (Figure 4.56, left). Small particles can also be embedded by the same approach. If sol–gel processing of alkoxides is performed in the solution of an organic polymer, the inorganic network (formed by sol–gel processing) and the organic network interpenetrate but are not bonded to each other (Figure 4.56, right).

An interesting application for this type of sol–gel material are biosensors. A biosensor is a sensor in which a biological species, for example, an enzyme or antibody, is used to detect certain other species. Biomolecules are very sensitive molecules that often react by denaturation to changes of the pH, higher temperatures, and organic solvents. The mild reaction conditions during sol–gel processing, particularly the low temperatures and the possibility of using buffered

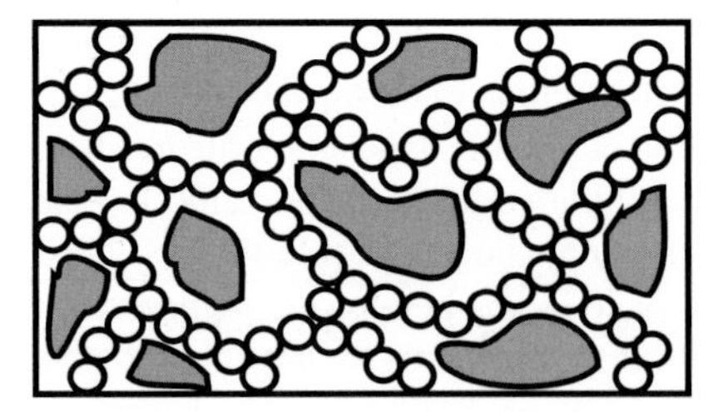
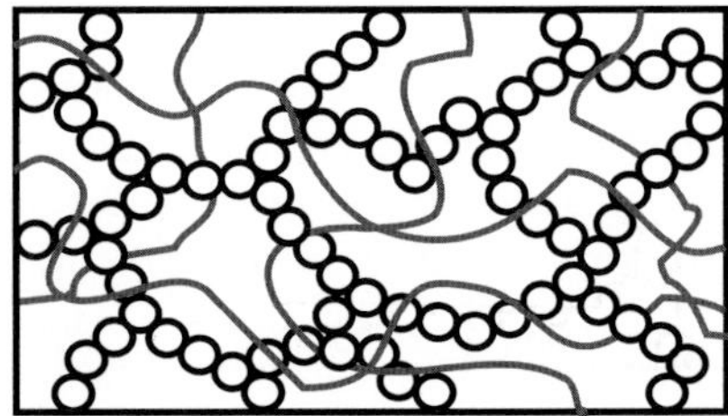

Figure 4.56 Embedding of molecules or particles (left) into an inorganic gel matrix; interpenetrating inorganic–organic polymer networks not covalently connected with each other (right). The pearls-on-a-string features represent the inorganic (sol–gel-derived) network.

aqueous solutions, allow the entrapment of enzymes and other biological species such as antibodies (or even whole cells) in inorganic materials. The biomolecule is stabilized against denaturation due to the adsorbed water on the hydrophilic surface inside the gel. Additionally, the fine porosity of the gel protects the biomolecule against influences from outside, but permits nutrients and reactants to penetrate.

Stabilization of entrapped molecules is a general phenomenon that is also observed for sensing molecules or catalytically active species.

Incorporation of Organic Groups via Covalent Bonding Very important modifications of sol–gel materials are based on a covalent linkage of the organic groups (Figure 4.57) ("class II" hybrid materials). In silicate systems it is possible to use organotrialkoxysilanes [$R'Si(OR)_3$], diorganodialkoxysilanes [$R'_2Si(OR)_2$] or even $(RO)_3Si–R'–Si(OR)_3$ as precursors for sol–gel processing in the same way as tetraalkoxysilanes. In the organically substituted derivatives, the group R′ is bonded through a Si–C link to the network-forming inorganic part of the molecule [Eq. (4.29)]. Since Si–C bonds are hydrolytically stable, the organic groups are retained in the final material.

$$R'Si(OR)_3 + {}^3/_2H_2O \longrightarrow R'SiO_{3/2} + 3ROH \qquad (4.29)$$

The (partial) replacement of $Si(OR)_4$ by $R'Si(OR)_3$ in the precursor mixture has several consequences (which are similar as already discussed for metal alkoxides substituted by → bidentate ligands):

- the crosslinking density of the inorganic network is decreased,
- the organically substituted trialkoxysilane exhibits a different hydrolysis and condensation behavior,
- the inorganic network is modified and/or functionalized, and
- organic substituents change the polarity of the system.

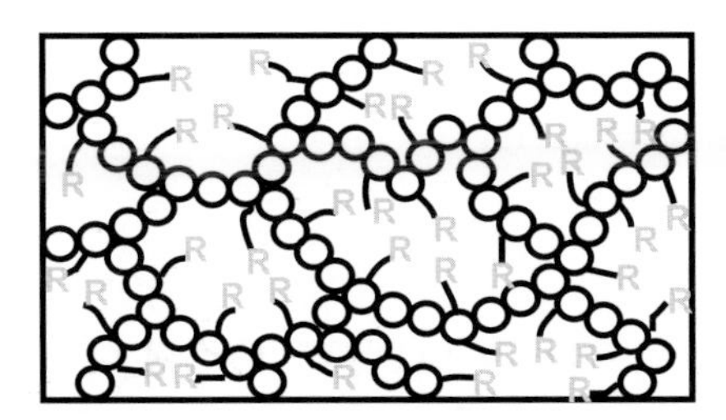
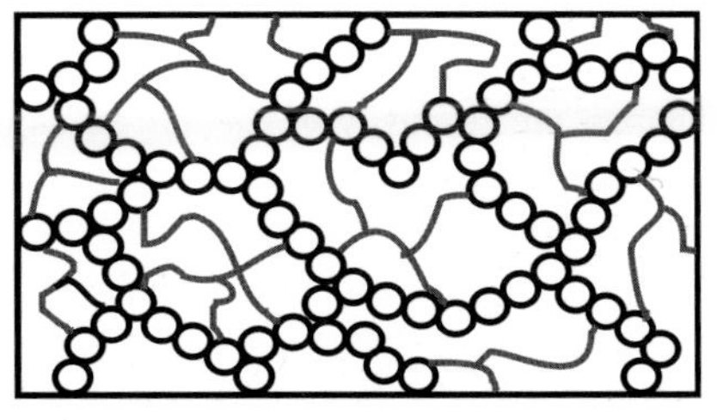

Figure 4.57 Incorporation of organic groups by covalent linkage into an inorganic gel matrix (left); dual inorganic–organic hybrid polymer networks connected with each other by covalent bonds (right). The pearls-on-a-string features represent the inorganic (sol–gel-derived) network.

Connectivity The connection of $[SiO_4]$ tetrahedra by four Si–O–Si bonds per silicon allows the formation of a three-dimensional network (see quartz or silica). Each substitution of an oxygen atom by a carbon atom lowers the crosslinking density of the gel network. For organo*di*alkoxysilanes, only linear polymer-like products can be obtained, as in silicones (see Section 5.2). Organo*tri*alkoxysilanes are typically coprocessed with tetraalkoxysilanes to obtain the mechanical properties characteristic of highly crosslinked networks.

An alternative to the formation of networks from three-connected building blocks are closed structures. Thus, cage compounds $(RSiO_{1.5})_n$, the so-called POSS (polyhedral oligomeric silsesquioxanes), can be obtained from $R'Si(OR)_3$ under certain reaction conditions. Similar cage compounds, the so-called spherosilicates, can also be formed during sol–gel processing of tetralkoxysilanes, with X = OH or OR. Both POSS and spherosilicates are interesting building blocks for materials syntheses in themselves. Examples are shown in Figure 4.58. Transition-metal equivalents to the spherosilicates are polyoxometallate clusters.

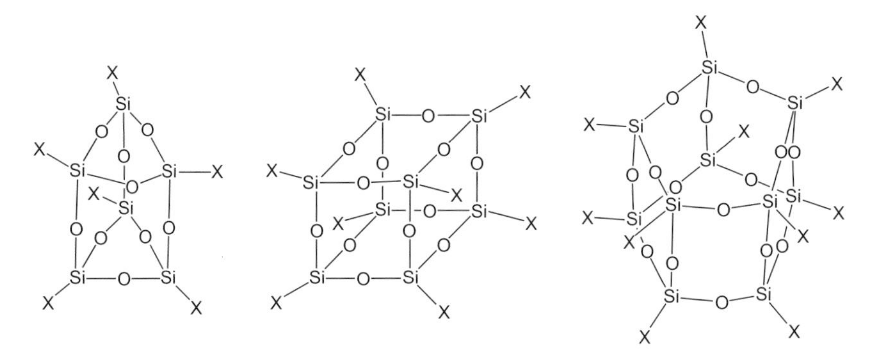

Figure 4.58 Molecular structures of $(XSiO_{1.5})_n$ cages for n=6, 8, 10; X=R, H, OH, and so on.

Hydrolysis and condensation behavior When (organically) modified alkoxides are reacted instead of the parent tetraalkoxides, the hydrolytically stable groups R change the reactivity of the silicon atom to which they are directly bonded. As discussed in Section 4.5.2, the electron density at the central silicon atom is typically increased due to the presence of the organic substituent (+I-effect).

If only the organotri- and dialkoxysilanes are employed as precursors, these factors just influence the overall rate of the hydrolysis and condensation reactions. However, if mixtures of the organically substituted alkoxysilanes and tetraalkoxysilanes are used, the reaction conditions must be chosen very carefully to obtain homogeneous materials and avoid phase separation. The organically substituted compounds react faster than the tetraalkoxysilane in acid-catalyzed systems and slower in base-catalyzed systems (Figure 4.59). Therefore, materials with a different structure will be obtained by working in solutions of different pH.

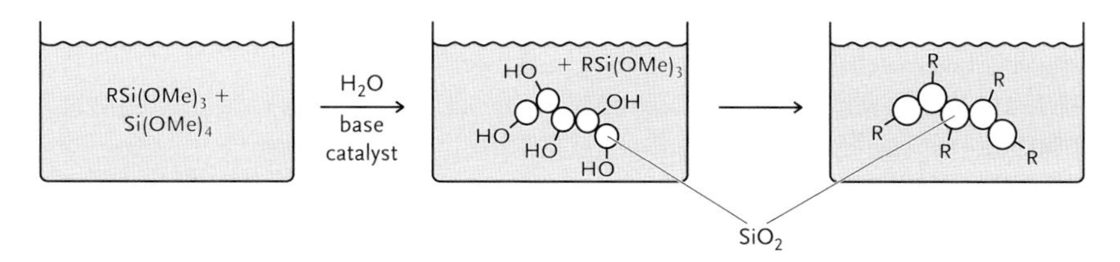

Figure 4.59 When a $Si(OMe)_4/R'Si(OMe)_3$ mixture is reacted under base-catalyzed conditions, the gel network is predominantly formed from $Si(OR)_4$ because it reacts faster. The $RSiO_{3/2}$ units then condense onto the network. Performing the same reaction in an acidic medium would reverse everything.

Modification and functionalization The choice of the organic group R′ for modification of the inorganic network is nearly unlimited. Table 4.6 lists some typical organotrialkoxysilanes, their properties, and some potential applications. Their synthesis will be outlined in Section 5.2.3.1.

While Si–C bonds are hydrolytically stable and thus allow introducing organic substituents into the gel materials, metal–carbon bonds in transition metals are usually cleaved by water. However, the organic groups can be introduced via the → bidentate ligands [Eqs. (4.27) and (4.28)] already discussed in Section 4.5.3. Usually, these ligands are used to moderate the reactivity of the alkoxide precursors, but they can also be used for an organic functionalization.

Generally speaking, precursors of the type $(RO)_nM$-Y-X-A are required, where $M(OR)_n$ is a metal alkoxide moiety, Y a bidentate group, X an inert spacer and A the functional organic group. For example, in the methacrylate derivative in Figure 4.60 (top), $M(OR)_n$ is a $Ti(OR)_3$ group (dimerized through OR bridges), Y a β-diketonate group, and the methacrylate double bond the functional organic group. For materials syntheses, this compound would be the titanium-equivalent to the methacrylate-substituted silane $(MeO)_3Si(CH_2)_3OC(O)CMe{=}CH_2$ in Table 4.6.

These examples show that the modification of inorganic gel networks by covalent linkage of all kinds of organic groups (left side of Figure 4.57) offers a huge variety of possibilities to modify or functionalize sol–gel materials.

4.5.4.2 Inorganic–Organic Hybrid Polymers

A related possibility is the formation of dual inorganic–organic hybrid polymer networks (Figure 4.57, right), in which cluster or polymer-type inorganic structures are linked by organic groups or polymer fragments. There are three principal different approaches to prepare such materials (Figure 4.61):

1. Formation of hybrid polymers from compounds of the type $[(RO)_nM]_xY$, in which Y is an organic group linking two ($x = 2$) or more ($x > 2$) metal alkoxide units. The structure of the preformed organic building block is retained in the final material. For example, organic groups of variable length (e.g., saturated or unsaturated hydrocarbon chains, or polyaryls) substituted with $Si(OR)_3$

Table 4.6 Some organofunctional trialkoxysilanes used for the preparation of inorganic–organic hybrid materials by sol–gel processing, and their function.

$R'Si(OR)_3$	Function
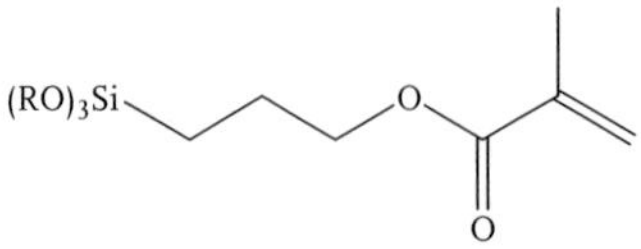 $(RO)_3Si$ O O	Polymerizable group for the preparation of hybrid polymers
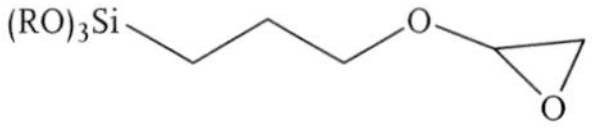 $(RO)_3Si$ O O	Group for organic polyaddition reactions; generation of hydrophilic diols by opening of the epoxide ring
 $(RO)_3Si$ NH_2	Hydrophilicity, coupling sites, coordination sites for metals
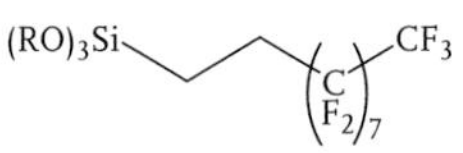 $(RO)_3Si$ $(CF_2)_7$ CF_3	Hydrophobicity
 $(RO)_3Si$ SH	Crosslinking site in thiol-ene UV cure systems, coordination site for metals
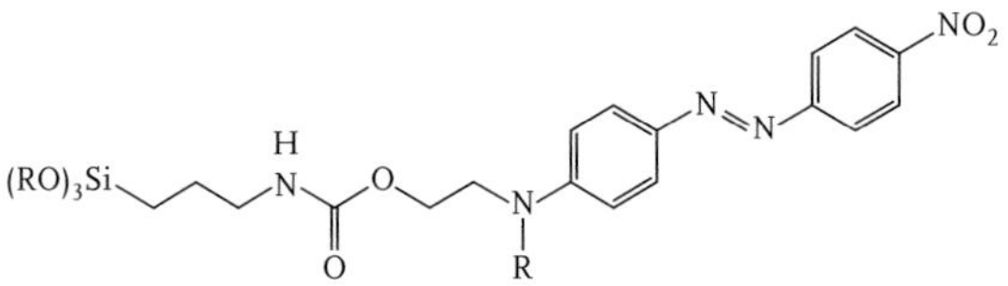 $(RO)_3Si$ H N O O N R N N NO_2	Chromophoric substituent (non-linear optic NLO dye)
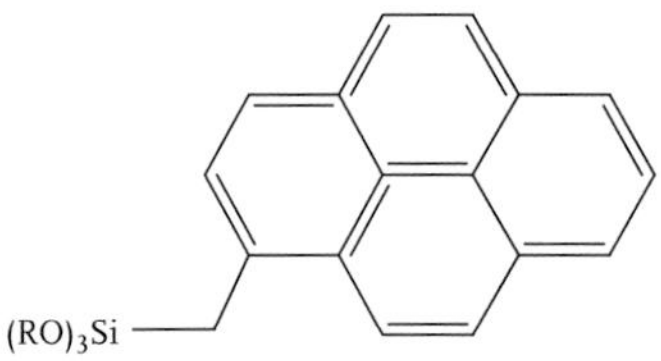 $(RO)_3Si$	Fluorescent substituent (pyrene derivative)
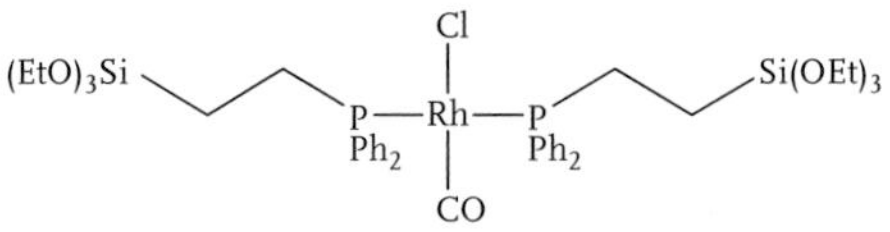 $(EtO)_3Si$ PPh_2 Cl Rh CO PPh_2 $Si(OEt)_3$	Coordination of transition metal complexes, catalysis

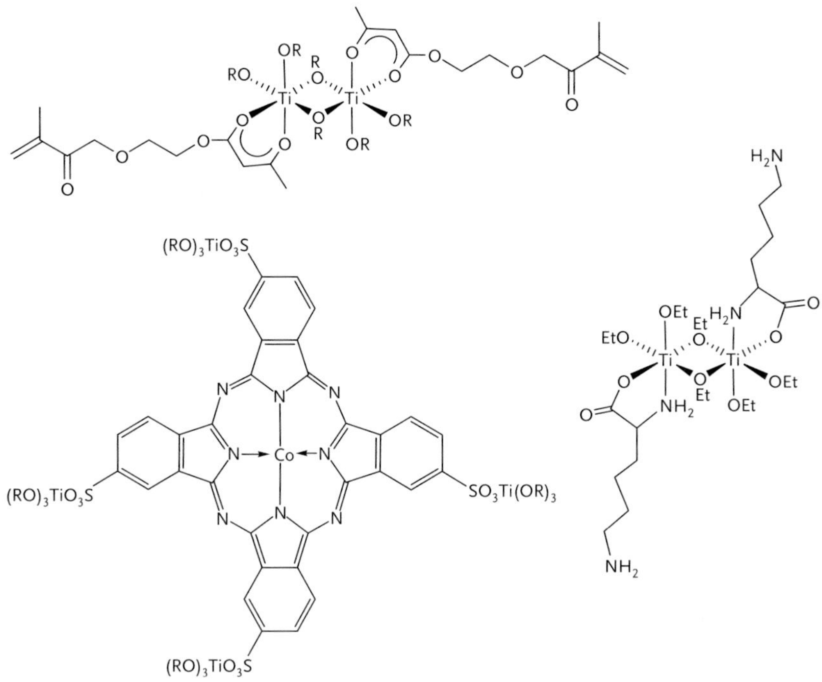

Figure 4.60 Some examples of organically modified titanium-based precursors.

groups at either ends or polymers with grafted $Si(OR)_3$ groups can be used (Figure 4.62).

2. Formation of hybrid polymers from bifunctional molecular precursors $(RO)_nM$-Y-X-A bearing an inorganic $(RO)_nM$, and an organic (A) functionality. The latter must be capable of undergoing polymerization or crosslinking reactions. The methacrylate-substituted silane $(MeO)_3Si(CH_2)_3OC(O)CMe{=}CH_2$ is often used for this purpose. After sol–gel processing, the organic network is formed by → polymerization of the methacrylate substituents.
3. Formation of hybrid polymers from functionalized inorganic building blocks. Here, the preformed inorganic structures are crosslinked by polymerization reactions of the organic functions. For example, methacrylate substituted metal oxide clusters or vinyl-modified POSS (Figure 4.58, X = vinyl) can be used.

An example of a material prepared by this approach is shown in Figure 4.63. A scratch-resistant coating with good adhesion on polymeric organic substrates is obtained by sol–gel processing of a mixture of vinyl and mercapto-substituted trialkoxysilanes (see Table 4.6). A sol is first formed by reaction of the silane mixture with water. After coating, the film is photochemically cured. In this step an organic link is formed by addition of the SH group to the double bond [Eq. (4.30)].

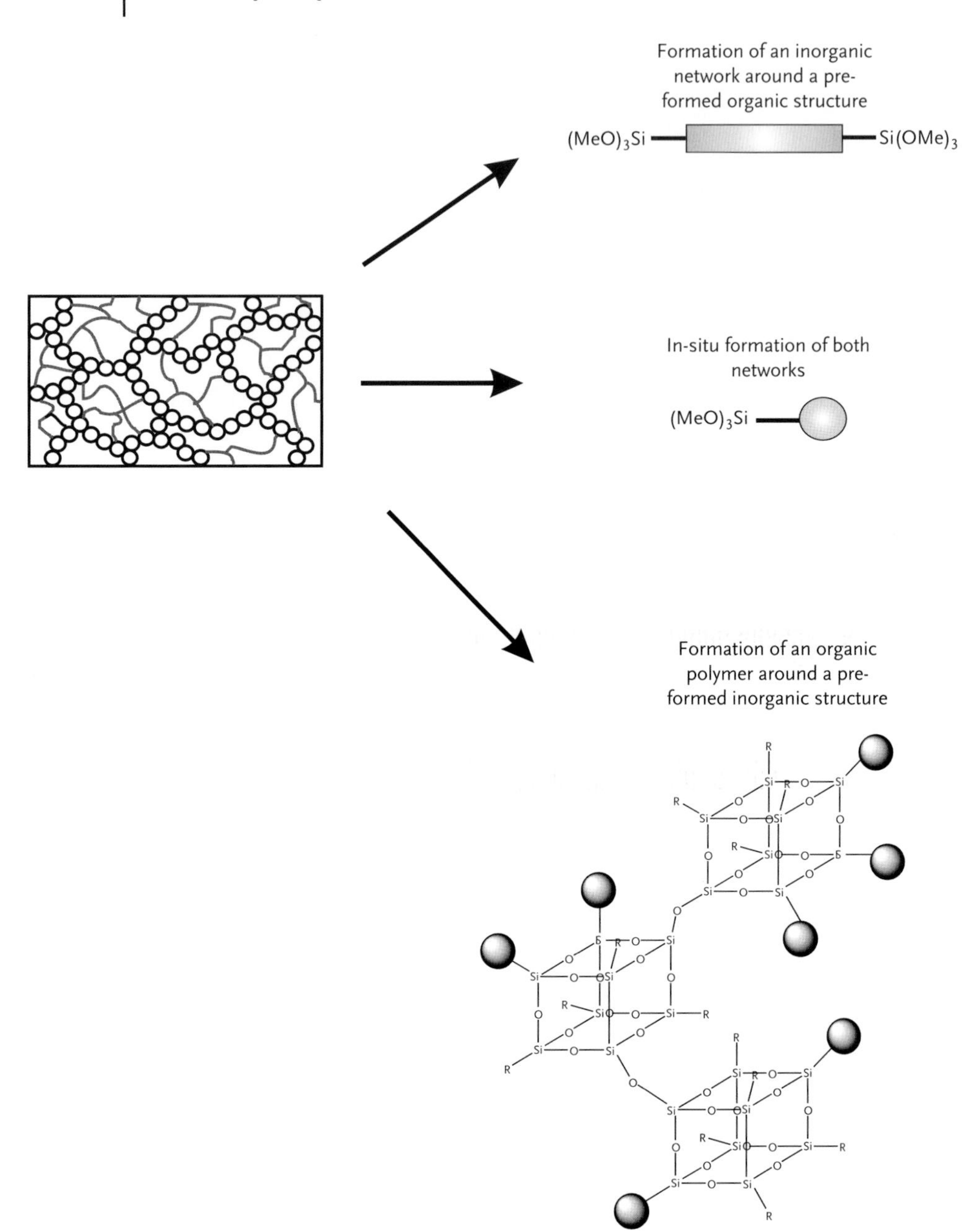

Figure 4.61 Approaches to dual inorganic–organic networks.

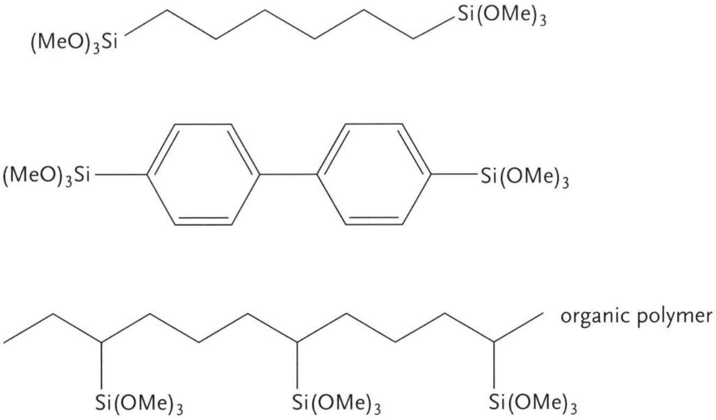

$(MeO)_3Si$ ⁄⁄⁄ $Si(OMe)_3$

$(MeO)_3Si$–C_6H_4–C_6H_4–$Si(OMe)_3$

organic polymer

$Si(OMe)_3$ $Si(OMe)_3$ $Si(OMe)_3$

Figure 4.62 Precursors with preformed organic linkages.

Figure 4.63 Scratch-resistant coating on polycarbonate (left half of slide coated, right half uncoated).

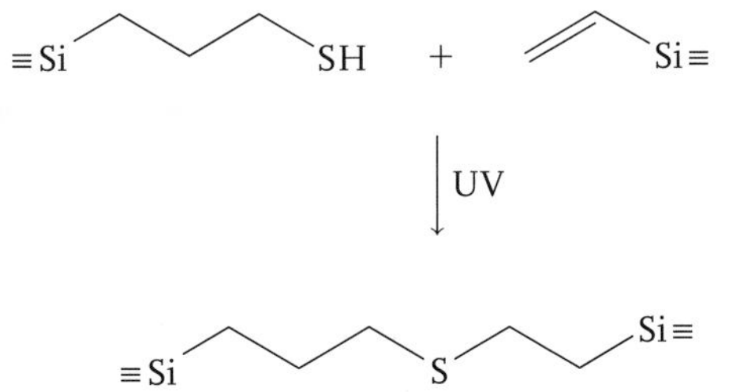

$$\equiv Si\text{–}CH_2CH_2CH_2\text{–}SH + CH_2{=}CH\text{–}Si\equiv \xrightarrow{UV} \equiv Si\text{–}CH_2CH_2CH_2\text{–}S\text{–}CH_2CH_2\text{–}Si\equiv \qquad (4.30)$$

Films and coatings represent the earliest commercial applications of sol–gel materials, since most of the disadvantages of the sol–gel technology such as large shrinkage, high costs of the starting materials, and so on can be overcome or are less important for thin films. Typical applications for inorganic–organic hybrid sol–gel materials are also coatings or films. For example, they are used for the protection of surfaces (e.g., against corrosion, abrasion, scratching, etc.), for optical applications (e.g., absorption, emission, or reflection of radiation), as chemically active layers (e.g., for sensing or catalysis), as membranes, as diffusion barriers, or to modify the surface properties.

The special properties of a surface originate from the fact, that atoms on a surface have a higher energy than those within the bulk. For a liquid on the surface of a substrate typically three different interfaces and thus three different interfacial energies have to be distinguished: γ_{SV}, the energy of the surface/vapor interface, γ_{SL}, the energy of the surface/liquid interface; and γ_{LV}, the energy of the liquid/vapor interface. If a layer of liquid spreads on a smooth and planar solid surface, the change in energy is given by Eq. (4.31):

$$\Delta E = \gamma_{SL} + \gamma_{LV} - \gamma_{SV} \tag{4.31}$$

If $\Delta E < 0$, the energy of the system is reduced so the liquid will spread spontaneously (Figure 4.64a); otherwise, the solid/liquid/vapor interface will be characterized by a contact angle, Θ (Figures 4.64b and c). The balance of tensions at the

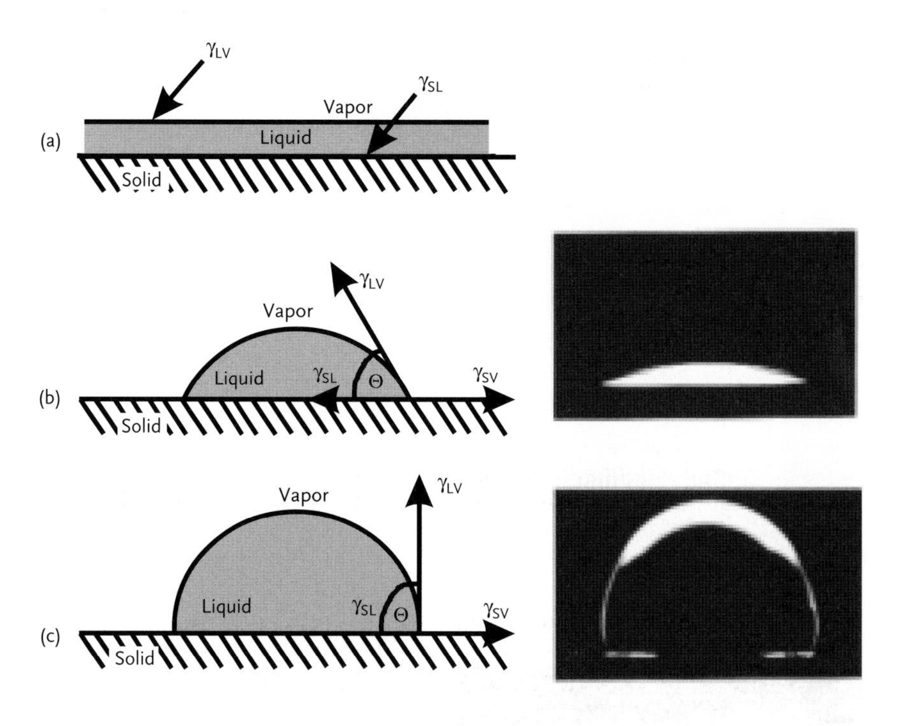

Figure 4.64 Different wetting behavior on a surface: (a) the liquid spreads easily over the surface; (b) the liquid wets the surface partly; and (c) the liquid does not wet the surface. Corresponding to the drawings, a water droplet on a hydrophilic (b) and a hydrophobic surface (c) is shown.

point of intersection leads to a relationship between the surface tensions that is known as Youngs equation [Eq. (4.32)]:

$$\gamma_{SV} = \gamma_{SL} + \gamma_{LV} \cos\Theta \quad (4.32)$$

If $\Theta \cong 0$ (as in Figure 4.64a), the liquid spreads easily and the solid will be covered with a liquid film. If $\Theta > 0$, the liquid does not really wet the surface and the contact angle is determined by the balance of forces at the intersection of solid/vapor and liquid/vapor interfaces.

Surfaces can be modified deliberately to control their polarity. This allows not only an achievement of hydrophobicity/hydrophilicity, but even oleophobicity/oleophilicity can be obtained for coatings for antisoiling and antifogging applications.

Imagine you are asked to modify a surface or substrate to render it scratch resistant, and at the same time insure that it is not wetted by organic solvents. According to the building block approach, it should be possible to add an oleophobizing component to the precursor solution used for making scratch resistant coatings (see above). This is indeed possible. The antisoiling coating shown in Figure 4.65 was prepared by adding fluoro-substituted alkoxysilanes (see Table 4.6) to the mixture of vinyl and mercapto-substituted trialkoxysilanes used for the scratch-resistant coating in Figure 4.63.

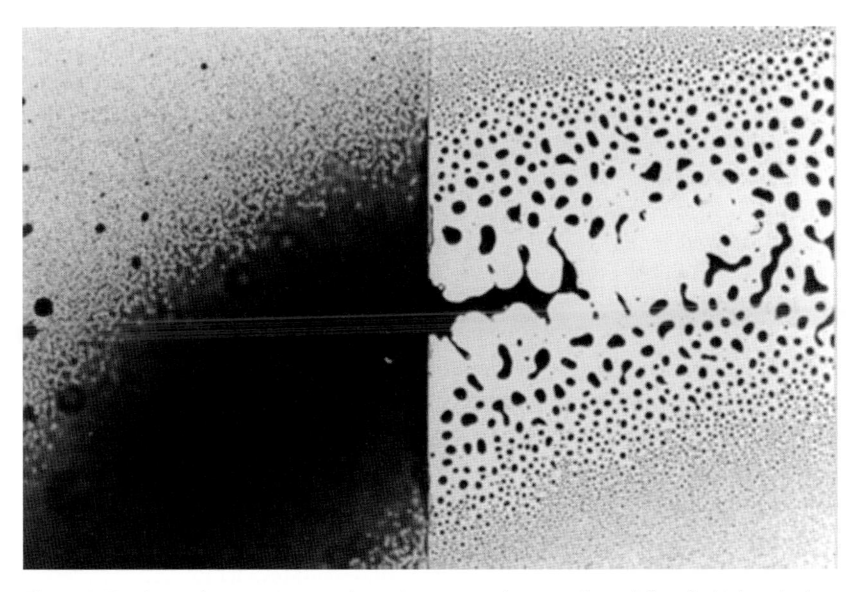

Figure 4.65 Antisoiling coating on glass; the picture shows a glass slide, of which only the right half is coated. The coated part is not wetted by the sprayed paint.

4.5.5 Non-hydrolytic Sol–Gel Processes

The conventional sol–gel process is based on the hydrolysis and condensation of molecular precursors, leading to oxide networks. The oxo ions originate from water that is added as a reagent or may be formed *in situ* by water-producing reactions. Variations of these reactions have been developed in which the oxo groups are formed by alkyl chloride or, less often, ester elimination instead of water or alcohol elimination [Eq. (4.17)] in the traditional sol–gel process. This process has been termed "non-hydrolytic sol–gel," because no water is involved. The base reactions are given in Eq. (4.33).

$$\begin{array}{c} \equiv \text{M-OR} + \equiv \text{M'-X} \\ \Updownarrow \\ \equiv \text{M-X} + \equiv \text{M'-OR} \end{array} \longrightarrow \begin{array}{c} \equiv \text{M-O-M'} \equiv + \text{R-X} \\ (\text{X} = \text{Cl, OOCR'}) \end{array} \tag{4.33}$$

Condensation occurs at temperatures between 20 and 100 °C; sometimes a catalyst is needed ($FeCl_3$ being most commonly used). A noteworthy feature of this method is that non-hydrated oxides without residual hydroxyl groups are obtained, due to the aprotic conditions. Redistribution between the OR and X ligands can take place before the actual reaction. For example, $Ti(O^iPr)_{4-x}Cl_x$ species are formed rapidly when $TiCl_4$ and $Ti(O^iPr)_4$ are mixed at room temperature. The chloro-alkoxo compounds are the actual precursors. The metal alkoxide species can often be generated *in situ* by reaction of metal chlorides with ethers or alcohol [Eq. (4.34)].

$$\begin{aligned} &\text{M-Cl} + \text{R-O-R} \longrightarrow \text{M-OR} + \text{RCl} \\ &\text{M-Cl} + \text{ROH} \longrightarrow \text{M-OR} + \text{HCl} \end{aligned} \tag{4.34}$$

The kinetics of non-hydrolytic sol–gel processes depends on the nature of the metal, the nature of the oxygen donor, electronic effects of the group R, and the composition of the initial metal alkoxide/metal chloride (carboxylate) mixture.

Various metal oxides as well as binary and even ternary oxides were prepared by non-hydrolytic routes. Various possibilities by which SiO_2–TiO_2 has been prepared are summarized in Eq. (4.35) as examples. In mixed-oxide systems, this route mostly leads to better control over the distribution of the elements because the problem of different precursor reactivities in hydrolytic sol–gel processes is avoided. However, the reaction [Eq. (4.34)] requires an equal number of alkoxo and chloro or carboxylate groups, and therefore the M/M′ ratio in mixed oxides is fixed to certain stoichiometries. The approach was also extended to the preparation of inorganic–organic hybrid materials by using $R'\,SiCl_3$ or $R'Si(OR)_3$ as precursors.

$$\begin{aligned} &TiCl_4 + Si(O^iPr)_4 \longrightarrow SiO_2/TiO_2 + 4\,^iPrCl \\ &Ti(O^iPr)_4 + SiCl_4 \longrightarrow SiO_2/TiO_2 + 4\,^iPrCl \\ &Ti(O^iPr)_4 + Si(OOCMe)_4 \longrightarrow SiO_2/TiO_2 + 4\,MeCOO^iPr \\ &TiCl_4 + SiCl_4 + 4\,^iPr_2O \longrightarrow SiO_2/TiO_2 + 4\,^iPrCl \end{aligned} \tag{4.35}$$

4.5.6
Aerogels

Drying of a gel body is a complex process in which different stages during the evaporation of the pore liquid can be distinguished (Section 4.5.2.4). Drying by means of evaporation of the pore liquid typically results in cracking and collapse of the gel network, which leads to powders or strongly shrunken monoliths (Figure 4.66). This section concentrates on special drying methods by which the wet gels can be transformed into aerogels in which the filigrane network structure is maintained even after drying. It is remarkable that a variety of inorganic, inorganic–organic or organic gels can be converted into aerogels despite a typical solid content of only 1–15 vol%.

The pore structure of aerogels is difficult to describe, but it is characterized by well-accessible, cylindrical, interconnected pores with pore sizes from the microporous to the macroporous regime. However, the majority of the pores fall in the mesoporous range with diameters between 2 and 50 nm and a broad pore-size distribution. The typical structure of a silica aerogel is shown in Figure 4.67. The inorganic network is built up from nanometer-sized silica particles, forming a

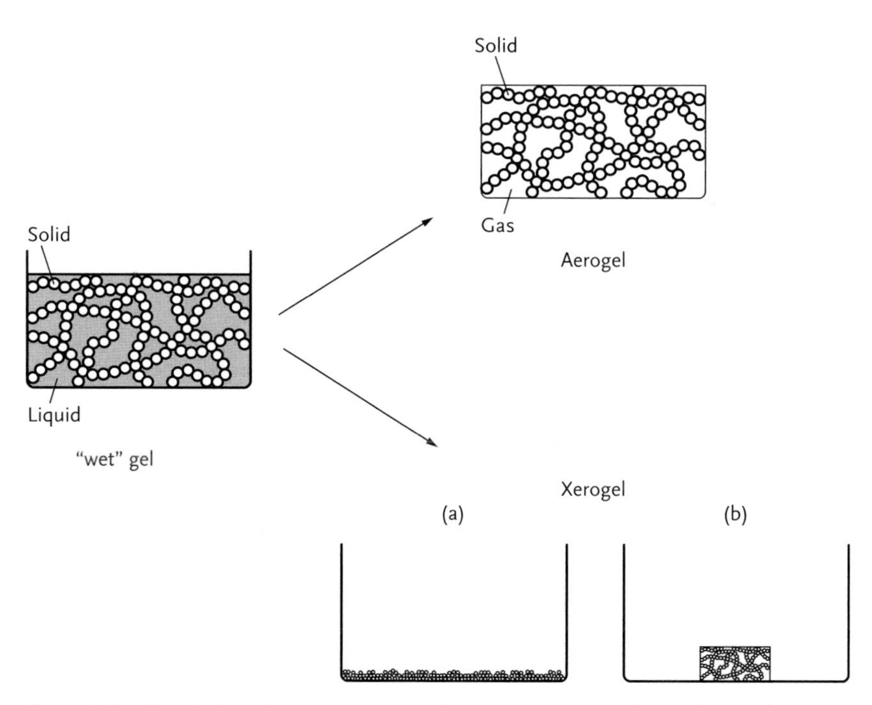

Figure 4.66 Top: Drying of a wet gel body to give an aerogel (the volume of the body remains approximately constant). Bottom: Conventional drying to give a xerogel powder (a) or monolith (b), associated with large shrinkage of the gel body.

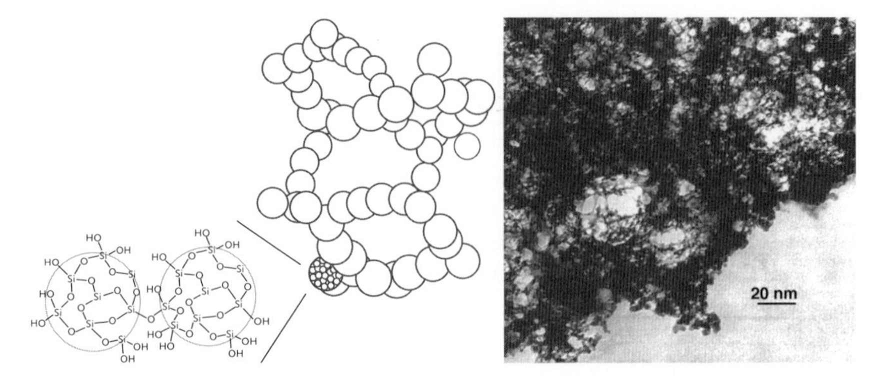

Figure 4.67 Silica aerogel; left: Schematic of the structure; right: a transmission electron microscopy (TEM) image.

three-dimensional network in which all architectural features (particle size, pore size) are in the nanometer regime. Other mesoporous materials are fumed silica, Vycor glass, carbon soot, or M41S materials (see Section 6.3.2). Among them, aerogels are unique owing to their extremely high porosity (low density) and high specific surface area, and the possibility of shaping the material into different morphologies, not only powders and granules, but also large monoliths. The bulk density of aerogels is in the range of 0.004–0.5 g cm^{-3} owing to the high porosity (for comparison: the density of air is 0.00129 g cm^{-3}). Aerogels definitely belong to the lightest inorganic solids available today.

4.5.6.1 **Drying Methods**

For the production of aerogels, different ways have been developed in order to preserve the pore structure of the wet gels, that is, to avoid irreversible shrinkage. The most important ones are the supercritical drying process and the so-called "ambient pressure-drying" process.

Supercritical drying In this procedure, the pore liquid is put into the supercritical state, in which no liquid/gas interfaces exist (see Section 4.4.1 and Figure 4.68). The critical point marks the end of the liquid–vapor coexistence curve in the phase diagram. It is noteworthy that the melting curve extends over the supercritical region. For example, the pressure required to solidify CO_2 at its critical temperature is 5.7 kbar, but that for water is an enormous 140 kbar.

Supercritical fluids exhibit simultaneously properties associated with both gases and liquids (see also Table 4.7). Thus, they are compressible like gases, but they also display solvencies similar to those of liquids. A unique feature of supercritical fluids may be demonstrated by starting with a subcritical liquid at point 1 in Figure 4.68. If the liquid is depressurized isothermally from point 1 to point 5, the presence of a liquid/vapor interface meniscus is observed as the vapor pressure line is

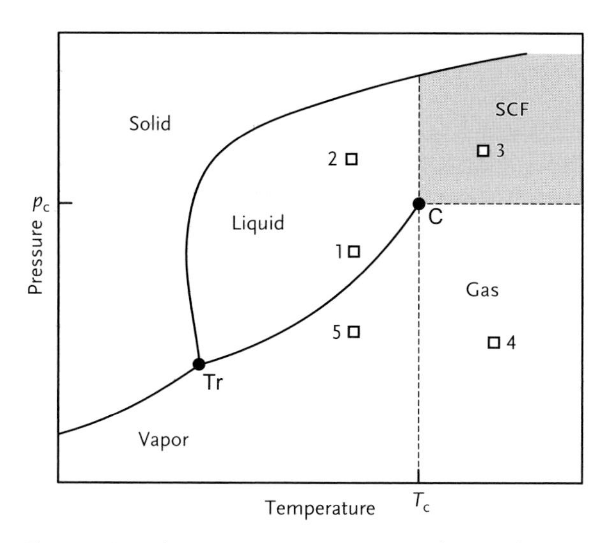

Figure 4.68 Schematic pressure–temperature diagram for a pure compound. The shaded area represents the supercritical fluid region (SCF), where C is the critical point. Tr represents the triple point, and 1 to 5 are random points in the phase diagram (see text).

Table 4.7 Comparison of some physical properties of gases, liquids, and supercritical fluids.

Property	Gas	Liquid	Supercritical fluid
Density [g cm^{-3}]	10^{-4}–10^{-3}	0.6–1.4	0.1–1
Diffusion coefficient [cm^2 s^{-1}]	10^{-1}	10^{-5}	10^{-3}–10^{-4}
Solvency	No	Yes	Yes
Compressibility	Yes	No	Yes

crossed. However, if the path 1–2–3–4–5 is taken, the fluid then passes from a liquid phase to a gas phase without crossing the liquid–vapor interface at any point. This pathway is used in supercritical drying to avoid collapse of delicate microstructures by the strong capillary forces that arise at liquid/vapor interfaces.

Table 4.4 in Section 4.4.1 shows some typical critical pressure and temperature data for different solvents. For the drying of aerogels, the most convenient way is to put the pore liquid, which is usually an alcohol or acetone, into the supercritical state. However, problems may arise from the combination of high pressures with high temperatures as well as the flammability of these solvents. Alternatively, liquid carbon dioxide has been used. Supercritical CO_2 is of particular interest due to its low critical temperature (31 °C). Additionally, it is non-flammable, nontoxic and – especially when used to replace organic solvents – environmentally friendly. However, a time-consuming solvent exchange (organic solvent versus liquid CO_2) is required prior to supercritical drying.

How is supercritical drying of a wet gel performed in practice? The wet gel is placed into an autoclave and covered with additional solvent. Partial drying of the samples (which would lead to the formation of cracks) is thus avoided. After the autoclave is closed, the temperature is slowly raised, resulting in a pressure increase. Both the temperature and the pressure are adjusted to values above the critical point of the corresponding solvent (T_c, p_c) and kept there for a certain period of time. This ensures that the autoclave as well as the sample, for example, a monolith, is completely filled with the supercritical fluid. The fluid is then slowly vented at constant temperature, which results in a pressure drop. When ambient pressure is reached, the vessel is cooled to room temperature and then opened. When organic solvents are used as medium, drying is often performed in a way that the vessel is pre-pressurized with nitrogen to avoid evaporation of the solvent. The phase boundary between the liquid and the gas must not be crossed during drying at any time.

Ambient-pressure drying To make aerogels interesting for large-scale commercial applications, one should avoid supercritical drying as the most expensive and risky part of the preparation. Therefore, the interest in alternative ways for exchanging the pore liquid in the gels by air and keeping the structure of the gel is very great.

The capillary forces exerted by the meniscus of the pore liquid and the pressure gradient upon the large shrinkage of the network (see Section 4.5.2) are the main reasons for the collapse of the filigrane structure. To obtain aerogels at ambient conditions, the network must be strengthened in order to avoid its collapse, that is, its irreversible shrinkage. Additionally or alternatively, the contact angle (see also Section 4.5.4) between the pore liquid and the pore walls must be influenced by deliberate modification of the inner surface and by variation of the solvent properties to minimize the capillary forces. For silica gels, the following route was developed that allows drying under ambient conditions.

The water/alcohol mixture in the pores of the gel is first exchanged for a water-free solvent, and the Si–OH groups at the surface are silylated (e.g., by reaction with chlorotrimethylsilane). The reactivity of the gel surface is thus reduced, and the surface hydrophobized. The actual drying process is performed after removal of excess chlorotrimethylsilane by washing with a water-free solvent. As expected, the gel shrinks strongly during evaporation of the solvent from the pores. However, no irreversible narrowing of the pores via the formation of Si–O–Si bonds is possible because no siloxane bonds can be formed due to silylation. The gel therefore expands to nearly its original size after reaching the maximum value for the deformation. This is called the spring-back effect. However, the network of the gel must be stable enough to tolerate a reversible shrinkage to 28% of its original volume (Figure 4.69).

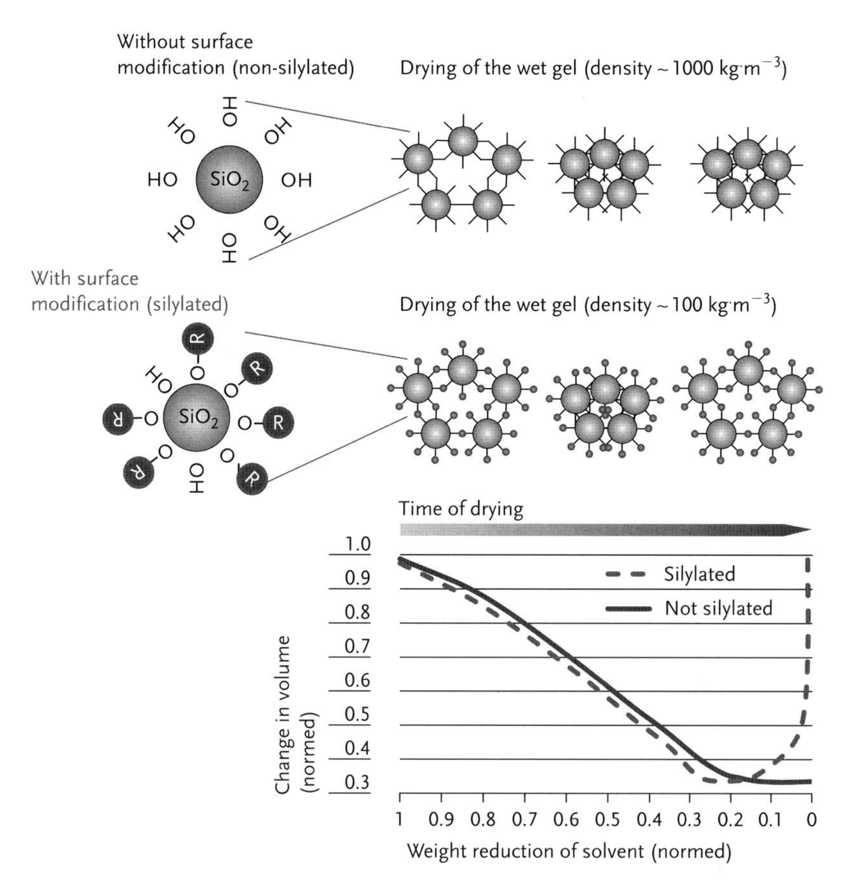

Figure 4.69 Ambient-pressure drying; top: irreversible shrinking of an unmodified silica gel; middle: reversible shrinking of a silica gel modified by organic groups; bottom: volume change of unmodified and modified gels upon ambient-pressure drying.

4.5.6.2 **Properties and Applications**

Aerogels possess a unique combination of properties due to their high porosity combined with low densities. Non-silicate aerogels are interesting materials for catalytic applications due to their high surface area. Silica aerogels are transparent, have excellent heat insulation properties, show a very good impact energy absorption capacity, and unusual acoustic properties. The sound velocities in SiO_2 aerogels are in the range of 100–300 m s^{-1}, which are among the lowest for inorganic solids and worth mentioning, since sound velocities of 5000 m s^{-1} were measured for quartz glass.

The thermal conductivity of silica aerogels is extraordinarily low. The passage of thermal energy through a material occurs through three mechanisms: solid conductivity, gaseous conductivity, and radiative (infrared) transmission. The sum of these three components gives the total thermal conductivity of the material. Solid conductivity is an intrinsic materials property, which is very high for dense silica. However, the volume fraction of silica in aerogels is very low, and additionally the solid is composed of very small particles linked to a three-dimensional network with many dead ends. Therefore, thermal transport through the solid phase in not particularly effective. The open pores allow the passage of gases (air) that can transport thermal energy through the material. The final mode of thermal transport through silica aerogels involves infrared radiation. At low temperatures, radiative transport is very low, but at higher temperatures radiative transport becomes the dominant process of thermal conduction. Evacuation of the aerogel system and addition of an infrared opacifier such as carbon soot or TiO_2, greatly improves the high-temperature insulation properties. With these improvements there is no question that silica aerogels are among the best heat-insulation materials currently available. Additionally, they are non-flammable and, in the absence of an opacifier, transparent. Aerogels can be utilized for the passive use of solar energy, for example for paneling house walls or for coating solar cells.

Silica aerogels are used by NASA for the collection of cosmic dust. This is quite a difficult task because these hypervelocity particles travel at nine times the speed of a bullet fired from a rifle. Although each of the captured particles is smaller than a grain of sand, high-speed capture could alter their shape and chemical composition – or vaporize them entirely.

To collect the particles without damaging them, a project called STARDUST was established, which used silica aerogels to perform the task. When a particle hits the aerogel, it will bury itself in the material, creating a carrot-shaped track up to 200 times its own length, as it slows down and comes to a stop. The situation is similar to that of an airplane setting down on a runway and braking to reduce its speed gradually. Since silica aerogels are transparent, scientists will use these tracks to find the tiny particles.

The dust-collector device consists of blocks of 1 and 3 cm thick silica aerogel tiles mounted in modular aluminum cells (Figure 4.70). For the STARDUST mission,

Figure 4.70 Aerogel collector for cosmic dust.

cells were mounted on both sides of a two-sided, grid-shaped array that deploys from the sample-return capsule. After exposure, the cells assembly folded up to a compact configuration for stowage into the earth return capsule.

Further Reading

1 Baccile, N., Babonneau, F., Thomas, B., and Coradin, T. (2009) Introducing ecodesign in silica sol–gel materials. *J. Mater. Chem.*, **19**, 8537–8559.
2 Brinker, C.J. and Scherer, G.W. (1990) *Sol–Gel Science: The Physics and Chemistry of Sol–Gel Processing*, Academic Press, Boston.
3 Evered, D. and Harnett, S. (eds) (1988) *Cell and Molecular Biology of Vertebrate Hard Tissue (Ciba Foundation Symposium 136)*, John Wiley & Sons Ltd, Chichester.
4 Fendler, J.H. and Meldrum, F.C. (1995) The colloid chemical approach to nanostructured materials. *Adv. Mater.*, **7**, 607–632.
5 Hüsing, N. and Schubert, U. (1998) Aerogels–airy materials: chemistry, structure, and properties. *Angew. Chem. Int. Ed. Engl.*, **37**, 22–45.
6 Iler, R.K. (1979) *The Chemistry of Silica*, John Wiley & Sons Ltd, New York.
7 Kickelbick, G. (ed.) (2007) *Hybrid Materials – Synthesis, Characterization and Applications*, Wiley-VCH, Weinheim.
8 Krampitz, G. and Graser, G. (1988) Molecular mechanisms of biomineralization in the formation of calcareous shells. *Angew. Chem. Int. Ed. Engl.*, **27**, 1145–1160.
9 Kumar, S. and Nann, T. (2006) Shape control of II–VI semiconductor nanomaterials. *Small*, **2**, 316–329.
10 Laudise, R.A. and Kolb, E.D. (1969) Hydrothermal synthesis of single crystals. *Endeavour*, **28**, 114–117.
11 Livage, J., Henry, M., and Sanchez, C. (1988) Sol–gel chemistry of transition metal oxides. *Prog. Solid State Chem.*, **18**, 259–341.
12 Livage, J., Sanchez, C., and Babonneau, F. (1998) Molecular precursor routes to inorganic solids, in *Chemistry of Advanced Materials – an Overview* (eds L. V. Interrante and M. Hampden-Smith), Wiley-VCH, New York, pp. 389–448.
13 Mann, S. (2000) The chemistry of form. *Angew. Chem. Int. Ed. Engl.*, **39**, 3392–3406.
14 Mann, S. (2001) *Biomineralization*, Oxford University Press, Oxford.
15 Mann, S., Webb, J., and Williams, R.J.P. (eds) (1989) *Biomineralization, Chemical and Biochemical Perspectives*, Wiley-VCH, Weinheim.
16 Matijevic, E. (1981) Monodispersed metal (hydrous) oxides – a fascinating field of colloid science. *Acc. Chem. Res.*, **14**, 22–29.
17 Mullin, J.W. (2001) *Crystallization* (4th ed.), Butterworth Heinemann, Boston.
18 Pan, D., Wang, Q., and An, L. (2009) Controlled synthesis of monodisperse nanocrystals by a two-phase approach without the separation of nucleation and growth processes. *J. Mater. Chem.*, **19**, 1063–1073.
19 Perry, C.C. (1998) Biomaterials, in *Chemistry of Advanced Materials: An Overview* (eds L.V. Interrante and M.J. Hampden-Smith), Wiley-VCH, New York, pp. 499–562.
20 Pierre, A.C. (1998) *Introduction to Sol–Gel Processing*, Kluwer, Boston.
21 Rabenau, A. (1985) The role of hydrothermal synthesis in preparative chemistry. *Angew. Chem. Int. Ed. Engl.*, **97**, 1026–1040.
22 Sanchez, C. and Livage, J. (1990) Sol–gel chemistry from metal alkoxide precursors. *New J. Chem.*, **14**, 513–521.
23 Sanchez, C., Julián, B., Belleville, P., and Popall, M. (2005) Applications of hybrid organic–inorganic nanocomposites. *J. Mater. Chem.*, **15**, 3559–3592.
24 Schubert, U. (2003) Silica-based and transition metal-based inorganic–organic

hybrid materials – a comparison. *J. Sol-Gel Sci. Technol.*, **26**, 47–55.

25 Schubert, U. (2007) Organically modified transition metal alkoxides – Chemical problems and structural issues on the way to materials syntheses. *Acc. Chem. Res.*, **40**, 730–737.

26 Schubert, U., Hüsing, N., and Lorenz, A. (1995) Hybrid inorganic–organic materials by sol–gel processing of organofunctional metal alkoxides. *Chem. Mater.*, **7**, 2010–2027.

27 Shelby, J.E. (2005) *Introduction to Glass Science and Technology* (2nd edition), The Royal Society of Chemistry, Cambridge.

28 Stávek, J., Šípek, M., Hirasawa, I., and Toyokura, K., (1992) Controlled double-jet precipitation of sparingly soluble salts. A method for the preparation of high added value materials. *Chem. Mater.*, **4**, 545–555.

29 Sugimoto, T. (1987) Preparation of monodispersed colloidal particles. *Adv. Colloid Interf. Sci.*, **28**, 65–108.

30 Suryanarayana, C. (ed.) (1999), *Non-equilibrium Processing of Materials*, Pergamon, Amsterdam.

31 Tao, A.R., Niesz, K., and Morse, D.E. (2010) Bio-inspired nanofabrication of barium titanate. *J. Mater. Chem.*, **20**, 7916–7923.

32 Telford, M. (2004) The case for bulk metallic glass. *Materials Today*, **7** (3), 36–43.

33 Vallet-Regi, M. and Gonzalez-Calbet, J. M. (2004) Calcium phosphates as substitution of bone tissues. *Prog. Solid State Chem.*, **32**, 1–31.

34 Vioux, A. (1997) Non-hydrolytic sol–gel routes to oxides. *Chem. Mater.*, **9**, 2292–2299.

35 Vogel, W. (1992) *Glaschemie*, Springer, Berlin.

36 Xu, A.W., Ma, Y., and Cölfen, H. (2007) Biomimetic mineralization. *J. Mater. Chem.*, **17**, 415–449.

37 Zarzycki, J. (Ed.) (1991) Materials science and technology, *Glasses and Amorphous Materials*, Vol. **9**, Wiley-VCH, Weinheim.

5
Preparation and Modification of Inorganic Polymers

5.1
General Aspects

Silicones (polysiloxanes, Section 5.2) are the commercially most successful types of "inorganic polymers," that is, polymers that have a backbone containing elements other than carbon, nitrogen, and oxygen, to which organic or organometallic substituents are attached. Other inorganic polymers (Figure 5.1) with a high commercial potential as specialty polymers are polyphosphazenes (Section 5.3) and polysilanes (Section 5.4). The technical interest in polycarbosilanes (Section 5.5) and polysilazanes (Section 5.6) is their use as *preceramic polymers,* that is, polymers that give ceramic materials upon pyrolysis. This option is of great interest, because the use of polymer technology offers certain processing advantages over conventional ceramic processing routes (see Section 2.1.5). Section 5.7 will provide an overview on polymers with other main-group elements in the inorganic backbone. New materials are needed in all fields of modern technology. Because of the almost limitless combination of elements, there should be other inorganic polymers with useful properties. A class of polymeric materials with very interesting potential applications are coordination polymers, which will be treated in Section 5.8.

Polymers are chemically characterized by the sequence of atoms forming the main chain ("backbone"), such as... Si–O–Si–O... in silicones, and by the nature of the groups ("side groups") attached to (all or some) atoms of the backbone (the groups R and R′ in Figure 5.1). The main chain may be branched or crosslinked. In this sense, sol–gel hybrid materials (see Section 4.5.4) with three-dimensional inorganic networks to which organic groups are attached, may also be considered inorganic polymers.

Before turning to the chemistry of the particular classes of inorganic polymers (Sections 5.2–5.8), some common aspects will be discussed.

There is a variety of methods for the preparation of *organic* polymers, such as polymerization of multiple bonds, polyaddition, or → polycondensation reactions. Since there are no thermodynamically stable double bonds for the elements of the third and higher periods, preparation methods involving double bonds are not applicable. There are only two general methods to synthesize inorganic polymers

Synthesis of Inorganic Materials, Third Edition. Ulrich Schubert and Nicola Hüsing.

Published 2012 by WILEY-VCH Verlag GmbH & Co. KGaA

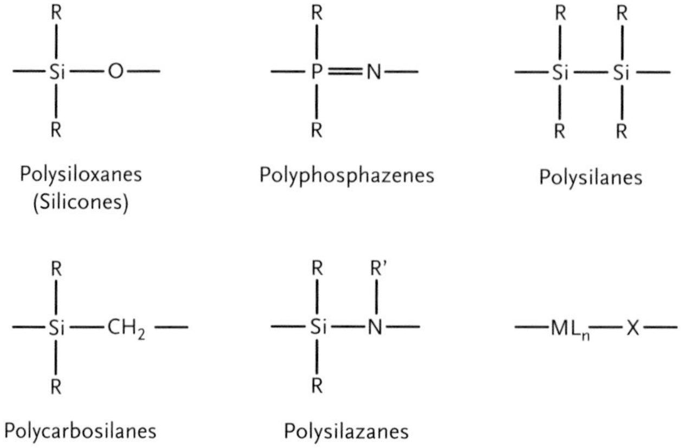

Polysiloxanes (Silicones)

Polyphosphazenes

Polysilanes

Polycarbosilanes

Polysilazanes

Figure 5.1 The repeat units in silicones, polyphosphazenes, polysilanes, polycarbosilanes, polysilazanes, and coordination polymers (ML_n = metal complex fragment; X = any coordinating group).

in which the elements in the backbone are connected to each other by covalent bonds. Examples for both types of reactions will be given in the following sections. Substitution, or chemical modifications of the substituents are options to modify the properties of inorganic polymers.

- *Polycondensation reactions* [Eq. (5.1)]

$$Y\text{–}A\text{–}B\text{–}X \longrightarrow [A\text{–}B\text{–}]_n + X\text{–}Y \tag{5.1}$$

where A and B are the elements forming the inorganic backbone, and X and Y are the eliminated groups. The spectrum of X–Y ranges from H_2O or HCl to Me_3SiOR, H_2 and NaCl. A major drawback of polycondensation reactions is that small cyclic oligomers (mainly six- and eight-membered rings) are thermodynamically more favorable than polymers. One major reason for this is the steric hindrance generated by the side groups. Irrespective of the chain conformation of the polymer, the substituents are always closer to their neighbors along a chain than they are in a small cyclic oligomer. The polycondensation reactions must therefore be performed under kinetically controlled conditions to obtain polymers instead of cyclic oligomers.

- *Ring-opening polymerization* (ROP) [Eq. (5.2)]

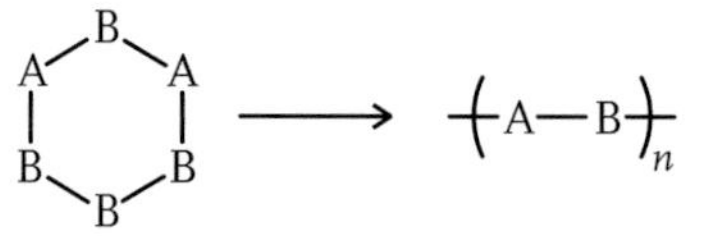

$$(AB)_3 \text{ ring} \longrightarrow \left[A\text{—}B\right]_n \tag{5.2}$$

Although rings are thermodynamically more favorable, there are methods to open the rings and attach the segments to each other. The obtained polymer may be stable at temperatures where the rate of depolymerization to the cyclic oligomers is very slow (although it would depolymerize when heated to higher temperatures). The cyclic oligomers will not ring-open and polymerize when the repulsion of the substituents is too large.

Coordination polymers (Section 5.8) are obtained when → bidentate molecules or ions of the type X–Y–X are used to bridge two or more metal centers.

$$\cdots X{-}ML_n{-}X{-}Y{-}X{-}ML_n{-}X{-}Y{-}X{-}\cdots$$

(ML_n is a metal complex fragment containing n → coligands L not integrated in the polymer backbone; X is any coordinating group which provides a stable link to the ML_n fragment; and Y is an organic or inorganic spacer). Both two- and three-dimensional coordination polymers have been prepared; this will be elaborated in Section 5.8. Coordination polymers are different to the polymers discussed in Sections 5.2–5.7 because the M–X link is usually a dative bond, and the polymers are formed by interaction of Lewis acidic metal complex fragments and the Lewis-basic ligands X.

5.1.1 Polymeric Materials

The main interest in inorganic polymers is that they may combine the advantages and minimize the disadvantages of biological or carbon-based polymers and inorganic materials. Organic polymers are light, tough, and easy to fabricate, but in general they cannot be exposed to high temperatures, strongly oxidizing conditions, or high-energy radiation.

In inorganic polymers, combinations of properties can be achieved and tuned by the combination of the properties of the inorganic backbone and the organic substituents. The backbone of inorganic polymers can provide heat, fire, or radiation resistance, or electrical conductivity, as well as flexibility. The substituents control properties such as solubility, → liquid crystallinity, optical properties, hydrophobicity or hydrophilicity, adhesion or biological compatibility. They also provide possibilities for generating crosslinks between the polymer chains (see below).

The question whether a polymer is an oil, an elastomer or a grease, and so on, depends not only on the nature of its repeating unit(s) (backbone plus side groups), but also on the three-dimensional structure of the macromolecule, the degree of polymerization (its molecular mass), the degree of branching, and its crystallinity. Many polymers are amorphous, that is, their three-dimensional structure exhibits no long-range order as in a crystal. However, a polymer may partially crystallize by packing of chains into a stereoregular (ordered) arrangement.

The flexibility of the main chain determines the degree of either random coiling or "rigid rod" character in solution. Highly flexible, long-chain molecules become entangled in solution and markedly increase the viscosity. In the solid state, non-crosslinked macromolecules interact strongly with their neighbors, either through entanglements or through the formation of ordered (micro)crystalline domains owing to interchain packing. The presence of branched structures and any molecular feature that eliminates symmetry or linear order disfavors microcrystallite formation that have a profound influence on materials properties.

An important characteristic of any polymeric material is the temperature dependence of its physical properties. An amorphous polymer will be a glass at temperatures below its glass-transition temperature (T_g) (see Section 4.1), and an elastomer above it.

Factors that affect the T_g and other physical properties of polymers are:

- Elements and bond types in the main chain determine the torsional barrier and hence the intrinsic mobility of the chain. For example, some of the lowest T_g values known (-100 to $-130\,^{\circ}$C) are obtained with polymers that possess Si–O or P–N backbone bonds. These bonds have very low barriers to torsion. Furthermore, only every second atom along the polymer chain in these polymers bears substituents.
- Shape and rigidity of the side groups affect chain flexibility by steric interactions between them. Bulky substituents which sterically interfere when the backbone bonds are twisted, will have high T_gs.
- The size and shape of the substituents may create free volume between themselves and their counterparts on neighboring chains. The greater the free volume, the easier it will be for macromolecular motion to occur. Hence, systems with large free volumes are expected to have low T_gs.
- Side-group regularity and symmetry along the chain favor efficient chain packing that leads to reduced free volume and the formation of microcrystallites.

5.1.2
Crosslinking

The initially obtained macromolecules are very often linear or weakly branched polymeric chains. For many uses, this structure needs to be changed after some initial processing steps by crosslinking reactions, by which bonds are formed between different chains. The process of crosslinking is also called vulcanization or curing. Crosslinking has several important consequences for the materials properties, and is thus one of the most important aspects of polymer science.

- The macromolecules in an amorphous uncrosslinked polymer at a temperature above T_g have considerable freedom of motion and are thus in a quasiliquid state. An average of only 1.5 crosslinks per polymer chain severely restricts these motions. Crosslinks prevent chains from sliding past each other when the material is stretched. This raises the tensile strength and generates rubber–elastic properties from materials that would otherwise be gums. Heavy crosslinking provides *rigidity* and *strength* to the system. When a crosslink exists for every two or three chain atoms, the material becomes inflexible and infusible.
- An average of only 1.5 crosslinks per chain is sufficient to prevent *dissolution* of soluble polymers, although the polymer may swell. Increasing the number of crosslinks per chain will progressively reduce the degree of swelling. Swelled polymers may contain up to 95% of their mass as solvent, yet have definite size and shape.

- Polymers in their non-crosslinked state are easily fabricated but are prone to decompose at moderately high temperatures by evolution of volatile products of low molecular mass. Crosslinking *improves thermal stability*.
- The proper degree of crosslinking is essential for the → rheological properties of polymers and polymer melts.

Some general concepts for the crosslinking of inorganic polymer chains are exemplarily shown in Figure 5.2; we will repeatedly come back to this issue in the following sections.

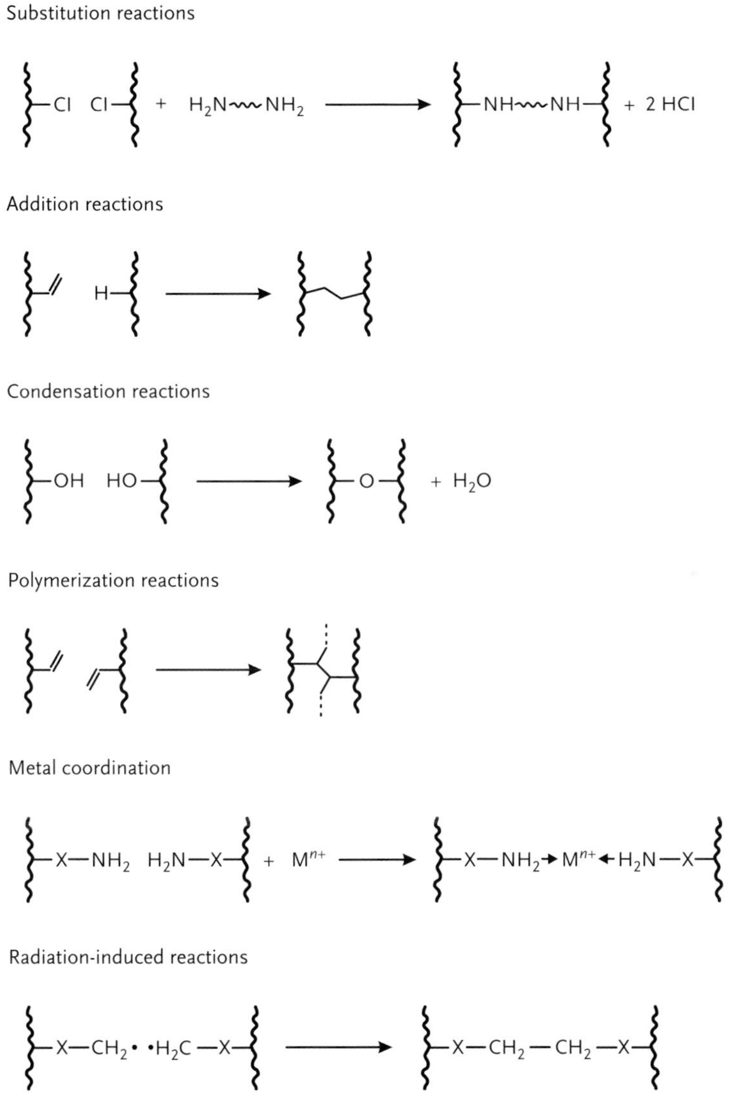

Substitution reactions

$$—Cl \quad Cl— + H_2N{\sim\sim}NH_2 \longrightarrow —NH{\sim\sim}NH— + 2\,HCl$$

Addition reactions

Condensation reactions

$$—OH \quad HO— \longrightarrow —O— + H_2O$$

Polymerization reactions

Metal coordination

$$—X—NH_2 \quad H_2N—X— + M^{n+} \longrightarrow —X—NH_2 \rightarrow M^{n+} \leftarrow H_2N—X—$$

Radiation-induced reactions

$$—X—CH_2\bullet \quad \bullet H_2C—X— \longrightarrow —X—CH_2—CH_2—X—$$

Figure 5.2 General methods for the linking of inorganic polymer chains.

Substitution reactions These are only rarely used, because polymers with displaceable substituents are not readily prepared and handled. This method is sometimes used in polyphosphazene chemistry. A variation of this method is the → metathetical exchange of groups. An example is the crosslinking of OCH_2CF_3-substituted polyphosphazene chains by treatment with $NaOCH_2(CF_2)_xCH_2ONa$ (by which P-$OCH_2(CF_2)_xCH_2O$-P links are formed and $NaOCH_2CF_3$ is released).

Addition reactions These are more widely used for all classes of inorganic polymers. The most general option is the addition of reactive element–hydrogen bonds (SH, SiH, etc.) to unsaturated groups. An industrially used example is the crosslinking of polysiloxanes by reacting Si–H and Si–$CH{=}CH_2$ entities in different chains (see Section 5.2.3.4). This results in the formation of Si–CH_2CH_2–Si links. A related example is given in Eq. (4.30).

Condensation reactions These are very important for silicon-based materials by which the very stable Si–O–Si bonds are formed by elimination of water starting from Si–OH groups. Another option is the formation of Si–NR–Si or P–NR–P bonds by elimination of amines starting from Si–NHR or P–NHR substituted polymer chains.

Polymerization → or polyaddition reactions These have proven especially effective. Almost any unsaturated organic group in the side chain can be used for polymerization, but vinyl crosslinking is most widely employed.

Crosslinking via metal coordination This is restricted to rather special applications and is nearly exclusively used in polyphosphazene chemistry.

Radiation-induced crosslinking This includes irradiation with UV, X-ray, γ or electron radiation, and has many advantages, mainly because most inorganic backbones are stable to irradiation and transparent to visible and near-UV light (polysilanes are different, see Section 5.4). Furthermore, no functional side groups are required. The underlying chemistry is radiation-induced homolytic cleavage of C–H or C–C bonds. The thus-generated carbon radicals combine to give covalent bonds.

5.1.3 Preceramic Polymers

The fabrication of ceramics from preceramic polymers involves the steps shown in Figure 5.3.

The main advantages of this route, compared to conventional syntheses of ceramics, are the lower reaction temperatures and the fact that polymer processing technologies can be used for shaping (molding, fiber drawing, coating, infiltration, etc.). The shaped preceramic polymer is pyrolyzed to the final ceramics. The → green body shrinks during pyrolysis, but must retain its shape. The shrinkage is mainly due to the change in density from polymer ($\sim 1\,g\,cm^{-3}$) to ceramic

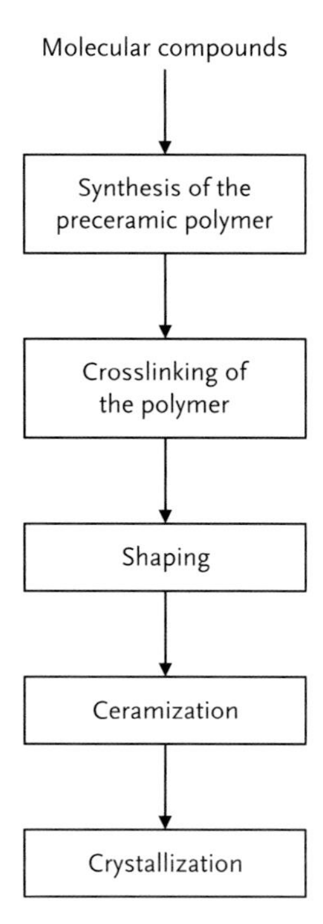

Figure 5.3 Processing steps in the preparation of ceramics from preceramic polymers.

($\sim$2–3 g cm^{-3}) and to the mass loss that occurs during polymer pyrolysis. Pyrolysis usually results in amorphous ceramics, which optionally can be converted into thermodynamically stable ceramic phases by heat treatment at high temperatures.

Typical applications of preceramic polymers include the following:

- For the fabrication of *ceramic fibers* a polymer is needed, which can be processed by melt-spinning or dry-spinning techniques (see Section 5.5).
- Preceramic polymers can be used for the preparation of *ceramic films*. After coating, the polymer layer is decomposed to a ceramic material. Typical applications for ceramic films are the improvement of the oxidation resistance or the → tribological properties of the substrate.
- The pyrolysis of large-volume polymeric → green bodies is very difficult, because of the shrinkage problem and the release of gases during pyrolysis. However, preceramic polymers are used to form the *ceramic matrices in composite materials*. The polymer is infiltrated in porous structures such as fiber filaments, and the infiltrated part is subsequently pyrolyzed.
- As *low-loss binders* the preceramic polymers could replace the widely used organic binders in ceramics processing (see Section 2.1.5.3). The result would be a stronger part that has far fewer defects, because the binder is for the most part converted to ceramic.
- The gases formed during pyrolysis allow production of *porous materials*, which can be used as ceramic membranes, for example.

Not every inorganic polymer is a useful preceramic polymer. The requirements depend somewhat on the intended application and are partially contradicting. The chemical challenge is to design a preceramic polymer with a good compromise of properties.

- The starting monomers should be commercially available and reasonably cheap.
- The polymer synthesis should be simple and proceed in high yields.
- The preceramic polymer must be processable by conventional polymer-processing techniques. It therefore should be liquid, or a fusible or soluble solid. It must have the correct → rheological properties for the intended shaping process.
- Pyrolysis of the preceramic polymer should give a high yield of ceramic residue: → ceramic yields of 60–75% are acceptable, but higher yields are desirable. For this reason, hydrogen and methyl groups are preferred substituents owing to their low mass.
- The polymer should have a high molecular mass to avoid evaporation during pyrolysis. The structure should be branched, because many linear inorganic polymers decompose thermally by evolution of volatile products of low molecular mass.
- To retain its shape during pyrolysis, the polymer must be infusible or rendered infusible by extensive → crosslinking during the initial stages of the pyrolysis. This requires appropriate reactive substituents.
- There should be minimal gas evolution during pyrolysis. Excessive formation of gaseous compounds lowers the ceramic yield, increases shrinkage of the ceramic body, creates porosity (which may not be wanted), and may even cause cracking or rapture of the ceramic part.
- The volatile products of pyrolysis should be non-hazardous and non-toxic.

The correct ratio of the elements in the preceramic polymer does not guarantee that a stoichiometric ceramic phase will be obtained on pyrolysis, because one of the elements may be partially lost as volatile products. One approach to achieve the desired elemental composition is to use chemical or physical combinations of two different polymers in the appropriate ratio, for example a polycarbosilane giving silicon-rich SiC in combination with another giving carbon-rich SiC. The stoichiometry can also be controlled by an appropriate atmosphere during pyrolysis. The gas used may be either inert or reactive. Important reactive gases are, for example, ammonia or nitrogen for the formation of nitrides, hydrogen for SiC, or water for oxides.

An approach to overcome the shrinkage problem in the fabrication of large-volume ceramic bodies is the active filler – controlled polymer pyrolysis process. The polymer is loaded with an "active filler" that undergoes a volume expansion upon reaction with the decomposition products of the polymer or a reactive gas atmosphere. This allows the near-net-shape manufacturing of components with complex geometries. Suitable active → fillers are elements or compounds forming carbides, nitrides or oxides, such as Al, B, Si, Ti, $CrSi_2$, or $MoSi_2$, which exhibit a high specific volume increase upon reaction.

5.2 Polysiloxanes (Silicones)

Silicones have been used technologically for almost 50 years, and may be produced as oils, elastomers (rubbers), or resins. About 60% of the produced silicones worldwide are oils (including → emulsions and greases), 30% elastomers and 10% are resins.

5.2.1 Properties and Applications of Silicones

The many applications are based on a number of interesting properties:

- Silicones exhibit excellent thermal stability up to 200–300 °C, and have low T_gs, for instance −123 °C of pure poly(dimethylsiloxane). Their physical properties change only slightly on changing the temperature.
- Silicones have a very good oxidation stability. They are only oxidized by air at $T > 200$–300 °C, the final inorganic product being silica. They also show prolonged resistance to ultraviolet irradiation and weathering.
- Most silicones act as very good electrical insulators. The volume → resistivity (10^{14}–$10^{16}\,\Omega$ cm), → dielectric strength (>15 kV mm^{-1}) and → dielectric loss tan δ ($<10^{-5}$ at 100 Hz) are nearly unchanged between 20 and 200 °C.
- The excellent water repellence is mainly due to methyl substituents. The contact angle (see Figure 4.64) for water droplets can reach 110° (comparable to paraffins). Closely connected with the hydrophobic properties are antistick properties against oils, fats, or organic polymers.
- Silicone oils have a very high spreadability on nearly all organic and inorganic materials, except fluoropolymers. Their surface tension (21 mN m^{-1} for poly(dimethylsiloxane)) is in the same range as that of tensides (≈30 mN m^{-1}).
- Silicones are chemically and physiologically inert.
- Liquids and gases permeate more rapid through silicone rubbers than through other polymers. For example, silicones are 10 times more permeable for oxygen than natural rubber or low-density polyethylene. A good oxygen → permeability is important for medical applications (e.g., contact lenses).
- Silicones retain their flexibility even at low temperatures, and they do not become brittle.

5.2.1.1 Silicone Oils

Silicone oils are $SiMe_3$-terminated poly(dimethylsiloxanes), Me_3SiO-$[SiMe_2O]_n$-$SiMe_3$. Their properties depend only on the chain length (n). There is a linear dependence between the logarithm of the viscosity and the logarithm of the average chain length. Viscosity at room temperature for $n = 50$ is about 70 mPa s, and about 10^5 mPa s for $n=1000$. This changes only slightly with temperature, and even silicone oils with a molecular mass of 10^5 are still fluids. Organic polymers with a much smaller molecular mass are already solid. The reason for the low viscosity is the weak intermolecular interactions between poly(dimethylsiloxane) chains.

Silicone oils are used as capacitor and transformer fluids, hydraulic oils, compressible fluids for liquid springs, and lubricants. Other uses are as heat-transfer media in heating baths and as components in car polish, sun-tan lotion, lipstick, shampoo, and other cosmetic formulations. Their low surface tension leads to their use as release agents, antifoaming agents, and hydrophobizing agents, such as water-repelling impregnations for textiles or buildings. Their complete non-toxicity allows them to be used in cooking oils and the processing of fruit juices, for example.

5.2.1.2 **Silicone Elastomers**

Silicone *elastomers* (silicone rubbers) are crosslinked poly(dimethylsiloxanes) of high molecular weight (10^4–10^7). Most silicones are reinforced by a → filler (mostly silica, but also silicates, $CaCO_3$, $BaSO_4$, etc.) to increase their modulus, tensile strength, tear strength and abrasion resistance. The reinforcing effect depends on the size, structure and surface properties (type and amount of surface OH groups, adsorbed water, surface modification) of the particles and the filler loading.

Silicone elastomers obtain their final properties only after curing → ("vulcanizing"). There are four main classes: *room-temperature vulcanizing* elastomers (RTV), which can consist of one (RTV-1) or two components (RTV-2), *liquid rubber* (LR), and *high-temperature vulcanizing* formulations (HTV; also called *heat curing* formulations, HC). Silicone rubbers in general are unmatched by any other synthetic or natural rubbers in retaining their inertness, flexibility, elasticity, and strength up to 250 °C (heat-stabilized silicone rubbers) and down to −120 °C. General aspects of → crosslinking have already been discussed in Section 5.1; the specific reactions employed in the production of silicones will be treated in Section 5.2.3.4 after the chemistry of silicones has been developed.

RTV-1 silicones These have good adhesive properties, because they wet most substrates. They are therefore mainly used as adhesives or coatings in cars, buildings, industrial plants, and household items, and in mechanical and electrical engineering. Adhesion can be improved, if necessary, by adding an adhesion promoter or by pretreating the substrate surface with a "primer." Both are mostly silanes with reactive organic groups. Each substrate/silicone combination requires a special primer. → Vulcanization of RTV-1 silicones occurs on contact with humidity.

RTV-2 silicones These find use for sealing and encapsulating electrical and electronic components, for elastic damping (vibration absorption), plugs, rubber stamps, rams for screen printing, and so on. They are also often used for making accurate molds and to provide rapid, accurate, and flexible impressions for dentures and inlays. Silicone foams are, for example, used for fire protection or for the isolation of cable passages. In contrast to RTV-1 formulations, RTV-2 silicones adhere to almost no material. However, reactive organic groups can be incorporated into the silicones to provide them with adhesion properties.

The two components of RTV-2 silicones have to be intimately mixed in a certain ratio before use. The mixture then must be used within a certain period of time, typically between half an hour and several hours. The viscosity of the mixture steadily increases after mixing until the formulation is no longer processable.

A representative application is the making of accurate molds, not only for the replication of art objects (Figure 5.4) and decorations, but also for the production of prototypes and models in various industries. For all these applications it is important that a negative of the object is made that reproduces every detail, particularly the texture of its surface. RTV-2 silicones are easy to apply, have a high elasticity, do not stick, and provide a precise copy of the object.

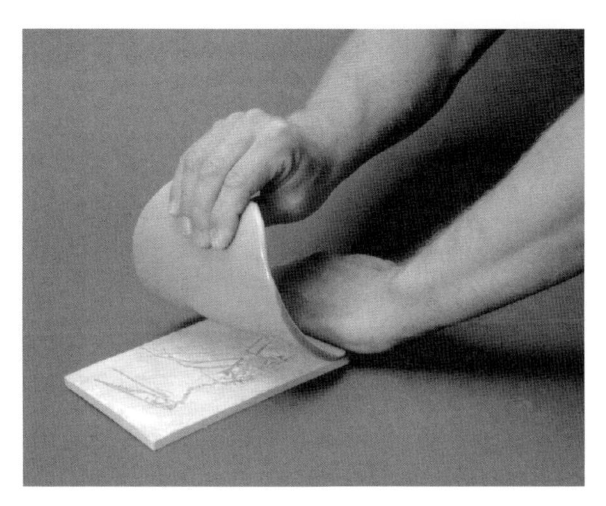

Figure 5.4 RTV-2 silicones allow accurate copies of objects to be made because they do not adhere to inorganic and organic materials.

LR formulations These were developed particularly for the quick, automated and cheap production of small elastic objects in a high number of pieces by injection molding. LR formulations also consist of two components that are mixed before injection into the mold. → Vulcanization occurs in the heated mold (180–200 °C) within seconds. Typical objects prepared from LR silicones are O-rings, plugs, contact mats, nipples, membranes, diaphragms, tube couplings, corrugated bellows, or gaskets.

A typical example for the application of such silicones is shown in Figure 5.5. Keyboard switches made of silicone rubber have become an indispensable part of phones, pocket calculators, remote controls, and so on. The mat under the push button has to establish electrical contact at a certain pressure, and in many cases the rubber contact itself is touched. This application requires the material to endure a continuous mechanical → stress ($>10^7$ switching cycles). It must have a high specific volume → resistivity, pigmentability, and printability (for lettering the contact).

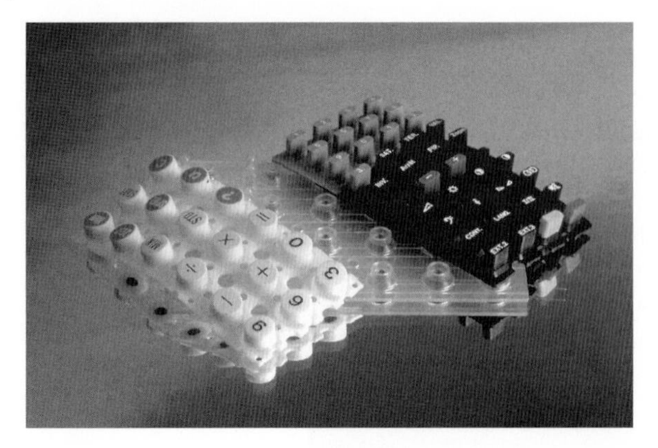

Figure 5.5 Silicone rubber contact.

Keyboard switches can be made as integrated composition of silicone rubber and conducting silicone rubber (Figure 5.6), the latter being produced by filling the silicone with carbon black.

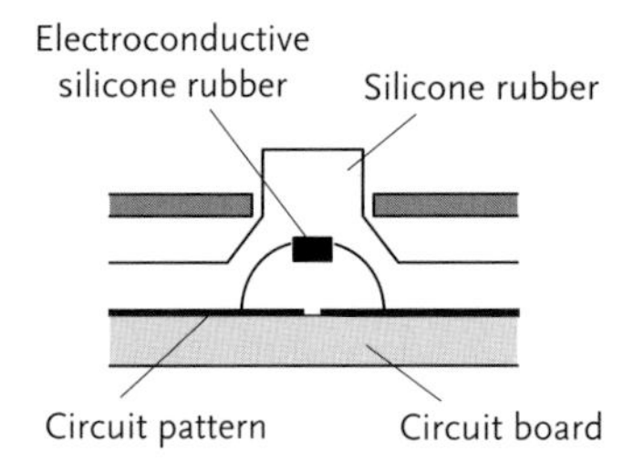

Figure 5.6 The working principle of rubber contacts.

HTV silicone rubbers These have found widespread applications for many uses. These include transparent tubings in food industry and medicine, electrical insulations, gaskets, belting, rollers, plug-and-socket connectors, keyboards, oxygen masks, soft-tissue prostheses, and so on. The uncured silicones are rubbery materials. Their forming processes are the same as in the rubber industry, such as compression or injection molding, extrusion, calandering, and so on.

5.2.1.3 **Silicone Resins**

Silicone resins are used in the insulation of electrical equipment and machinery, as laminates for printed circuit boards, and for the encapsulation of components such as resistors and integrated circuits. Non-electrical uses include high-temperature binders in paints and coatings for industrial plants, as well as household items, such as cooking ware, ovens, and so on.

5.2.2
Structure of Silicones

Silicones are composed of four structural units (Figure 5.7). There are 15 possible combinations: QT, QD, QM, TD, TM, DM, TT, DD, MM, QQ, QTD, QTM, QDM, TDM, and QTDM. MM represents disiloxanes, and QQ SiO_2 or silicates. Mainly DM, QM, TD, TM, TDM, and TT are realized in technical products. Materials of the QT and TT type are the domain of sol–gel chemistry (see Section 4.5).

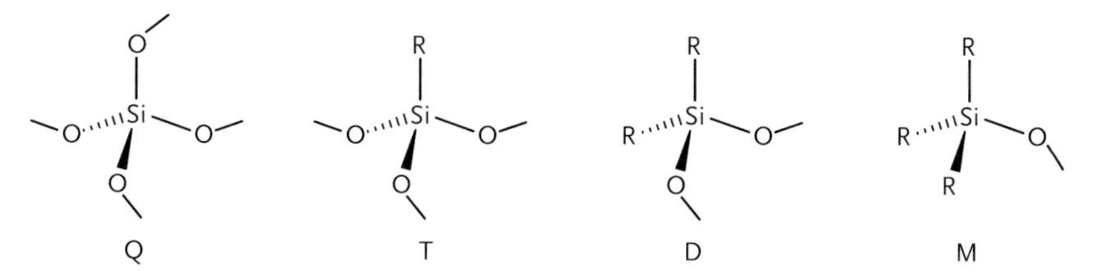

Figure 5.7 Structural units of silicones (R = organic group).

The properties of silicones are determined by:

- the type and proportion of the structural units,
- the type of the organic groups R,
- the distribution of the organic groups if there is more than one kind of R, and
- the chain length and chain-length distribution.

The properties of phosphazene polymers (Section 5.3) are to a very large extent determined by the properties of the organic substituents. In contrast, the properties of silicones are mostly influenced by the distribution of the structural units. Each of them has different functions. M units terminate chains or three-dimensional entities. A higher proportion of M units therefore results in a lower molecular mass of the silicone. The combination of D units results in chains, while each Q and T unit is a branching point. Q units are only employed in silicone resins and for some → crosslinking reactions of silicone elastomers (see below). A small number of T units in a TD silicone results in light crosslinking of the polysiloxane chains, but no three-dimensional network is formed.

The high thermal stability of silicones is mainly due to the strength of the Si–O bonds (420–500 kJ mol^{-1}). The Si–O–Si backbone of silicones has a unique dynamic flexibility. Compared to organic polymers (with a C–C–C backbone) polysiloxanes have longer bond distances (Si–O 164 pm, C–C 154 pm), and every second atom along the chain (the oxygen atom) has no substituent. This renders the chains more flexible. There are only low intermolecular forces between the polysiloxane chains. This is, *inter alia*, the reason for the very high spreadability and the low viscosity of silicone oils, and also explains the small temperature dependency of the physical properties of silicone oils and elastomers.

Methyl groups are predominant in silicones. They can be partially replaced by other groups to achieve special materials properties:

- Phenyl groups disrupt crystallinity and therefore improve flexibility at low temperatures. For example, cured dimethylsiloxane elastomers lose their elasticity at about −50 °C. Formulations containing about 8 mol% PhMeSiO groups instead of Me_2SiO, or 5 mol% Ph_2SiO, retain their elasticity down to −110 °C. Phenyl groups increase the already high thermal stability and weatherability of silicones, raise their refractive index, are less prone to oxidation, and additionally strengthen the Si–O bonds. In silicone resins, they improve the processability and the compatibility with organic polymers.
- Vinyl groups and hydrogen substituents are important for organic crosslinking reactions (see below).
- $CH_2CH_2CF_3$ (and other fluorinated alkyl) groups reduce swelling of silicone rubbers in many solvents ("fluorosilicones").

A variety of other organofunctional groups is used for the preparation of specialty silicones, the most important groups are of the type $(CH_2)_nX$ with

$$X = –SH, –NR_2, –O–H_2C–HC\overset{O}{\frown}CH_2, –OH, –O–\overset{O}{\overset{\|}{C}}–R$$

(see also Section 4.5 on sol–gel materials). In silicone chemistry, these groups mainly act as coupling sites, for example for providing links to organic polymer structures. However, organofunctional substituents can also determine physical properties of the silicone. This aspect will be discussed in more detail for polyphosphazenes (Section 5.3), where it plays a major role. For example, many →liquid-crystalline silicone polymers have been prepared by substituting the silicon atoms in polysiloxanes with →mesogenic side groups.

5.2.3 Preparation of Silicones

5.2.3.1 Precursor Syntheses

Silicones are prepared from the corresponding silanes $R_{4-n}SiX_n$ as precursors, where X almost always is Cl. The methyl derivatives are produced in the Müller–Rochow process ("direct process") in which methyl chloride is reacted with elemental silicon at 280–300 °C and 200–400 kPa in the presence of copper as the catalyst and promoters [Eq. (5.3)] in a fluidized-bed reactor. Zinc is mainly used as a promoter, in addition to small amounts of other elements. The actual catalyst (or at least the precursor to active surface species) is η-Cu_3Si highly dispersed at the silicon surface. A mixture of methyl(chloro)silanes is formed. The process has been optimized so that 80–90% Me_2SiCl_2 is obtained, the most important precursor for the preparation of silicones, particularly for silicone elastomers.

The second largest product is $MeSiCl_3$ (3–15%), followed by Me_3SiCl (2–5%), $Me(H)SiCl_2$ (0.5–4%) and a variety of other mono- and disilanes. The spectrum of products and the reproducibility is influenced by the processing conditions such as temperature, pressure, catalyst and promoters, inhibitors, silicon purity, particle size distribution of the solid reactants, homogeneity of the fluidized bed, dust removal from the reactor, and purity of the methyl chloride. About 1.4 million tons of silanes is produced annually by the Müller–Rochow process.

$$Si + CH_3Cl \xrightarrow{Cu} (CH_3)_x SiCl_{4-x} \tag{5.3}$$

The Müller–Rochow process can also be used for the preparation of phenyl-substituted silanes. Silanes with functional organic groups of the general type $Cl_3Si–X–A$ with a great variety of groups A connected to the silicon atom by an inert spacer X are easily prepared by indirect routes. The spacer X usually is a $(CH_2)_n$ ($n = 2$, 3) chain. Chlorosilanes $Cl_3Si–X–A$ or the derived alkoxysilanes $(RO)_3Si–X–A$ (prepared from the chlorosilanes by reaction with alcohols) have found widespread industrial applications, not only as components in silicones, but also as adhesion promoters, for derivatizing surfaces, or for the immobilization of substrates.

The more general preparation routes for these compounds are summarized in Eqs. (5.4) to (5.6) (Y = Cl or OR).

$$Cl_3SiH + CH_2{=}CH{-}X{-}A \longrightarrow Cl_3Si{-}CH_2{-}CH_2{-}X{-}A \tag{5.4}$$

$$Y_3Si{-}(CH_2)_n{-}A + A' \longrightarrow Y_3Si{-}(CH_2)_n{-}A' + A \tag{5.5}$$

$$Y_3Si{-}(CH_2)_n{-}A + Y{-}X{-}A' \longrightarrow Y_3Si{-}(CH_2)_n{-}A(Y){-}X{-}A' \tag{5.6}$$

The hydrosilylation of alkenes or alkynes, that is, the addition of Si–H bonds to double or triple bonds, is a rather general method for the formation of Si–C bonds. In practice, $HSiCl_3$ (obtained by reaction of HCl gas with elemental silicon) is added to unsaturated organic compounds [Eq. (5.4)]. For example, the technically very important silanes $Cl_3Si(CH_2)_3Cl$, $Cl_3Si(CH_2)_2CN$ or $Cl_3Si(CH_2)_3OC(O)$–$CMe{=}CH_2$ are prepared in this way, starting from allylchloride, acrylonitrile, or allylmethacrylate, respectively.

Substitution of a group A by a group A′ [Eq. (5.5)] or addition reactions of functional groups [Eq. (5.6)] is often the easier way to introduce more complex functional groups. For example, the technically important thio-substituted compound $[(EtO)_3Si(CH_2)_3]_2S_4$ (used as an adhesion promoter between rubber and silica fillers) is prepared by reaction of $(EtO)_3Si(CH_2)_3Cl$ (A=Cl) with Na_2S_4 ($A'{=}S_4$). An example of the modification by addition reactions to functional groups A [Eq. (5.6)], such as (meth)acrylate, isocyanate or epoxide groups, is given in Eq. (5.7).

EtC(–O–C(=O)–CH=CH$_2$)$_3$ + (RO)$_3$Si–CH$_2$CH$_2$CH$_2$–SH

↓

(RO)$_3$Si–CH$_2$CH$_2$CH$_2$–S–CH$_2$CH$_2$–C(=O)–O–CH$_2$–C(Et)(–O–C(=O)–CH=CH$_2$)$_2$ (5.7)

5.2.3.2 **Hydrolysis, Methanolysis, and Polycondensation Reactions**

The conversion of the chlorosilanes to polysiloxanes is exemplarily discussed for Me_2SiCl_2. Processing of other chlorosilanes proceeds analogously.

Reaction of Me_2SiCl_2 with water results in the formation of cyclic and OH-terminated linear dimethylsiloxane oligomers [Eq. (5.8)], which are in equilibrium with each other.

$$(n+m)\,Me_2SiCl_2 \xrightarrow[-HCl]{H_2O} [Me_2Si(OH)_2] \xrightarrow[-H_2O]{} HO[-Me_2SiO]_n-H + (Me_2SiO)_m \quad (5.8)$$

The prime reaction is the hydrolysis of the Si–Cl groups to Si–OH groups. However, the reaction cannot be stopped at this point. As has already been discussed in Section 4.5, hydrolysis and condensation reactions [Eq. (5.9)] usually cannot be decoupled from each other, particularly in the presence of HCl (formed as a byproduct) which acts as a catalyst.

$$\begin{aligned} &\equiv Si-Cl + H_2O \rightleftharpoons \equiv Si-OH + HCl \\ &2 \equiv Si-OH \rightleftharpoons \equiv Si-O-Si\equiv + H_2O \end{aligned} \quad (5.9)$$

Both the hydrolysis and the condensation reactions are equilibria. Thus, cyclic and linear products, and linear products with different chain lengths are in equilibrium, and their ratio is influenced by the reaction conditions. A very important parameter is the HCl concentration and its contact time with the siloxane. Rapid removal or neutralization of the HCl results in siloxanediols $HO[SiMe_2O]_nH$ with short chain lengths (n). The cyclic products are dominated by $(Me_2SiO)_4$ ("D_4").

Me_2SiCl_2 is completely converted into OH-terminated linear siloxanes in the technically used continuous hydrolysis process. The cyclic products are continuously distilled off and fed back to the starting silane. Acid-catalyzed ring-opening polymerization (ROP) of the cyclosiloxanes $(SiMe_2O)_n$ (see Eq. (5.15) below) and reaction with Me_2SiCl_2 gives α,ω-dichlorsiloxanes, $Cl[SiMe_2O]_mSiMe_2Cl$, which are then hydrolyzed. A flow chart of the continuous hydrolysis process is shown in Figure 5.8.

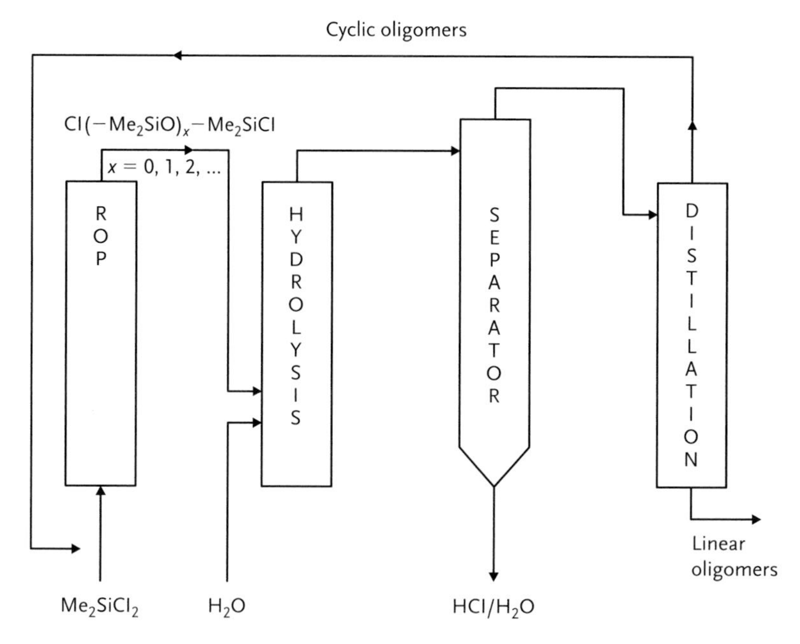

Figure 5.8 Continuous hydrolysis process.

The HCl obtained at the hydrolysis of Me_2SiCl_2 is reacted with methanol to give methyl chloride, which is fed back in the Müller–Rochow process. Chlorine is thus recycled.

An option in which chlorine from the Si–Cl groups is directly converted in methyl chloride is the reaction of Me_2SiCl_2 with methanol ("methanolysis," [Eq. (5.10)]). The cyclic products can be recycled in a continuous process similar to that in Figure 5.8.

$$(n+m)\mathrm{Me_2SiCl_2} \xrightarrow[-\mathrm{MeCl}]{\mathrm{MeOH}} \mathrm{HO[-Me_2SiO]}_n\mathrm{-H} + (\mathrm{Me_2SiO})_m \tag{5.10}$$

To obtain highly polymeric silicones, the OH-terminated dimethylsiloxane oligomers obtained by the hydrolysis or methanolysis reaction have to undergo → polycondensation reactions, in which the oligomer units are coupled, [Eq. (5.11)]. The polycondensation can be catalyzed by strong acids or bases, or by solid catalysts. The presence of precursors of chain-terminating [Me_3SiO–] or cross-linking groups [$MeSi(O–)_3$] must be carefully avoided for the production of linear polysiloxanes.

$$2 \sim\!\mathrm{O-Si(Me)_2-OH} \xrightarrow[-\mathrm{H_2O}]{\mathrm{cat.}} \sim\!\mathrm{O-Si(Me)_2-O-Si(Me)_2-O}\!\sim \tag{5.11}$$

5.2.3.3 **Equilibration and Ring-Opening Polymerization of Cyclosiloxanes**

When cyclic oligosiloxanes or linear polysiloxanes are exposed to a base or acid catalyst, an equilibrium mixture between cyclic and linear species is obtained (upper part of Eq. (5.12)), as mentioned above. The process of simultaneous cleavage and re-formation of siloxane bonds is called equilibration.

Acid- or base-catalyzed ROP reactions are performed under non-equilibrium conditions. In the *acid-catalyzed reaction*, the Si–O–Si bond of a cyclic or linear siloxane is *broken* by the electrophilic attack of the Lewis acid (proton) to the oxygen atom. This generates species terminated by an OH group at one end and by an active electrophilic center at the other. The group A in Eq. (5.12) may be the counterion of the acid (X) or an oxonium ion ($\equiv$Si–O$^+$H$_2$ generated by substitution of X by water, which is present in the system owing to the subsequent condensation reactions). Two Si–O–Si bond-*forming* mechanisms can occur simultaneously. The active center can add to another Si–O–Si bond (addition polymerization mechanism) or hydrolysis/condensation reactions can take place. All these reactions are basically equilibria.

$$D_4 + H^+ X^- \rightleftharpoons HO{-}(SiMe_2{-}O)_3{-}SiMe_2{-}A \quad A = X^- \text{ or } H_2O$$

$$HO{-}(SiMe_2{-}O)_3{-}SiMe_2{-}A \xrightarrow{+D_4} HO{-}(SiMe_2{-}O)_7{-}SiMe_2{-}A \longrightarrow \longrightarrow$$

$$HO{-}(SiMe_2{-}O)_3{-}SiMe_2{-}A \underset{-HA}{\overset{+H_2O}{\rightleftharpoons}} HO{-}(SiMe_2{-}O)_4{-}H \underset{-H_2O}{\overset{\times 2}{\rightleftharpoons}} HO{-}(SiMe_2{-}O)_8{-}H \longrightarrow \longrightarrow \tag{5.12}$$

KOH is the most widely used catalyst for the *anionic* → *polymerization*. The OH$^-$ ion attacks a silicon atom and thus generates a $\equiv$Si–OH and a $\equiv$SiO$^-$

(silanolate) terminated end [Eq. (5.13)]. The silanolate group than attacks another Si–O–Si group, and so on (living anionic polymerization). Before the residual cyclosiloxanes are removed from the reaction mixture, the catalyst must be destroyed. Otherwise the equilibria would be shifted towards the cyclic products.

$$D_4 + OH^- \longrightarrow HO\left(\begin{matrix}Me\\|\\Si\\|\\Me\end{matrix}-O\right)_3\begin{matrix}Me\\|\\Si\\|\\Me\end{matrix}-O^- \tag{5.13}$$

Both equilibration and acid- or base-catalyzed ROP reactions play an important role for the regulation of chain lengths, and the production of copolymers from two differently substituted polysiloxanes. They allow the very reproducible production of silicone products. Technically important examples are:

- Silicone oils with very defined chain lengths and viscosities are made by shaking suitable proportions of $Me_3Si–O–SiMe_3$ (as the precursor for end-stopping $–OSiMe_3$ units) and either a cyclo(dimethylsiloxane) or a linear poly (dimethylsiloxane) with a small quantity of 100% H_2SO_4 [Eq. (5.14)]. The reason why this reaction proceeds towards polymers (although all reactions are equilibria) is that the reaction rate of Si–O–Si bond cleavage in acidic media decreases in the following order: $D_4 > MM > MDM > MDDM$. Thus, cleavage of a Si–O–Si bond in D_4 is preferred to that of an oligomeric or polymeric chain.

$$Me_3Si-O-SiMe_3 + \left[\begin{matrix}Me\\|\\Si\\|\\Me\end{matrix}-O\right]_n \xrightarrow{\text{cat.}} Me_3Si-O\left[\begin{matrix}Me\\|\\Si\\|\\Me\end{matrix}-O\right]_n SiMe_3 \tag{5.14}$$

- Endblocks with reactive groups (X) such as Cl, OR, OAc, $–ON{=}CR_2$ or $–NR'C(O)R$ can be introduced by a related reaction [Eq. (5.15)]. Polymers with such groups are needed for the production of silicone elastomers (see below). The acid-catalyzed ring opening of cyclosiloxanes in the presence of Me_2SiCl_2 in the continuous hydrolysis process (Figure 5.8) is another example.

$$Me_2SiX_2 + m(Me_2SiO)_n \longrightarrow X–(Me_2SiO)_{nm}–SiMe_2X \tag{5.15}$$

- Polysiloxanes containing a certain number of Si–H groups are needed for the production of silicone elastomers (see below). Their synthesis by equilibration of $SiMe_3$-terminated poly- or oligo(methylsiloxanes) and cyclic or linear *di*methylsiloxanes is shown in Eq. (5.16). The proportion of Si–H groups in the equilibrated polymer determines the later → crosslinking density, and thus the physical properties of the elastomers. The clearly defined incorporation of these groups is therefore a very important issue. Other functional groups can be incorporated by the same method.

$$Me_3Si-O\left[\begin{matrix}Me\\|\\Si\\|\\H\end{matrix}-O\right]_m SiMe_3 + \left[\begin{matrix}Me\\|\\Si\\|\\Me\end{matrix}-O\right]_n \xrightarrow{\text{acid}} Me_3Si-O\left[\begin{matrix}Me\\|\\Si\\|\\H\end{matrix}-O\right]_m\left[\begin{matrix}Me\\|\\Si\\|\\Me\end{matrix}-O\right]_n SiMe_3 \tag{5.16}$$

- When the cyclosiloxanes are distilled off from an equilibrium mixture between cyclic and linear species, the equilibrium is continuously shifted, and the polymer is eventually completely degraded. This depolymerization reaction can be used for the recycling of silicone products. The thus-obtained cyclosiloxanes can again be converted to linear polysiloxanes using the processes discussed above.

5.2.3.4 **Curing ("Vulcanizing")**

→ Crosslinking of the polysiloxane chains is necessary to obtain silicone materials with a higher viscosity, such as elastomers, rubbers, or resins (see above). This requires functional groups bonded directly to silicon, or as part of the organic substituents. General methods for crosslinking were already discussed in Section 5.1; the following paragraphs will describe specifically which of these processes are used industrially to obtain the various silicone varieties.

Silicones of the RTV-1 type (see Section 5.2.1.2) are prepared by reaction of OH-terminated polysiloxanes with silanes of the type $RSiX_3$ [Eq. (5.17)] or by the ring-opening reaction shown in Eq. (5.15). Curing occurs upon exposure to humidity [Eq. (5.18)]. Therefore, fast surface curing is obtained, but hardening of unexposed material can be quite long, especially for thick samples. This is favorable to adhesion, because there is a long period of time for spreading and wetting of the elastomer on the substrate. Cure accelerators have been developed to applications that require shorter curing periods. Depending on the cleaved HX molecule, basic (cleavage of an amine, $HX = NH_2R$), neutral (cleavage of an oxime or an amide, $HX = HON{=}CR_2$ or $R'NHC(O)R$) or acidic (cleavage of a carboxylic acid, $HX = HOOCR$) systems are distinguished.

$$HO{-}[Si(Me)_2{-}O]_n{-}H + 2RSiX_3 \longrightarrow R{-}SiX_2{-}O{-}[Si(Me)_2{-}O]_n{-}SiX_2{-}R + 2HX \quad (5.17)$$

$$\sim\!Si(Me)_2{-}X + X{-}Si(Me)_2{-}O\!\sim + H_2O \xrightarrow{-2HX} \sim O{-}Si(Me)_2{-}O{-}Si(Me)_2{-}O\!\sim \quad (5.18)$$

The two main crosslinking reactions in RTV-2 formulations are shown in Eqs. (5.19) and (5.20).

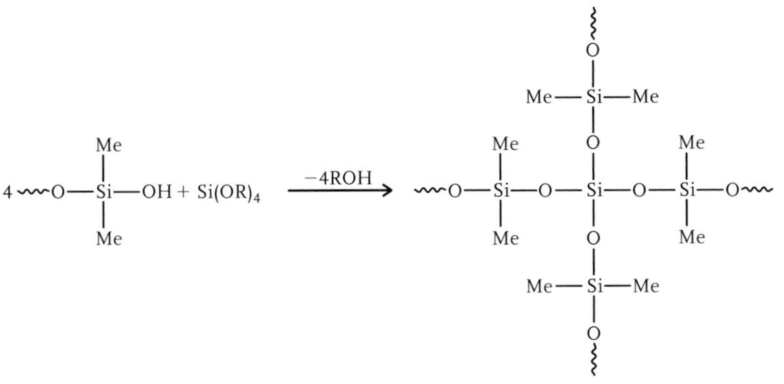

$$4 \sim\!O\!-\!Si(Me)_2\!-\!OH + Si(OR)_4 \xrightarrow{-4ROH} \sim\!O\!-\!Si(Me)_2\!-\!O\!-\!Si[\!-\!O\!-\!Si(Me)_2\!-\!O\!\sim]_2\!-\!O\!-\!Si(Me)_2\!-\!O\!\sim \quad (5.19)$$

In the condensation–crosslinking process [Eq. (5.19)], the first component is an OH-terminated polysiloxane, and the second a tetraalkoxysilane. The latter is mixed with a catalytically active tin compound in one package, and the former with water in another. Curing begins upon intimate mixing. Since no atmospheric water is required, the surface and the bulk cure at the same rate. The eliminated alcohol must be removed to avoid the backreaction. The main advantage of this crosslinking method is the variable ratio of the two components.

In the addition-crosslinking process [Eq. (5.20)], Si–H groups in one polysiloxane chain are catalytically added to Si–CH$=$CH$_2$ groups in another chain. In contrast to Eq. (5.19), this crosslinking is irreversible. There is typically one crosslink for every 200–2000 Si atoms. The → crosslinking is strongly influenced by the temperature. → Vulcanization of the *liquid rubber (LR) formulations* proceeds by the same process at higher temperature (in the heated mold). There is almost no shrinkage upon curing, because the reaction proceeds without the loss of mass. The main disadvantage of the addition–crosslinking method is that the catalyst (usually 5–50 ppm of a Pt compound) can be poisoned by materials that are in contact with the silicone.

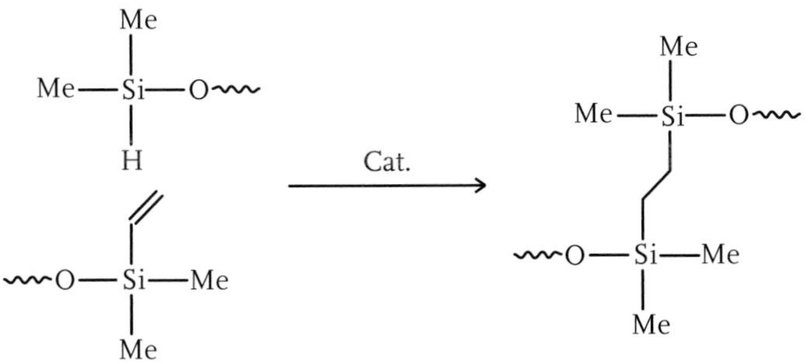

$$Me_2Si(H)\!-\!O\!\sim \; + \; CH_2\!=\!CH\!-\!Si(Me)_2\!-\!O\!\sim \xrightarrow{\text{Cat.}} \sim\!O\!-\!Si(Me)_2\!-\!CH_2\!-\!CH_2\!-\!Si(Me)_2\!-\!O\!\sim \quad (5.20)$$

In *HTV formulations*, mainly vinyl groups are crosslinked. The solid rubbers are vulcanized by radical reactions at 150–250 °C using 1–3% of an aroyl or alkyl peroxide [Eq. (5.21)]. Crosslinking by coupling of Si–CH$_2^{\bullet}$ radicals is also possible.

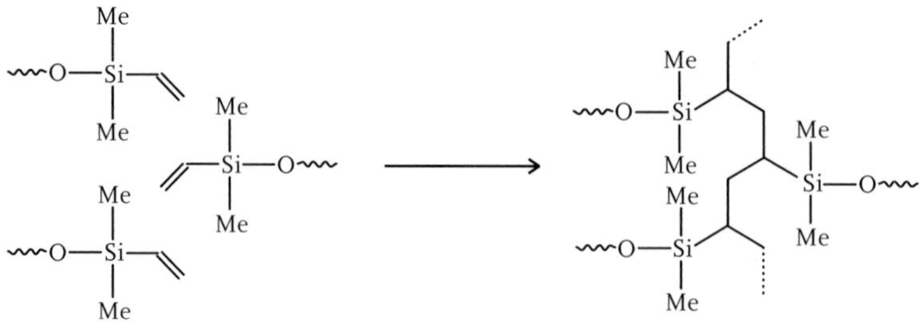

(5.21)

While the chain-forming Me_2SiCl_2 is the main component for the preparation of silicone oils and elastomers, *silicone resins* are prepared by using phenyl- or methyltrichlorosilane as the base component. The rigid skeleton is made more flexible by addition of Me_2SiCl_2, $PhMeSiCl_2$, and Ph_2SiCl_2. $Me(H)SiCl_2$, $CH_2{=}CHSiCl_3$ and $CH_2{=}CH(Me)SiCl_2$ enable hardening by crosslinking reactions (as discussed above), and Me_3SiCl serves to provide end groups. The hydrolyzed mixture of chlorosilanes is washed with water to remove HCl and partly polycondensed to a stage at which the resin is still soluble. The resins are normally applied in this form. After evaporation of the solvent, the resin is crosslinked to a three-dimensional siloxane network by heating above 150 °C in the presence of a catalyst to condense the silanol groups. Additional → crosslinking via vinyl groups, either by radical polymerization or by addition of Si H groups, as for the elastomers, is also possible.

5.2.3.5 **Polysiloxane Copolymers**

Totally different materials properties are obtained when polysiloxane structures are combined with organic polymer structures.

The most general method to make such copolymers is to react silicones with appropriate organofunctional substituents with organic monomers (or oligomers containing the reactive end groups). The preparation of a silicone–polymethacrylate copolymer [Eq. (5.22)] may serve to illustrate the general approach.

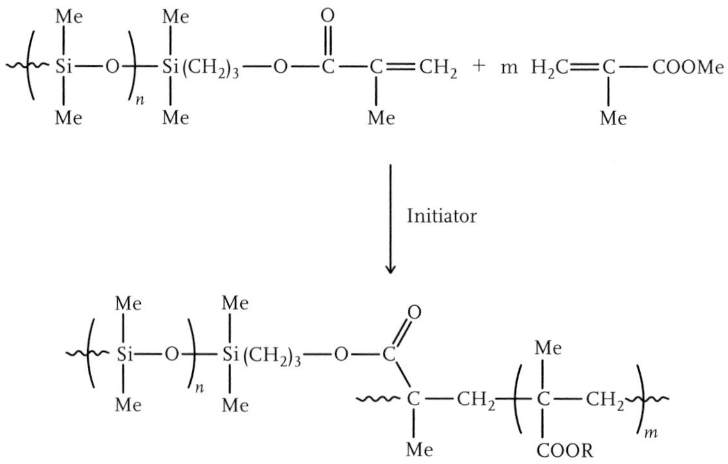

$$\sim\!\!\left[SiMe_2{-}O\right]_n{-}SiMe_2(CH_2)_3{-}O{-}C({=}O){-}C(Me){=}CH_2 \;+\; m\, H_2C{=}C(Me){-}COOMe \xrightarrow{\text{Initiator}} \sim\!\!\left[SiMe_2{-}O\right]_n{-}SiMe_2(CH_2)_3{-}O{-}C({=}O){-}C(Me)({\sim}){-}CH_2{-}\left[C(Me)(COOR){-}CH_2\right]_m\!\sim \tag{5.22}$$

Copolymers between polysiloxanes and nearly all types of organic polymers have been made, and some are applied industrially. A detailed discussion of the preparation and properties of these copolymers is beyond the scope of this book.

5.3 Polyphosphazenes

Several hundred different polyphosphazenes have been synthesized, with a wide range of physical and chemical properties and with high molecular weights. Polyphosphazenes exhibit a very broad spectrum of useful chemical, mechanical, optical, and biological properties. Their properties vary from fluids to vulcanizable elastomers and glasses, from crystalline to amorphous, from water-soluble to hydrophobic, from bioinert to bioactive, and from electrical insulators to conductors.

5.3.1 Properties and Applications of Polyphosphazenes

High-performance elastomers, which often contain → fillers, are the technically most advanced applications, and their utility parallels that of silicones. For example, a commercially available polyphosphazene with OCH_2CF_3 and $OCH_2(CF_2)_nCF_3$ substituents (and a small amount pendent allyl groups for curing) is a soft gum that can be processed by conventional rubber technology. The polymer is cured with peroxides or by radiation. The cured polymer has a tensile strength of 7–14 MPa (depending on the → filler), can be elongated by 75–200%, adheres very well to metals and fabrics, has excellent vibration-damping properties and a hydrophobic surface, is resistant to fungi, ozone and a broad range of fluids, is non-flammable, shows an excellent weatherability, and is resistant to liquid oxygen. It maintains its properties between −65 °C and +175 °C. Because of these properties the elastomer is used for seals, gaskets, and shock mounts in demanding automotive, aerospace, and petroleum applications.

The backbone of polyphosphazenes consists of (NPR_2) or ($NPRR'$) repeat units. The most common substituents R (or R′) are alkoxy (OR) or amino groups (NHR or NR_2), rarely directly bonded organic moieties. Although the bonding in the backbone is formally represented as a series of alternating single and double bonds [Eq. (5.23)], all the bonds along the chain are equal in length. There is no conjugation comparable to organic unsaturated polymers because of the participation of d-orbitals at phosphorus in the π-bond.

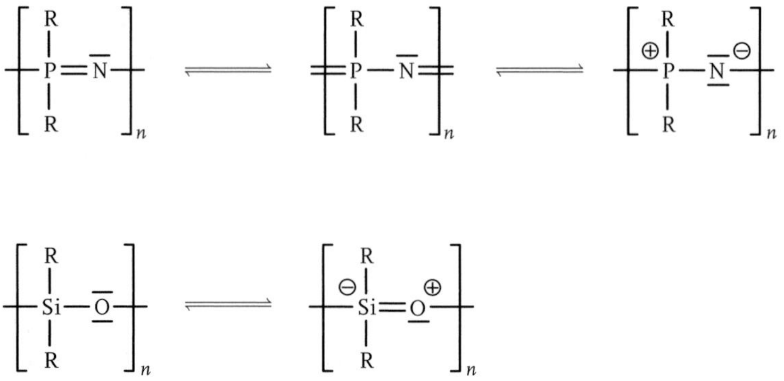

$$\left[\text{P}(R)_2{=}\text{N} \right]_n \rightleftharpoons \left[{=}\text{P}(R)_2{-}\text{N}{=} \right]_n \rightleftharpoons \left[{}^{\oplus}\text{P}(R)_2{-}\text{N}^{\ominus} \right]_n$$

$$\left[\text{Si}(R)_2{-}\text{O} \right]_n \rightleftharpoons \left[{}^{\ominus}\text{Si}(R)_2{=}\text{O}^{\oplus} \right]_n \qquad (5.23)$$

Polyphosphazenes are isoelectronic to polysiloxanes (see Eq. (5.23) and Section 5.2). This explains the similarity of some of their physical properties, for example, their elasticity down to low temperatures. In both polyphosphazenes and polysiloxanes the substituents are only attached to every second backbone atom, which results in the high flexibility of the polymer chain. There is no torsional barrier of the skeleton, because there are no p_π–p_π bonds. The conformation of a polyphosphazene chain is therefore mainly determined by interactions between the substituents at neighboring phosphorus atoms. It can be seen from Figure 5.9 that the repulsion is minimized in the *cis–trans* planar conformation compared to the *trans–trans* planar conformation or any other conformation between. The repeating distance found in nearly all polyphosphazenes studied so far is about 50 pm, which is the value expected for a configuration near *cis–trans* planar.

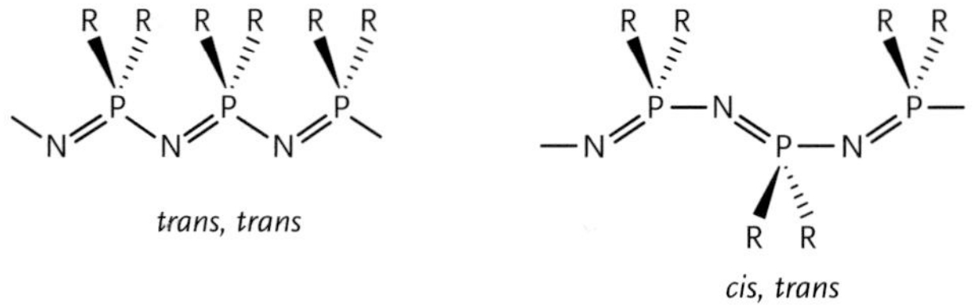

Figure 5.9 *Trans, trans*, and *cis, trans* conformation of a polyphosphazene chain.

While the properties of silicones are largely determined by the type and proportion of the structural units, the properties of polyphosphazenes are mainly varied by the types of substituents, and their sequencing along the chain if there is more than one type. The influence of a particular substituent on the properties of a polymer is probably quite general. However, the substituents are more easily varied in polyphosphazenes than in any other polymer.

Enough is known about the effect of the substituents to allow, to some extent, the prediction of the polymer properties. For example, polyphosphazenes have low glass-transition temperatures (T_g) when the substituents are small (R = F: −96 °C, Cl: −66 °C, OMe: −74 °C) or when the substituents themselves are very flexible ($OCH_2CH_2OCH_2CH_2OMe$: −84 °C, OPr: −100 °C). Large and inflexible substituents sterically interfere with each other as the P N skeleton undergoes twisting motions and therefore impose restrictions on the flexibility of the macromolecule. For example, if the substituents are OPh or OC_6H_4-*p*-Ph the T_g is −8 °C or +93 °C, respectively.

A selection of substituents and their influence on other materials properties is given in Table 5.1. The refractive index is increased by increasing the number of electrons in the substituents, for instance by biaryl substituents. Rigid, → mesogenic substituents can give rise to liquid crystallinity. Other properties, such as solubility, chemical stability, hydrophobicity or hydrophilicity, ionic conductivity, → non-linear optical activity, photochromic properties and biological behavior can also be introduced by appropriate substituents. Membranes have been developed by amphiphilic polymers with both NHMe (hydrophilic) and OCH_2CF_3 (hydrophobic) substituents.

Most polyphosphazenes are stable to water. However, if a P–R bond is cleaved by hydrolysis, a P–OH group is generated. The proton of this group can migrate to a nitrogen atom to give a saturated phosphazane (–P(O)R–NH– repeating units) that is susceptible to hydrolytic cleavage of its skeleton. If hydrolysis is very slow, as for R=NHCH(R)COOEt, hydrolytic sensitivity is a useful property, because it allows the use of such polyphosphazenes as biodegradable materials in medicine.

Table 5.1 The influence of selected substituents R on the properties of polyphosphazenes $[NPR_2]_n$.

R	Property/Application
NHMe, glucosyl, glyceryl	Hydrophilicity
$OCH_2CH_2OCH_2CH_2OCH_3$	Li^+ conductivity in the solid state, water solubility, elasticity, low T_g
OR (R = alkyl)	Elasticity
$NHCH_2CH{=}CH_2$	→ Crosslinking sites
OC_6H_4–p–OC(O)CH=CHPh	Photosensitivity
$OCH_2(CF_2)_xCF_3$	Low T_g, increased solvent resistance, hydrophobicity
OCH_2CF_3	High transmission coefficient for O_2, solubility in organic solvents, bioinert materials
OR (R = aryl)	Thermoplasticity, solubility in organic solvents
$[OC_6H_4\text{–p–}COO^-]_2Ca^{2+}$	Microencapsulation of bioactive species
$OCH_2CH_2N(Et)C_6H_4\text{–}N{=}N\text{–}C_6H_4\text{–p–}NO_3$	χ^2 → NLO activity
$O(CH_2CH_2O)_3C_6H_4\text{–p–}C_6H_4\text{–p–}OMe$	→ Liquid crystallinity
OC_6H_4Ph	High refractive index
$OC_6H_4\text{–}PPh_2$	Coordination of metal complexes
$OC_6H_4\text{–p–}N{=}CH(CH_2)_3CHO$	Immobilization of enzymes
NHCH(R)COOEt	Biodegradable materials

5.3.2
Preparation and Modification

The most general and most widely used method for the preparation of polyphosphazenes is the ROP/substitution sequence shown in Eq. (5.24).

$$(NPCl_2)_3 \xrightarrow{250\,^\circ C} \left[-P(Cl)_2{=}N- \right]_n \xrightarrow[\text{substitution}]{\text{nucleophilic}} \left[-P(R)_2{=}N- \right]_n \qquad (5.24)$$

Hexachlorocyclotriphosphazene, $(NPCl_2)_3$, is prepared on an industrial scale by the reaction of PCl_5 with NH_4Cl in an organic solvent such as chlorobenzene or tetrachlorethane. The purified, molten compound is heated in a sealed vessel in the absence of moisture (which would convert the P–Cl bonds to P–O–P bonds) for several days at 200–250 °C to induce ROP. The polymerization can also be conducted in an organic solvent at lower temperatures in the presence of a Lewis acid catalyst. It is assumed that ROP is induced by ionic species generated, for example, by heterolytic cleavage of a P–Cl bond or by abstraction of Cl^- by the Lewis acid. ROP of $(NPCl_2)_3$ results in linear poly(dichlorophosphazene), $[NPCl_2]_n$, soluble in benzene, toluene, or tetrahydrofuran. The control of the molecular weight of this polymer is difficult, and the → polydispersity is large.

The chlorosubstituted polymer $[NPCl_2]_n$, known since the end of the 19th century as "inorganic rubber," is still very moisture sensitive. It also reacts rapidly with a variety of nucleophiles such as metal alkoxides, amines, or organometallic reagents by substitution of the P–Cl groups. A wide range of groups can thus be introduced to give polymers that are stable towards water. With some reactive reagents, all the chloride atoms of the polymer are replaced within minutes at room temperature! In any case, the substitution reaction has to be driven to completion, because residual P–Cl bonds would result in moisture-sensitive polymers.

Certain bulky or insufficiently nucleophilic organic groups do not replace all chlorine atoms under mild reaction conditions. However, after a bulky substituent has partially substituted the P–Cl groups, the controlled introduction of a second (smaller) type of substituents is possible. For example, when $[NPCl_2]_n$ is reacted with diethylamine (even in excess), only one chlorine atom per phosphorus is replaced. The remaining chlorine can then be substituted by treatment with methylamine or short-chain alkoxides, for example.

$[NPCl_2]_n$ can be deliberately reacted with a deficiency of the reagent. The remaining chlorine substituents can then be reacted with another nucleophile, resulting in mixed-substituted polymers $[NPRR']_n$. An example is shown in Eq. (5.25). Alternatively, both reagents may be allowed to compete simultaneously for the available P–Cl groups. The disposition of substituents along the polyphosphazene chain depends on the directing characteristics of the substituents already present. The order in which two or more different substituents are introduced will also affect the substitution pattern. However, there is no strict control of the substitution pattern in most cases. Polymers with two or more different substituents generally have a better elasticity because they crystallize less easily.

$$\left[P(Cl)_2{=}N \right]_n \xrightarrow[-\,NaCl]{+\,NaOR} \left[P(OR)(Cl){=}N \right]_n \xrightarrow[-\,HCl]{+\,R'NH_2} \left[P(OR)(NHR'){=}N \right]_n \qquad (5.25)$$

The length of the polymer chain is usually unaffected by the replacement of Cl by OR or NRR′ groups. Organometallic reagents result in more complicated reactions. For example, in the reaction of $[NPCl_2]_n$ with RMgX or LiR, replacement of chlorine by the group R is accompanied by cleavage of P–N bonds in the skeleton. Coordination of the organometallic molecules to the nitrogen atoms of the polyphosphazene backbone is necessary for the cleavage reaction. Thus, any feature that increases electron density at the nitrogen atoms and thus favors coordination will also favor cleavage of the backbone. Chlorine atoms at phosphorus lower the electron density due to their electron-withdrawing properties and thus protect the skeleton against cleavage. As an organometallic substitution reaction proceeds, this protective effect is successively lost, and there is an increased probability of P–N bond cleavage.

Due to the restrictions in the reactions with organometallic reagents, a modification of the ROP process was investigated. In this approach, the organic substituents are introduced at the cyclic trimer level, followed by ROP. However, the tendency for ROP declines as more and more halogen atoms in the trimer are replaced by organic groups. Cyclic trimers that bear only one or two organic substituents usually polymerize almost as easily as $(NPCl_2)_3$. The remaining chlorine atoms in the partially organo-substituted polymers can be replaced by nucleophiles as discussed before. A typical reaction sequence is shown in Eq. (5.26). $(NPR_2)_3$ with two organic substituents per phosphorus atom only undergo equilibration to $(NPR_2)_4$ when heated, but no ROP.

$$(NPCl_2)_3 \xrightarrow{+\,3LiMe} [NP(Cl)(Me)]_3 \xrightarrow{T\uparrow} \left[P(Cl)(Me){=}N \right]_n \xrightarrow[-\,NaCl]{+\,NaOR} \left[P(OR)(Me){=}N \right]_n \qquad (5.26)$$

The first intermediate in the synthesis of cyclic $(NPCl_2)_3$ from PCl_5 and NH_4Cl is $Cl_3P{=}NH$. The reaction then proceeds stepwise by elimination of HCl. In a promising new approach, Me_3SiX (X = OR, Cl) is eliminated instead of HCl. This method makes use of the high tendency to form the strong Si–O or Si–Cl bonds. When phosphoranimines of the type $R_2(X)P{=}NSiMe_3$ (R = alkyl, aryl, OR′) are heated, polymeric $[NPR_2]_n$ is obtained by elimination of Me_3SiX, [Eq. (5.27)]. This reaction provides a direct access to alkyl- and aryl-substituted polyphosphazenes that are difficult to obtain by the ROP/substitution approach, as discussed above.

$$n\ \mathrm{X{-}P(R)(R'){=}N{-}SiMe_3} \xrightarrow[\text{or cat.}]{T\uparrow} \left[\mathrm{P(R)(R'){=}N}\right]_n + n\ \mathrm{Me_3SiX} \quad (5.27)$$

Polycondensation of $X_3P{=}NSiMe_3$ ($X{=}OR$, Cl) gives $[NPX_2]_n$. Thus, room-temperature polycondensation of $Cl_3P{=}NSiMe_3$, initiated by small amounts of PCl_5, is a new route for the synthesis of $[NPCl_2]_n$ with controllable molecular weights and narrow → polydispersities. Polycondensation of $(RO)_3P{=}NSiMe_3$ is an alternative route to $[NP(OR)_2]_n$ instead of the ROP/substitution route. The required temperature depends on the nature of OR (e.g., $R{=}CH_2CF_3$: 200 °C; R = aryl: 60 °C).

The stability of the polyphosphazene backbone also allows the modification of the substituents bonded to phosphorus. Groups can thus be introduced that are not accessible by substitution reactions. An example is shown in Eq. (5.28).

$$\left[\mathrm{P(O{-}C_6H_4{-}Br)_2{=}N}\right]_n \xrightarrow[-\,\mathrm{BuBr}]{+\,\mathrm{BuLi}} \left[\mathrm{P(O{-}C_6H_4{-}Li)_2{=}N}\right]_n \xrightarrow{CO_2} \left[\mathrm{P(O{-}C_6H_4{-}COOH)_2{=}N}\right]_n \quad (5.28)$$

Other examples are the tethering of metal complexes or even enzymes to suitable functional groups present in the organic side groups.

Methyl-substituted polyphosphazenes can be deprotonated. Substituted products are formed upon reaction of the thus-generated anionic $P{-}CH_2^-$ groups with electrophiles. Side-group reactivity can, for example, be utilized for the → crosslinking reactions shown schematically in Figure 5.2. The anionic sites can also be used to initiate anionic → polymerization reactions. A variety of graft polymers have been prepared in this manner. An example (polyphosphazene-graft-polystyrene) is given in Eq. (5.29).

+ BuLi

$$\left[\mathrm{P(Ph)(Me){=}N}\right]_n \xrightarrow[\text{THF}]{^n\mathrm{BuLi}} \left[\mathrm{P(Ph)(Me){=}N}\right]_x\left[\mathrm{P(Ph)(CH_2Li){=}N}\right] \xrightarrow[(2)\ \mathrm{H^+}]{(1)\ \mathrm{CH_2{=}CHPh}} \left[\mathrm{P(Ph)(Me){=}N}\right]\left[\mathrm{P(Ph)(CH_2{-}[CH_2{-}CHPh]_y{-}H){=}N}\right] \quad (5.29)$$

5.4 Polysilanes

Although the first polysilanes (polysilylenes) were prepared in the first half of the twentieth century, they have only attracted scientific and technical interest since the late 1970s. Polysilanes are now sold commercially as polymeric precursors to SiC ceramics by the Yajima process. This process will be discussed in Section 5.5. Polysilanes are promising polymeric materials mainly as UV → photoresists for microelectronics and materials for → waveguides, but also as photo- and charge-conducting materials, → thermochromic materials, NLO materials, or radical photoinitiators for organic → polymerization reactions.

5.4.1 Properties and Applications of Polysilanes

Polysilanes range from highly crystalline, hard, brittle solids to glassy amorphous materials and rubbery elastomers, from insoluble solids to those soluble in organic solvents or even in water. The T_gs are between −75 and 120 °C. When purified, most polysilanes are inert to oxygen at ambient temperatures, and reasonably stable towards hydrolysis. Some are thermally stable to almost 300 °C in an inert gas atmosphere.

More than in any other inorganic polymer, the physical and chemical properties of polysilanes are influenced by both the kind of the backbone and by the properties of the substituents.

5.4.1.1 Effects Originating from the Substituents

$[Me_2Si]_n$, $[MeEtSi]_n$, $[Et_2Si]_n$, and most $[aryl_2Si]_n$ are crystalline because of crystallization of the side chains. Such polysilanes are insoluble and infusible. Crystallinity is reduced by decreasing the symmetry or by longer alkyl chains. Thus, most poly(aryl,alkyl)silanes, the higher poly(methyl,alkyl)silanes, $[MeRSi]_n$, and the symmetrical poly(dialkylsilanes), $[R_2Si]_n$ with R = *n*-propyl or larger, are meltable solids or rubbery elastomers soluble in organic solvents. The minimum T_g for $[MeRSi]_n$ at about −75 °C is reached for R = *n*-hexyl. Crystallinity is also lower for copolymers than for the homopolymers with the same substituents. For example, the copolymer $[Me_2Si]_n[Ph_2Si]_m$ is soluble and meltable. In the technically important "poly(silastyrene)" copolymers, $[PhMeSi]_n[Me_2Si]_m$, crystallinity is at a minimum near a $n : m$ ratio of 1.

5.4.1.2 Effects Originating from the Backbone

One of the most remarkable properties of polysilanes is the extensive *σ-electron delocalization* along the Si–Si chain. Many of the technical uses and the remarkable properties of polysilanes result from this unusual mobility of the σ-electrons. As a consequence, their electronic and photochemical behavior is very different from that of most other inorganic or organic polymers.

Other polymers, such as polyacetylene and polythiophene, also show electron delocalization, but in these materials the delocalization involves π-electrons. The main reason for the delocalization of the σ-electrons in polysilanes is the relatively large interaction between adjacent σ-orbitals, because they are more diffuse than C–C-σ-orbitals. For the description of bonding in polysilanes the band model is therefore more appropriate than a model with localized bonds. The situation is comparable to the difference between diamond and elemental silicon.

The delocalization of σ-electrons is the reason for the conductivity of polysilanes. Pure polysilanes are insulators, but treatment with oxidizing agents renders them electrically conducting. Conductivities as high as $0.5\,S\,cm^{-1}$ were reached by doping crosslinked poly(silastyrene) with AsF_5. This value is in the range for → semiconductors. The conductivity of polysilanes arises from radical cations (a "hole") created in the polysilane chain by the oxidizing agent migrating through the polysilane.

When polysilanes absorb UV light, electrons are promoted from the σ valence band to the σ* conduction band. Because this transition is permitted, the electronic absorptions are intense, with extinction coefficients of 3×10^3 to 10^4 per Si–Si bond. As the number of silicon atoms in the chain increases, the σ/σ* energy gap becomes smaller, that is, there is a redshift for the σ→σ* transition. The absorption maximum of $Me_3Si–SiMe_3$ is at 220 nm, in poly(dialkylsilanes) it is shifted to 300–325 nm. Aryl substituents bonded to silicon cause a redshift of 20–30 nm per substituent due to the electronic interaction of π-orbitals with the orbitals of the Si–Si backbone (see Figure 3.2 for a wavelength scale).

The σ–σ* separation and thus λ_{max} also depends on the conformation of the polysilane chain. λ_{max} increases as the number of *trans*-Si–Si–Si–Si conformations increases. Therefore, many polysilanes show strong reversible UV → thermochromism both in solution or in the solid state. For example, poly(di-*n*-hexylsilane) undergoes a striking change in UV absorption from $\lambda_{max} = 372$ nm to 317 nm when the temperature is raised above 42 °C (an absorption maximum at 317 nm is also observed in solution at room temperature). The reason for this change is a profound conformational change at the transition temperature. The low-temperature form is highly crystalline with an all-*trans*, zigzag arrangement of the polymer chain. The high-temperature form is a hexagonal columnar liquid-crystalline phase (packing of the chains) in which the polymer chain adopts an all-*gauche* helix.

Both the extinction coefficient and the wavelength of the maximum absorption depend on the chain length in the all-*trans* conformation and reach a limit for about 40–50 silicon atoms (Figure 5.10). Therefore, polysilanes show spectral bleaching, which is the key to a number of applications. The reason is the photochemical scission of Si–Si bonds. Shorter oligosilanes are thus formed that absorb at lower wavelengths and have lower extinction coefficients.

The major reactions in photolysis reactions of polysilanes (Figure 5.11) are homolysis to give silyl radicals and silylene elimination. Bond homolysis occurs at all wavelengths absorbed by the polymer, while R_2Si is only eliminated at wavelengths <300 nm. Some chain scission with transfer of a substituent also takes place. The reactive intermediates generated by the photochemical reaction (silylenes, silyl radicals) undergo characteristic reactions, such as hydrogen

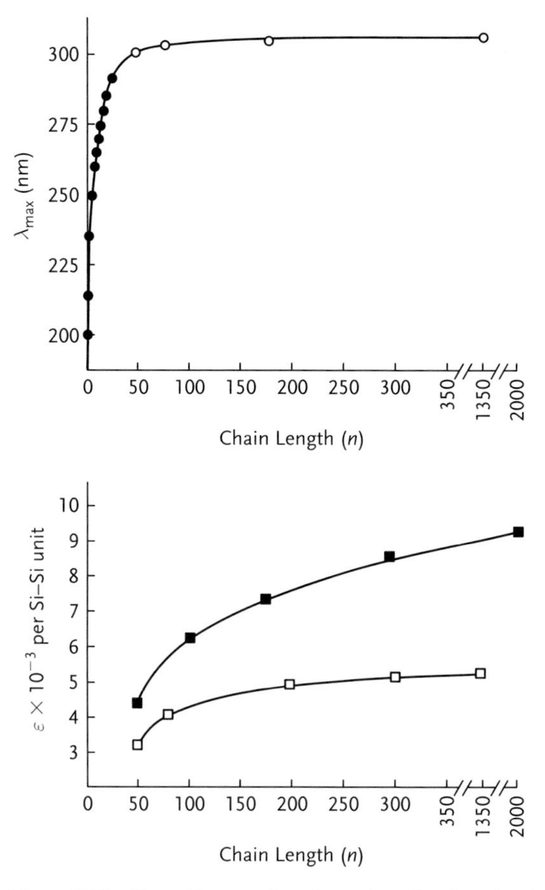

Figure 5.10 Absorption wavelength maxima (λ_{max}) and extinction coefficients (ε) as a function of the Si–Si chain length (n). • $(SiMe_2)_n$, ○ and □ (Si(Me)(n-dodecyl))$_n$, ■ $(SiMePh)_n$.

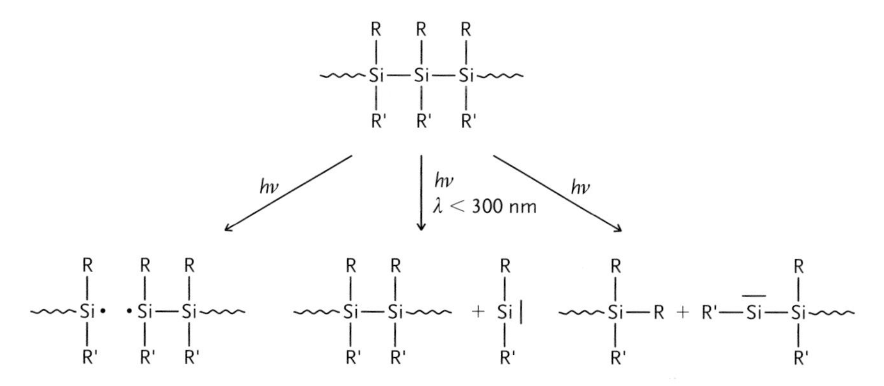

Figure 5.11 The most important photochemical reactions of polysilanes: homolysis (left), silylene extrusion (center), chain scission with transfer of a substituent (right).

abstraction, disproportionation, addition or insertion reactions. When 254-nm UV light is used, photolysis ceases at the disilane stage because UV radiation is no longer absorbed.

The photochemical degradation of polysilanes enables their use as photo-initiators in a variety of vinyl → polymerization reactions. The silyl radicals generated by scission of Si–Si bonds can add to C=C double bonds and start polymerization processes. Compared to conventional photoinitiators, the photo-initiation efficiency of polysilanes is lower by about a factor of ten. They are, nevertheless, interesting because they have the unique advantage that they are less susceptible to inhibition by oxygen than other photoinitiators. This makes industrial photoprocesses, such as photopolymerization of films and coatings, less expensive, because no protection from oxygen is required.

Another application of polysilanes is as → photoresists in microlithography. At present, visible light is used for microlithography (see Figure 3.2 for a wavelength scale). This limits the resolution to 0.5–1.5 μm due to diffraction effects. There is an increasing need for smaller resolutions on microprocessors, to increase the speed of operation, the device packing density and the number of functions that can reside on one single chip. This is only possible by using light of shorter wavelength in the structuring process. Polysilanes are well suited for this generation of positive → photoresists, since they are

- thermally and oxidatively stable, yet photochemically labile;
- strongly absorbing in the UV region, yet bleachable;
- soluble in organic solvents and thus can be coated as films;
- very resistant to etching under O_2 plasma conditions; and
- resistant to X-rays, electron beams, and so on.

A piezochromic behavior was observed for some polysilanes, which also originates from structural changes. For example, when pressure is applied to poly(di-*n*-hexylsilane) films at temperatures above 42 °C, λ_{max} changes reversibly from 317 to 372 nm as in the low-temperature conformation. Thus, the ordered *trans* structure is adopted when pressure is applied.

Polysilanes also show photoconductivity when irradiated with UV light, and are excellent charge-transport materials in electrophotography with high drift mobilities ($\sim 10^{-4}$ cm^2 V^{-1} s^{-1}). The activation energies associated with the charge hopping are very low for a polymeric material. Polysilanes are unique as charge-transport materials because the active sites are on the polymer backbone itself. In other electrophotographic materials, charged sites are on the substituents.

5.4.2
Preparation and Modification of Polysilanes

Dehalogenation of Diorganodichlorosilanes (Wurtz Coupling) Most polysilanes are synthesized by dehalogenation of diorganodichlorosilanes (R_2SiCl_2 or $RR'SiCl_2$) with finely divided sodium metal above its melting point (98 °C) in an inert, high-boiling solvent such as toluene, xylene, decane, diethyleneglycol dimethylether (diglyme), or mixtures of these [Eq. (5.30)]. This is also the technically applied method. Other reducing agents, such as potassium metal, Na/K → alloy or C_8K, have also been used, but the more reactive reagents often result in degradation of the polymer. The reaction is highly exothermic. Few functional groups can withstand such vigorous conditions; therefore, substituents are generally limited to alkyl, aryl, or silyl groups or other substituents that are intrinsically stable in a strongly reducing environment, such as fluoroalkyl or ferrocenyl.

$$RR'SiCl_2 \xrightarrow[T > 100\,°C]{Na/solvent} \left[-\overset{R'}{\underset{R}{Si}}- \right]_n + 2NaCl \qquad (5.30)$$

Cyclic oligomers (cyclosilanes) are the only products at equilibrium for the reasons discussed in Section 5.1. They cannot be converted to polymers. High molecular weight polymers are formed only when the reaction is kinetically controlled, but concomitant formation of cyclic oligomers cannot be avoided. The molecular-weight distributions are usually broad and → polymodal, and consist of a narrow low molecular weight fraction ($M_w < 10^3$), mostly cyclopentasilane and cyclohexasilane, a dominant intermediate fraction ($10^3 < M_w < 5 \times 10^4$) with a polydispersity index of about 10, and a lower high molecular weight fraction ($M_w \sim 10^6$) characterized by a polydispersity index of about 1.5. The oligomeric fraction is readily removed by solvent extraction, but the other two fractions are inseparable other than by fractionation. Their polymer yields depend upon the substituents on silicon and the reaction conditions, and are typically in the range of 10–50%. The molecular weight and the molecular-weight distribution also depend on the reaction conditions.

When mixtures of silanes are employed, either random or block-like copolymers are produced, depending on the relative reactivities of the chlorosilanes. In block-copolymers, the monomeric units are arranged in blocks (e.g., AAAABBBBAAAABBBB, etc.), in random copolymers their sequence is accidental (e.g., BAABABBAAABBABAB, etc.). Random copolymers are obtained for similar chlorosilanes, for example n-$BuMeSiCl_2$ and n-$HexMeSiCl_2$. If the chlorosilanes have rather different reaction rates, one will be consumed preferentially to generate block-like copolymers. For example, poly(silastyrene), $[PhSiMe]_n[Me_2Si]_m$, has a rather block-like structure.

The reactions leading to the formation of polysilanes are very complex. A simplified mechanism is shown in Figure 5.12. It more closely resembles a chain growth process initiated by the sodium surface than a classical condensation reaction. The reaction of chlorosilanes with sodium probably proceeds by initial formation of a silyl anion by two single-electron transfer steps via a silyl radical. The propagation step is the reaction of anion-terminated chains with dichlorosilane to add one R_2Si unit. This step is fast but nevertheless rate determining, and is very probably the major reaction for chain growth. The new terminating Si–Cl bond reacts again with

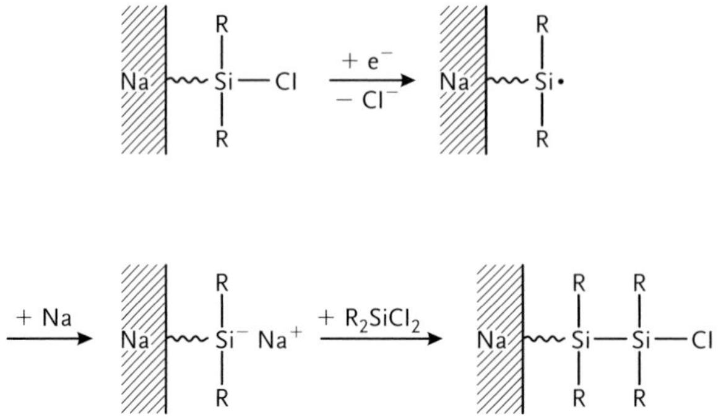

Figure 5.12 Mechanism for the dehalogenation of R_2SiCl_2 by sodium metal.

sodium, such that the polysilane chain grows away from the metal surface. The monomeric or polymeric radicals can also recombine.

In the presence of crown ethers or diglyme, the polymer yields are increased, and the molecular-weight distributions tend toward → monomodal as more polymer is formed within the intermediate M_w range. The ethers probably clean the sodium surface by solubilizing absorbed sodium salts and also complex the sodium cations of the $\equiv Si^- \ Na^+$ ion pairs. Ultrasound irradiation greatly accelerates the → polycondensation at temperatures lower than the melting point of the alkali metal. The reaction can then be carried out at about 60 °C. Sonification offers another possibility for achieving activation and regeneration of the metal surface than thermal activation.

Other intermediates than silyl radicals may also be involved. The balance among competing mechanisms may depend on solvent, temperature, sodium surface area, additives, and the nature of the substituents at silicon. It is therefore necessary to optimize the reaction conditions for each particular polysilane. Despite these drawbacks, this route is still the most practical.

Considering that alkali metals are expensive and relatively dangerous to handle on an industrial scale, several electroreductive routes have been proposed. The best results were obtained using a sacrificial magnesium or aluminum anode in an undivided cell at room temperature. The mechanism is very similar to that in Figure 5.12. The silyl anion is generated at the cathode by reduction of a Si–Cl group ($\equiv$Si–Cl + $2e^- \rightarrow \equiv Si^- + Cl^-$) and then reacts in solution with another chlorosilane. At the anode, the metal is oxidized.

Anionic Polymerization of Masked Disilenes In contrast to olefins, disilenes $R_2Si{=}SiR_2$ are only stable at room temperature if they are kinetically stabilized by bulky substituents. This excludes the possibility of preparing polysilanes from stable disilenes by → polymerization.

High molecular weight polymers are produced by reaction of alkyl anions with disilabicyclooctadienes, as for example shown in Eq. (5.31). These compounds can be considered to contain masked disilenes. However, the mechanism of polymerization does not involve disilenes. Instead, regiospecific attack of the alkyl anion occurs in the initiation step. The reaction then proceeds as a living anionic polymerization with elimination of the disilene fragment as a new silyl anion that can continue the reaction chain. This route is useful for making alternating polysilane copolymers or stereoregular polysilanes, and the first fully ordered polysilane of the ABAB type was obtained in this way. Ordered polysilanes may show unusual electronic behavior, and are hence of theoretical and perhaps practical interest.

$$Cl{-}SiMe_2{-}SiR_2{-}Cl + [C_6H_5{-}C_6H_5]^{\overline{\cdot}}\,Li^+ \xrightarrow[-78\,°C]{THF}$$

$$\text{(disilane-bridged Ph-substituted bicyclic adduct)} \xrightarrow[(2)\ +\ H^+]{(1)\ +\ LiR'} Ph{-}Ph + R'{-}[SiMe_2{-}SiR_2]_n{-}H \qquad (5.31)$$

Ring-Opening Polymerization Only *strained* cyclosilane rings undergo ROP under kinetically controlled conditions. The cyclotetrasilane $(Me_2Si)_4$ polymerizes to $[Me_2Si]_n$ on long-term standing. Poly(methylphenylsilane) has been synthesized by ROP of the strained cyclosilane $(PhMeSi)_4$ on reaction with elemental potassium, butyllithium, or silylcuprates. The ROP route allows for the preparation of well-defined polysilanes with controlled molecular weights and microstructures, low polydispersities, and defect-free structures.

Dehydrogenative Coupling of Diorganosilanes Metal-catalyzed dehydrogenation of diorganosilanes is an attractive alternative for polysilane synthesis [Eq. (5.32)]. Due to the mild reaction conditions, a greater variety of organic groups could probably be introduced directly. However, a sufficiently efficient catalyst is still lacking.

$$RR'SiH_2 \xrightarrow{catalyst} [SiRR']_n + H_2 \qquad (5.32)$$

Currently, the best catalysts appear to be dialkyltitanocenes and dialkylzirconocenes, Cp_2MR_2 (M=Ti, Zr). Truly polymeric compounds are only formed from primary silanes, $RSiH_3$. With diorganosilanes R_2SiH_2, the products are oligomers of different size, with a maximum chain length of about 20 silicon atoms. The polymerization is accelerated by the addition of olefins, which are hydrogenated and therefore act as "H_2 sponges." The main problem with the currently available catalysts, which operate by a σ-bond → metathesis mechanism, is that the metal center activates not only Si–H bonds, but also the Si–Si bonds of the oligomer; hence the metal complex also catalyzes depolymerization reactions. Another problem is the formation of cyclic products.

Redistribution Reactions Polysilanes can be prepared by the catalytic redistribution of Si–Si and Si–Cl bonds of disilanes at 250 °C as shown schematically in Eq. (5.33).

$$\text{Cl}\!\left[\begin{array}{c}\text{Me}\\ |\\ \text{Si}\\ |\\ \text{Cl}\end{array}\right]_n\!\text{Cl} + \text{Cl}-\begin{array}{c}\text{Me}\\ |\\ \text{Si}\\ |\\ \text{Cl}\end{array}-\begin{array}{c}\text{Me}\\ |\\ \text{Si}\\ |\\ \text{Cl}\end{array}-\text{Cl} \xrightarrow{\text{catalyst}} \text{Cl}\!\left[\begin{array}{c}\text{Me}\\ |\\ \text{Si}\\ |\\ \text{Cl}\end{array}\right]_{n+1}\!\text{Cl} + \text{MeSiCl}_3 \quad (5.33)$$

The most effective catalyst seems to be hexamethylphosphoric acid triamide (HMPA). This synthetic route uses mixtures of disilane byproducts from the "direct process" for the synthesis of methylchlorosilanes (Section 5.2.3.1). The approximate composition of this mixture is 55% $MeCl_2Si–SiCl_2Me$, 35% $Me_2ClSi–SiMeCl_2$, and 10% $Me_2ClSi–SiMe_2Cl$. Monosilanes are distilled off as well as $Me_2ClSi–SiMe_2Cl$, which does not react. The resulting yellow pyrophoric polymer is highly branched and still contains some chlorine.

Chemical Modification of Polysilanes The presence of functional groups in polysilanes is limited by the vigorous conditions of the usually employed sodium condensation reaction [Eq. (5.30)]. Any functions that react with molten sodium metal must therefore be introduced after the polymer is synthesized. Since the polysilane skeleton is rather sensitive towards acids or bases, the modification reactions must be well designed. Some examples are shown in Table 5.2. The thus-introduced functional groups can be further converted to other groups by conventional reactions. For example, chloro or triflate substituents on silicon are highly reactive and can be replaced by a variety of nucleophiles without disturbing the Si–Si bond to give substituted polysilanes.

Table 5.2 Examples of the modification of substituents in polysilanes.

$\equiv$Si–X–CH=CH_2	HBr/BBr_3	→ $\equiv$Si–X–CH_2CH_2Br
$\equiv$Si–X–$C_6H_4OSiMe_3$	MeOH	→ $\equiv$Si–X–C_6H_4OH
$\equiv$Si–Ph	$HCl/AlCl_3/C_6H_6$	→ $\equiv$Si–Cl
$\equiv$Si–Ph	CF_3SO_3H/CH_2Cl_2	→ $\equiv$Si–OSO_2CF_3
$\equiv$Si–H	CH_2=CHR/AIBN	→ $\equiv$Si–CH_2CH_2R

Crosslinking of Polysilanes → Crosslinking is essential if the polysilanes are used as precursors to SiC ceramics (Section 5.5). Without crosslinking, most of the polymer is volatilized before rearrangement to polycarbosilanes or thermolysis to SiC can take place. Common methods for the crosslinking of polysilanes are as follows:

- In the presence of vinyl or other alkenyl substituents, photolysis leads to crosslinking, which dominates over photochemical degradation of the polysilane chain. The crosslinking probably results by addition of initially formed silyl radicals to the C=C double bonds on neighboring chains.
- Related reactions can occur when organic molecules with two or more unsaturated groups are mixed with the polysilane. Typical examples are tetravinylsilane, $Si(CH{=}CH)_4$, or the hexavinyl-substituted compounds $(CH_2{=}CH)_3Si(CH_2)_nSi(CH{=}CH_2)_3$. The radicals necessary to cause crosslinking can be formed by photolysis of the polysilane or by heating the mixture with a free-radical initiator such as AIBN.
- Polysilanes that contain Si–H bonds can be crosslinked by hydrosilylation reactions by, for example, addition of trivinylphenylsilane in the presence of a catalyst.

5.5 Polycarbosilanes

Polycarbosilanes are a class of polymers with carbon and silicon atoms forming the backbone. The sequence of atoms in the industrially important carbosilanes is –Si–C–Si–C–. Polycarbosilanes are used exclusively as preceramic polymers for SiC; technical applications as polymeric materials are currently not at hand. However, new preparation methods and polycarbosilanes with other sequences of the Si and C atoms, such as $–Si–(C)_x–Si–$, $–Si–(C{=}C)_xSi–$, –Si–aryl–Si, and so on may change this situation in the future. Copolymers from polysilanes and organic polymers are not considered polycarbosilanes.

5.5.1 SiC Fibers from Polycarbosilanes (Yajima Process)

The preparation and properties of SiC, one of the most important high-performance non-oxide ceramic materials, have been discussed in Section 2.1.3. The traditional routes to SiC ceramic powders are not suitable for making ceramic films (at reasonable temperatures) or fibers.

SiC filaments are technically used in ceramic (CMC), metal (MMC), glass or polymer–matrix (PMC) high-performance → composites. Continuous-fiber reinforced materials have the advantage that they do not fail catastrophically, because after matrix failure, the fiber can still support a load. SiC filaments are particularly suitable for high-temperature use by incorporating them in BN, B_4C, AlN, SiC, Si_3N_4, TiC, TiO_2, TiB_2, carbon, and so on. For example, SiC fiber-reinforced SiC composites are suitable for structural components of motors or turbines. SiC filaments have also been used to reinforce a variety of organic polymers. For example, epoxy/SiC composites possess compressive strength twice that of epoxy/graphite. SiC fiber–Al matrix composites may serve as examples of MMCs. They are lightweight, high-strength materials used for components in aircrafts, helicopters, missiles, bridge structures, and so on.

SiC fibers are commercially available as Nicalon®, Tyranno® or Sylramic® fibers. SiC fibers of 10–15 μm diameter exhibit tensile strengths of ~2.0–3.4 GPa (depending on the diameter and the curing process), elastic moduli of 200–400 GPa and densities of 2.5–3.1 $g\,cm^{-3}$, which is less than for the bulk ceramic material. By comparison, SiC single crystals of 0.3–1.0 μm diameter (density ~3.25 $g\,cm^{-3}$ depending on the polymorph) exhibit tensile strengths of ~8 GPa and elastic moduli of ~580 GPa. Above 1200 °C, the mechanical strength of the SiC fibers is lost. Solid-state reactions between the oxygen introduced during the curing step (see below) and SiC generate gaseous CO and SiO. The gas evolution causes significant damage. Furthermore, rapid crystallization of β-SiC occurs above 1200 °C.

The properties of ceramic fibers are greatly influenced by the composition and structure of the preceramic polymer. The requirements for an ideal preceramic polymer have been discussed in Section 5.1. Polymers for fiber spinning must additionally be adjusted to the requirements of the fiber-making process (Figure 5.1.3).

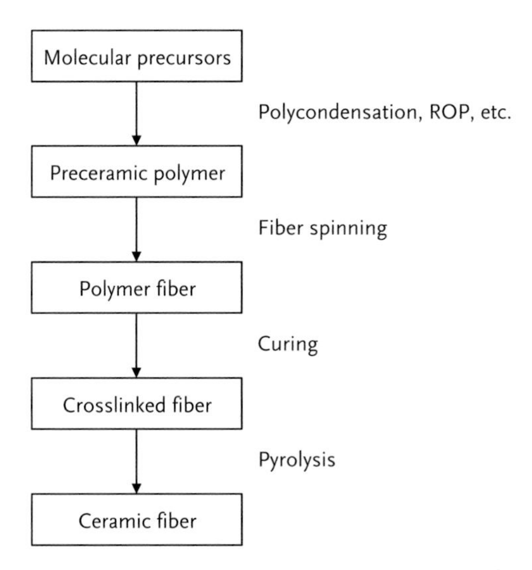

Figure 5.13 Processing steps in the preparation of ceramic fibers.

5.5.1.1 **Preparation of SiC Fibers**

This is carried out as a step-wise process:

- **Step 1**: *Spinning of the green fiber*. Spinning of SiC fibers is typically carried out by extruding the melted polymer through a spinneret ("melt spinning"), but dry spinning is also possible. In the latter process, the solution of the polymer is extruded into a heated drying chamber where the solvent is volatilized, leaving the solid polymer fiber. The polymer viscosity should exhibit non-Newtonian behavior during spinning, that is, it should be shear dependent. This behavior is obtained when the particles in the liquid strongly interact with each other. A compromise must be found between the rheological properties of the polymer and the requirements for obtaining high → ceramic yields. The viscosity should be sufficiently high that the green fiber retains its shape and supports itself. If the polymer is too highly crosslinked, the ceramic yield would probably be high, but melt spinning will not be possible.

- **Step 2**: *Curing.* The → green polymer fibers need to be cured (crosslinked) to render them infusible, so that the fiber form is retained on pyrolysis. The polymer therefore must have some latent reactivity for crosslinking.
- **Step 3**: *Pyrolysis.* After spinning, the polymer is pyrolyzed to give the desired ceramic material. The pyrolysis chemistry and the reaction conditions must be carefully controlled to ensure removal of the gaseous products and uniform densification. The polycarbosilane fibers are converted to β-SiC by firing at 800 –1400 °C in N_2 atmosphere. The weight loss during this process is 20–40%, and the volume decrease is 60–80%.

5.5.2 Chemical Issues of Polymer Preparation, Curing, and Pyrolysis

5.5.2.1 Polymer Preparation

Kumada rearrangement When poly(dimethylsilane) is heated under argon, the rearrangement shown in Eq. (5.34) ("Kumada rearrangement") takes place. The polycarbosilanes isolated after removal of volatile compounds are glassy resins that are meltable and soluble in organic solvents.

$$\left[\mathrm{Si(Me)_2} \right]_n \xrightarrow{450\,^\circ\mathrm{C}} \left[\mathrm{Si(H)(Me)}-\mathrm{CH_2} \right]_n \tag{5.34}$$

In reality, the structure of the obtained polycarbosilane is more complex. In particular, there are structural elements other than the $-Si(H)Me-CH_2-$ units in the idealized formula. A ladder-like structure with rings and chains was proposed from spectroscopic investigations, as shown schematically in Figure 5.14. There are three to four branchings per 10 silicon atoms. However, the carbon and silicon atoms are always alternating.

Considerable effort has been invested in optimizing the Kumada rearrangement, because this is a critical step in the generation of polycarbosilanes with the correct processing properties for fiber spinning. In the most common method, poly(dimethylsilane) is heated to 320 °C in flowing argon, melted, refluxed for 5 h,

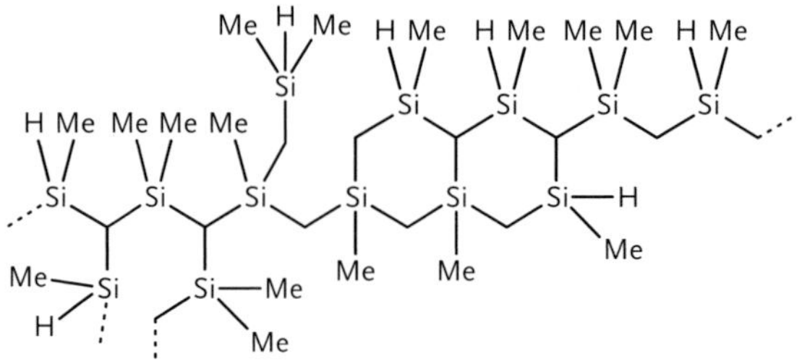

Figure 5.14 A structural model for the polycarbosilane obtained by Kumada rearrangement of poly(dimethylsilane).

and then heated to 470 °C to remove volatile materials. The resulting polycarbosilane is obtained in about 55% yield and has an average molecular weight $M_n = 1500$. Total conversion should lead to poly(silapropylene), $[Si(H)MeCH_2]_n$. However, the typical chemical analysis for Yajima polycarbosilanes prepared from poly(dimethylsilane) by the Yajima process is around $SiC_{1.8}H_{3.7}$ instead of SiC_2H_6.

The primary reaction in the Kumada rearrangement probably involves pyrolytic cleavage of Si–Si bonds. The thus-formed silyl radicals give $CH_2{=}SiMe$- and Me_2HSi-terminated fragments that subsequently combine to form $Si–CH_2–Si$ linkages. Alternative mechanisms involving the initial formation of silenes and silylenes are also consistent with the experimental results.

Lewis-acid catalysts improve the conversion of polysilanes to polycarbosilanes. The commercially used catalyst is poly(borodiphenylsiloxane) $[B(OSiPh_2O–)_{3/2}]_n$, prepared by condensation of boric acid, H_3BO_3, and Ph_2SiCl_2. Other boron compounds such as $B(OR)_3$ or $B(NEt_2)_3$ were equally successful. Polymers are typically obtained in about 60% yield at ambient pressure and ~350 °C by this modification of the process. Spectroscopic investigations indicate that the polymers also contain Si–Si bonds and are more branched than the polymers prepared without the use of the catalyst. A disadvantage is that oxygen is already introduced in this step.

C-Cl/Si-Cl dehalogenation There are several approaches to obtain the desired Si/C sequences in carbosilane chains by dehalogenative coupling of C–Cl and Si–Cl groups. The most obvious precursors are chloromethyl(chloro)silanes, $ClCH_2SiR_2Cl$, because they possess the desired CH_2–Si unit. Reaction of these precursors with elemental magnesium either gives the cyclic compounds $[SiR_2CH_2]_n$ ($n = 2$–4) (mostly the 1,3-disilacyclobutanes [$n = 2$], which can be ring-opened; see below) or polymers, depending on the electronic and steric properties of the substituents at silicon and the reaction conditions. For example, while $ClCH_2SiMeCl_2$ only gives oligomeric and polymeric products, a high portion of the disilacyclobutane is obtained when $ClCH_2Si(OR)Cl_2$ is similarly reacted with magnesium. A 53% yield of the polymer $[SiMe_2CH_2]_n$ and only 7% of cyclic $[SiMe_2CH_2]_2$ is obtained when $ClCH_2SiMe_2Cl$ is added to a suspension of Mg in THF, while the inverse addition, that is, addition of Mg to the silane solution, gives 13% of $[SiMe_2CH_2]_n$ and 69% of cyclic compounds.

The most direct route to carbosilanes is the coupling of CH_2X_2 and $R_2SiX'_2$. Several polycarbosilanes with moderate molecular weight were prepared by this route by condensation of dihalomethane derivatives, mostly CH_2Br_2, with dichlorosilanes [Eq. (5.35)]. The general approach can also be used to prepare polycarbosilanes with other Si/C sequences, as exemplarily shown in Eqs. (5.36) and (5.37).

$$nRR'SiCl_2 + nCH_2Br_2 + 4nNa \xrightarrow[-NaBr]{-NaCl} \left[Si(R)(R') - CH_2 \right]_n \quad (5.35)$$

$$Li\left[C\equiv C\right]_x Li + Cl-Si(R')(R)-Cl \xrightarrow{-LiCl} \left[Si(R')(R) \left(C\equiv C\right)_x \right]_n \quad (5.36)$$

$x = 1, 2$

$$Li-C\equiv C-C\equiv C-Li + ClMe_2Si-SiMe_2Cl \longrightarrow \left[SiMe_2-SiMe_2-C\equiv C-C\equiv C \right]_n \quad (5.37)$$

Ring-opening polymerization (ROP) Several types of polycarbosilanes can be prepared by metal-catalyzed ROP of 1,3-disilacyclobutane derivatives, [Eq. (5.38)]. Hexachloroplatinic acid, H_2PtCl_6, is a typical catalyst. This route provides strictly linear polycarbosilanes if there are no substituents at the silicon atoms that can undergo → crosslinking reactions (such as R=H or vinyl).

$$\text{cyclo-}[Si(R)(R')CH_2]_2 \xrightarrow{\text{catalyst}} \left[Si(R)(R') - CH_2 \right]_n \quad (5.38)$$

Linear $[Si(H)MeCH_2]_n$, prepared by ROP of 1,3-dichloro-1,3-dimethyl-1,3-disilacyclobutane (R=Me, R′=Cl in Eq. (5.39)) followed by $LiAlH_4$ reduction of the chloro-substituted polymer, $[Si(Cl)MeCH_2]_n$, has properties different from the polycarbosilane obtained by the Kumada rearrangement. This shows again that the latter has a different structure, as discussed above.

Unbranched $[SiR_2CH_2]_n$ is chemically related to polyolefins $[CR_2CH_2]_n$ and polysiloxanes $[SiR_2O]_n$ which have similar glass-transition temperatures. The asymmetrically substituted polymers $[Si(Me)(n\text{-alkyl})CH_2]_n$ exhibit very low T_gs down to −91 °C.

Chlorine substituents in the polymer can be replaced by other groups such as OEt or saturated and unsaturated organic groups. Likewise, Si–H groups can be added to olefins, which is another approach to introduce a specific side chain into the polycarbosilane.

ROP of 1,3-disilacyclobutanes gives polycarbosilanes with alternating silicon and carbon atoms in the polymer chain. Polycarbosilanes with other Si/C sequences can be prepared by anionic ROP of silacyclobutanes (giving $[Si–C–C–C]_n$ sequences) or silacyclopent-3-enes (giving $[Si–C–C{=}C–C]_n$ sequences).

Hydrosilylation and metathesis of alkenylsilanes Hydrosilylation allows another direct route to polycarbosilanes. For example, vinyldichlorosilane, $SiH(CH{=}CH_2)Cl_2$, contains both a Si–H and an olefinic group. Thus, the Si–H group of one molecule can add to the vinyl group of the next, and so on to give a polycarbosilane. The obtained chlorosilane polymer is then reduced by $LiAlH_4$, [Eq. (5.39)]. Hydrosilylation mainly affords linear $SiCH_2CH_2$ units, and only a minor portion of branched $–SiCH(Me)–$ units.

$$\text{H–SiCl}_2\text{–CH=CH}_2 \xrightarrow{\text{cat.}} \text{H}{+}\text{SiCl}_2\text{–CH}_2\text{CH}_2{+}_n\text{SiCl}_2\text{–CH=CH}_2 \xrightarrow{\text{LiAlH}_4} \text{H}{+}\text{SiH}_2\text{–CH}_2\text{CH}_2{+}_n\text{SiH}_2\text{–CH=CH}_2 \quad (5.39)$$

Allyl- or ethynyl-substituted silanes, $R_2SiH(CH_2CH{=}CH_2)$ or $R_2SiH(C{\equiv}CR')$ can be similarly polymerized by hydrosilylation, giving polymers with other Si/C sequences.

Other organic → polymerization reactions also allow the formation of polycarbosilanes. For example, high-molecular weight unsaturated polycarbosilanes were prepared by metal-complex-catalyzed → metathesis polymerization [Eq. (5.40)]. Any other organosilicon compound with two unsaturated substituents can be similarly used.

$$\text{H}_2\text{C=CH–SiMe}_2\text{–CH=CH}_2 \xrightarrow{\text{cat.}} {+}\text{CH}_2\text{–SiMe}_2\text{–CH}_2{+}_n + n\text{C}_2\text{H}_4 \quad (5.40)$$

5.5.2.2 Crosslinking

The polycarbosilanes obtained by one of the above-mentioned methods must be cured to give reasonable → ceramic yields of SiC upon pyrolysis. For example, the linear polycarbosilanes obtained by ROP of substituted disilacyclobutanes decompose with very low ceramic yields if they contain no latent functionality (Si–H or $Si–CH{=}CH_2$ groups) because cyclic or linear, low-boiling oligomers are formed that distill off. The polycarbosilanes obtained by the Kumada rearrangement, on the other hand, are of sufficiently low molecular weight ($M_n \sim 1–2 \times 10^3$) that they can be melt spun. However, the → green fiber must be cured. The commercially available SiC fibers are cured by heating the spun polycarbosilane fibers at ~200 °C in air for 30 min. This converts surface Si–H groups into Si–O–Si groups. Pyrolysis of the air-cured fibers gives → ceramic yields of 80–85%, compared to 60–70% without curing. Since significant amounts of oxygen are introduced into the fiber material, several alternative approaches were investigated.

- High molecular weight polycarbosilanes were prepared by thermal decomposition of poly(dimethylsilane) or poly(silastyrene) in an autoclave instead of the two-step process described before. They do not melt, but are soluble and can be dry spun. The thus-obtained fibers offer much better high-temperature stability as a consequence of their low oxygen content. The effect of the thermal pretreatment ("thermosetting") was also demonstrated for the completely linear and → atactic $[Si(H)MeCH_2]_n$ obtained by ROP of 1,3-dichloro-1,3-dimethyl-1,3-disilacyclobutane followed by $LiAlH_4$ reduction, [Eq. (5.38)]. The ceramic yield of the untreated polymer is quite low (20%). However, a pretreatment under nitrogen at 400 °C results in dehydrocoupling of Si–H bonds. Due to this → crosslinking reaction, the ceramic yield upon pyrolysis is improved to 66%.
- A variation of the Nicalon process consists in heating the polycarbosilane with $Ti(OBu)_4$. The titanium compound provides Ti–O–Si crosslinks. Ceramic fibers with higher ceramic yield (70%) are obtained after pyrolysis at 1200 °C in N_2 atmosphere. They are marketed under the trade name Tyranno.
- The mechanical properties of the fibers at high temperatures were improved by γ-ray, electron or neutron radiation instead of curing by oxidation due to the very low oxygen content. Such fibers retain a tensile strength of >2 GPa after long-time thermal exposure in argon at 1800 °C. However, this process makes the fibers more expensive.
- The incorporation of vinyl and Si–H groups into the starting polysilane allows crosslinking by hydrosilylation reactions. Yajima polycarbosilane, which contains Si–H bonds, can be also rendered infusible by hydrosilylation with appropriate amounts of an unsaturated compound prior to pyrolysis, (see Table 5.2). Low oxygen contents are obtained with this crosslinking method, but the carbon content is increased.

5.5.2.3 **Pyrolysis**

Pyrolysis of polycarbosilanes (obtained by the Kumada rearrangement) in an argon atmosphere proceeds in several steps. Low molecular weight carbosilanes distill off below ~350 °C. Between ~350 °C and 500 °C, the molecular weight of the polycarbosilane increases by formation of additional Si–C–Si linkages. The thermolysis reactions progressively end as the pyrolysis proceeds. In the next stage, from ~500 to 900 °C, the main transformation into an inorganic material occurs, with the loss of hydrogen and methane. As Si–H bonds are more reactive than C–H bonds, the SiC_4 tetrahedra found in SiC are formed first. Tetrasubstitution of the carbon atoms, that is, final conversion of $\equiv Si–CH_3$, $\equiv Si–CH_2–Si\equiv$ or $HC(Si\equiv)_3$ units in $C(Si\equiv)_4$ nodes, occurs at higher temperatures. At 900 °C the material is amorphous and can be described as a hydrogenated Si_xC_y ($y > x$). Crystallization of β-SiC starts at 1000–1200 °C. H_2 evolution continues in this temperature range, and represents the final stage in conversion of the polycarbosilane to the inorganic SiC network. SiC fibers obtained from the titanium-modified polycarbosilane remain amorphous on heating to 1200 °C and crystallize at higher temperatures into a mixture of β-SiC and TiC.

The final materials contain significant portions of oxygen and free carbon. Nicalon fibers have an overall composition of $Si_{0.58}C_{0.31}O_{0.11}$. About 40–50% of the silicon atoms are surrounded by four carbon atoms (as in crystalline SiC), about 6% by four oxygen atoms (as in crystalline SiO_2), and the remaining fraction is substituted by both oxygen and carbon atoms.

If the pyrolytic degradation of $[Si(H)MeCH_2]_n$ were to proceed optimally, then the overall reaction would produce $SiC + H_2 + CH_4$. However, the control of the Si : C ratio in the resulting fibers is still a crucial issue. They can contain an excess of elemental carbon or (less often) elemental silicon. For example, Nicalon fibers are black due to the presence of graphite crystallites. The nature of the pyrolysis atmosphere, and the reaction conditions in general, plays a very important role in the loss of carbon. For example, a hydrogen-rich atmosphere favors the formation of methane and thus decreases the carbon content of the ceramic material.

5.6
Polysilazanes and Polyborosilazanes

Polysilazanes are more difficult to handle, mainly due to their high reactivity towards water, particularly at low pH, and therefore have not achieved the same degree of importance. The main reason for developing polysilazanes is that they are preceramic polymers for silicon nitride (Si_3N_4).

Conventional routes for the preparation of Si_3N_4 ceramic powders have been discussed in Chapter 2, and the production of Si_3N_4 films by CVD in Section 3.2.5. Preceramic routes were mainly developed for the production of strong, high-modulus, oxidation-resistant continuous Si_3N_4 fibers that could be used in ceramic–matrix composites (CMCs).

Multicomponent nitride and carbide ceramics are highly interesting with respect to the "tailoring" of materials properties. However, they are difficult to synthesize by traditional high-temperature routes. The preceramic polymer route offers great advantages for the synthesis of metastable, single-phase multi-component systems because it operates at comparatively low temperatures. Multi-component systems consisting of main-group elements connected by covalent bonds (as in silicon and boron nitrides and carbides, for example) have high onset temperatures of crystallization and exhibit a high thermal stability. Below the crystallization temperature, these systems are amorphous with rigid, crosslinked networks. Although these solids are not in thermodynamic equilibrium, materials of high kinetic stability are accessible.

For example, thermal stability and mechanical performance of amorphous Si_3N_4 is limited by crystallization, which starts above 1000 °C (in the presence of elemental silicon the crystallization temperature is even lower). In carbon-containing systems the onset of crystallization is shifted to higher temperatures. Therefore, there are considerable efforts to prepare ternary Si/C/N (silicon carbonitride) from polycarbosilazanes with both Si–N–Si and Si–C–Si linkages in the main chain. Addition of boron to SiC, Si_3N_4 or Si/C/N greatly enhances ceramic properties, including a reduced crystallinity and an improved thermal and oxidative stability. Si/B/C/N ceramic materials, especially fibers, prepared from polyborosilazanes, have recently gained much interest due to their superior properties.

5.6.1 Preparation of Polysilazanes and Polyborosilazanes

Reaction of chlorosilanes with amines The most common way to form Si–N bonds is the reaction of ammonia or amines (ammonolysis or aminolysis) with halogenosilanes. Tetrachlorosilane, $SiCl_4$, reacts with ammonia under various conditions (in the liquid or gas phase) to give hydrolytically sensitive polymeric silicon diimide "$Si(NH)_2$" of unknown structure. This compound gives pure Si_3N_4 after pyrolysis at 1250 °C. Although the ammonolysis of $SiCl_4$ is useful for the production of Si_3N_4 powders or in the CVD process (Section 3.2.5), it is not suitable for obtaining fusible or soluble preceramic polymers.

Reaction of dihalogenosilanes R_2SiCl_2 with ammonia or primary amines gives oligomeric or polymeric silazanes, [Eq. (5.41)]. As discussed in the previous sections, the proportion of cyclic products (mainly cyclotrisilazanes and cyclotetrasilazanes) is largely determined by the steric properties of the substituents, both at silicon and nitrogen sites. The higher the steric bulk, the larger is the proportion of cyclic products.

$$R_2SiCl_2 + 3R'NH_2 \longrightarrow \left[-\underset{R}{\overset{R}{Si}}-\underset{R'}{N}- \right]_n + 2[R'NH_3]Cl \quad (5.41)$$

Polysilazanes with various substituents (including vinyl groups for → crosslinking) were prepared as preceramic polymers for silicon nitride or carbonitride ceramics. To obtain high → ceramic yields, methyl substituents are the second-best choice after hydrogen. Thus, methylated polysilazanes obtained from the ammonolysis reactions of $Me(H)SiCl_2$ (which is cheap and non-hazardous), Me_2SiCl_2, or $MeSiCl_3$, or from the reaction of H_2SiCl_2 with $MeNH_2$ were investigated for their use as preceramic polymers, especially for the production of ceramic fibers.

Simple polysilazanes (Si : N ratio = 1) are nitrogen-deficient with respect to the Si/N stoichiometry of Si_3N_4. Therefore, chlorosilanes were also reacted with hydrazine instead of ammonia or amines to increase the nitrogen content.

Polyborosilazanes were prepared by a related approach. When the mixed silicon/boron chloride derivate $Cl_3Si\text{-}NMe\text{-}BCl_2$ is reacted with methylamine [Eq. (5.42)], $(MeHN)_3Si\text{-}NMe\text{-}B(NHMe)_2$ is initially formed. As the temperature is increased, this compound transforms into a polymer of the composition $[SiN(Me)B](NMe)_{x/2}(NHMe)_{(5-2x)/2}$ by $MeNH_2$ elimination. The degree of condensation (x) mainly depends on the time, temperature and pressure. Controlling x allows adjusting the viscosity of the polymer, as the viscosity increases as condensation progresses.

$$\mathrm{Cl_3Si{-}N(Me){-}BCl_2} + \mathrm{MeNH_2} \xrightarrow{-\,[\mathrm{MeNH_3}]\mathrm{Cl}} \mathrm{(MeHN)_3Si{-}N(Me){-}B(NHMe)_2} \xrightarrow[\;T\uparrow\;]{-\,\mathrm{MeNH_2}} [\mathrm{SiN(Me)B}](\mathrm{NMe})_{x/2}(\mathrm{NHMe})_{(5-2x)/2} \tag{5.42}$$

Ammonolysis or aminolysis of chlorosilanes results in large amounts of ammonium chlorides [Eq. (5.41)] that are difficult to remove. This byproduct is undesirable since it introduces chlorine in the preceramic polymers and acts as a catalyst in the splitting of Si–N bonds. This problem has been overcome by the Si–Cl/Si–N → metathesis reaction using the readily available hexamethyldisilazane [Eq. (5.43)]. Volatile trimethylchlorosilane is the main byproduct. This reaction can be applied for the synthesis of a wide range of polymers.

$$\mathrm{R_2SiCl_2} + \mathrm{Me_3Si{-}NH{-}SiMe_3} \longrightarrow \left[\mathrm{SiR_2{-}NH}\right]_n + 2\mathrm{Me_3SiCl} \tag{5.43}$$

N-H/Si-H dehydrocondensation The formation of Si–N bonds by hydrogen elimination from Si–H and N–H groups is promoted by several catalysts. Various monomeric hydrogenosilanes R_2SiH_2 or $RSiH_3$ were reacted catalytically with amines and ammonia. However, the hydrogenosilanes particularly interesting for the preparation of preceramic polymers, $MeSiH_3$ or SiH_4, are difficult to handle. Therefore, it is easier to prepare oligomers by ammonolysis and modify them catalytically in order to obtain a suitable ceramic precursor.

Dehydrocondensation reactions are also catalyzed by strong bases such as KH, although mainly four-membered rings are obtained, [Eq. (5.44)]. This method is very convenient for polymerizing silazanes possessing –SiH–NH– groups. For example, ammonolysis of $Me(H)SiCl_2$ gives an oil that mainly contains six- and eight-membered silazane rings $[Si(H)Me–NH]_n$ ($n = 3$ or 4). KH-catalyzed → crosslinking of the cyclosilazanes gives a polymer of the approximate composition $[Si(H)Me–NH]_{0.4}[SiMeN]_{0.6}$, which is one of the more important starting polymers for the production of Si_3N_4 or Si/C/N ceramics.

$$2\ \text{—Si(H)—N(H)—} \xrightarrow{\text{base}} \text{>Si(μ-N)}_2\text{Si<} + H_2 \qquad (5.44)$$

ROP of cyclosilazanes High molecular weight linear polysilazanes can be prepared by anionic ROP of four- or six-membered cyclosilazanes, in some cases also by cationic or thermally induced ROP. The reactivity for the ROP process decreases with the steric bulk of the substituents. ROP of cyclosilazanes can also be promoted by transition metal complexes. Note the analogy of this reaction, [Eq. (5.45)], with that of Eq. (5.14).

$$n\ (Me_2SiNH)_4 + Me_3Si{-}NH{-}SiMe_3 \longrightarrow Me_3Si{-}NH{-}\left[SiMe_2{-}NH\right]_{4n}{-}SiMe_3 \qquad (5.45)$$

Transamination Transamination reactions play a very important role in the formation of the $N(Si{\equiv})_3$ nodes during pyrolysis of the polysilazanes (see below) or in CVD processes (see Section 3.2.5). This reaction can also be employed for the preparation of preceramic polymers with the advantage that no chlorosilanes are required. Tris(dimethylamino)silane, $HSi(NMe_2)_3$, is prepared from elemental silicon and dimethylamine under Müller–Rochow conditions (see Section 5.2.3.1). In the presence of ammonia or primary amines and a catalyst, this molecular precursor readily undergoes transamination reactions [Eq. (5.46)]. The intermediate aminosilanes with Si–NHR or Si–NH_2 groups can undergo condensation reactions to give Si–NR–Si or Si–NH–Si linkages by elimination of RNH_2 or NH_3 (similar to the formation of Si–O–Si bonds by water elimination from Si–OH groups).

$$(Me_2N)_3SiH + 2RNH_2 \xrightarrow{\text{catalyst}} \left[Si(NHR)(H){-}N(R)\right]_n + 3Me_2NH \qquad (5.46)$$

When tris(*N*-methylamino)methylsilanes, $CH_3Si(NHMe)_3$ are heated, transamination leads to the evolution of methylamine, $MeNH_2$, and the formation of highly crosslinked preceramic polysilazanes.

5.6.2
Curing and Pyrolysis Reactions

Latent reactivity in polysilazanes, leading to → crosslinking upon thermal treatment, mainly results from Si–H and N–H groups (SiH/NH dehydrocondensation; see above) or from vinyl groups present in the polymer (olefinic polymerization or hydrosilylation). The efficiency for crosslinking follows the order:

- hydrosilylation > SiH/NH dehydrocondensation > polymerization of vinyl groups ∼ SiH/SiH dehydrocondensation.

Similar results were obtained for polycarbosilazanes.

Pyrolysis of polysilazanes with organic substituents in an inert atmosphere at about 1000 °C results in single-phase, amorphous ceramic materials that may contain substantial portions of carbon. For example, pyrolysis of $[Si(H)Me–NH]_{0.4}[SiMeN]_{0.6}$ at 1000 °C under N_2 gives a material of the chemical composition $Si_{2.0}N_{1.8}C_{1.0}$. The → metastable amorphous material is than transformed into the thermodynamically stable phases Si_3N_4, SiC, and graphite.

As for the polycarbosilanes (see Section 5.5), the pyrolysis atmosphere plays a crucial role in the mechanisms and the final composition of the materials. Although nitrogen may act as a nitridating reagent above 1000 °C, ammonia is much more efficient. If pyrolysis is carried out under ammonia atmosphere, carbon-free ceramics are obtained. The nitridation reaction with ammonia is so efficient that silicon nitride can even be prepared by pyrolysis of poly*carbo*silanes at 500–1000 °C in an ammonia atmosphere.

Three well-resolvable reaction stages occur during pyrolysis of the polyborosilazane. From 200 to 600 °C, the polycondensation is completed while methylamine is removed. At 600 °C, the largest weight loss occurs and is accompanied by the evolution of molecular fragments. Above 1000 °C, H_2 and small amounts of N_2 are evolved. The ternary nitride $Si_3B_3N_7$ is obtained if ammonia is used as a reactive gas during pyrolysis, whereas pyrolysis in N_2 yields a ceramic material of the approximate composition $SiBN_3C$ with hardness comparable to that of sapphire. $SiBN_3C$ is stable up to 1800 °C, where it starts to decompose into Si, N_2, BN, and SiC. Exposing this material to air at 1400 °C results in the formation of a protective double layer (SiO_2/BN) that results in a maximum operating temperature for $SiBN_3C$ in air of 1500 °C. This is significantly higher than that for all non-oxidic materials reported so far.

5.7
Other Inorganic Polymers

Efforts to synthesize inorganic polymers with other non-metallic atoms in the main chain have been discussed in Sections 5.2–5.6 are motivated by the possibility of accessing new polymeric materials or preceramic polymers with interesting and useful properties. The intention of the non-comprehensive selection of examples in this section is to show the manifold possibilities, particularly when the polymer backbone consists of more than two elements. This section thus merely scratches the surface of what has already been investigated on a fundamental level, and what is feasible in terms of possible applications.

5.7.1
Preceramic Polymers for BN

α-Boron nitride, BN, is a commercially important non-oxide high-temperature ceramic. It is hard, has a high mechanical strength, a low thermal expansion coefficient, and a high corrosion and oxidation resistance. As in the case of Si_3N_4 and SiC, fibers, coatings, and foams cannot be prepared from BN powders.

The ammonolysis of BCl_3 or B_2H_6 is not suitable for obtaining fusible or soluble preceramic polymers. The precursor family with the greatest potential for the formation of preceramic polymers for BN are borazene derivatives $B_3N_3R_3R'_3$ (Figure 5.15). They have the correct B/N ratio, and the six-membered ring is the basic structural motif of α-BN. Borazene itself, $B_3N_3H_6$, is a low-boiling liquid that is easily prepared. Controlled thermolysis of liquid borazene *in vacuo* at 70 °C results in near-quantitative formation of soluble polyborazylene by H_2 elimination [Eq. (5.47)]. The molecular weight of the polymer depends on the time of heating. Analytical data indicate that the polymer is probably crosslinked with an average of three branchings per ring.

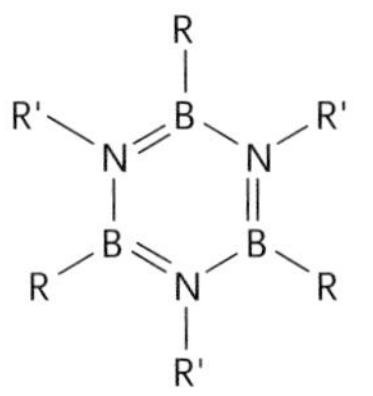

Figure 5.15 The structural formula of borazene derivatives $B_3N_3R_3R'_3$.

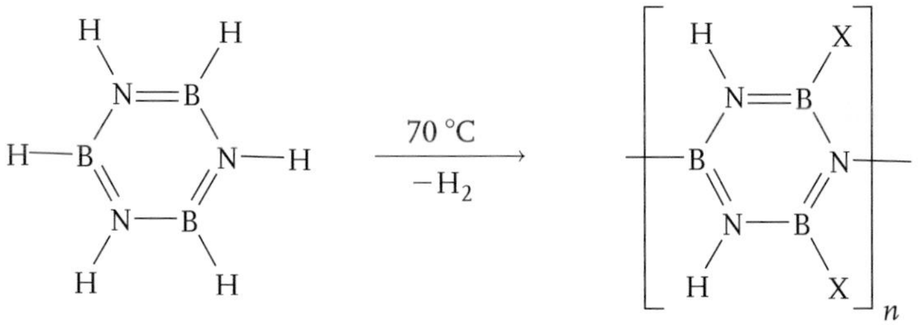

$$B_3N_3H_6 \xrightarrow[-H_2]{70\,^\circ C} [B_3N_3H_2X_2]_n \qquad (5.47)$$

X = H or link to another ring

Pyrolysis of the polymer under Ar or NH_3 to 1200 °C results in dehydrocoupling between adjacent borazene units and eventually in the formation of white, very pure BN powders with → ceramic yields up to 92%. High-quality BN coatings on alumina, Si_3N_4, SiC, and carbon fibers have been achieved using polyborazylene precursors.

A variety of substituted borazene derivatives have been investigated. The most promising is *B*-triamino-*N*-tris(trimethylsilyl)borazene ($R{=}NH_2$, $R'{=}SiMe_3$ in Figure 5.15). A fusible, soluble preceramic polymer is formed when this compound is heated at 200 °C. The polymer is melt spun into fibers that are slowly calcined in NH_3 to give BN at 1000 °C. However, the synthesis of this borazene monomer is a tedious multistep process.

A second class of borazene-based polymers are those in which the borazene units are not connected with each other but instead crosslinked by two ("two-point polymers") or three bridging groups per ring ("three-point polymers"). The most promising bridging groups are amino groups. The linkage of the borazene ring by amino groups is best achieved by reaction of B–Cl-substituted borazenes (R=Cl) with hexa- or heptamethyldisilazane, $(Me_3Si)_2NR$ (R=H, Me). The driving force is the formation of stable, volatile Me_3SiCl.

For example, the reaction of *B*-trichloroborazene and hexamethyldisilazane [Eq. (5.48)] can be performed in a variety of solvents. The solubility of the resulting three-point polymer in organic solvents appears to depend on the borazene/$(Me_3Si)_2NH$ ratio.

$$(ClBNH)_3 + (Me_3Si)_2NH \longrightarrow [(HNBNH)_3\text{-NH-bridged}]_n \qquad (5.48)$$

However, if the polymers are dissolved in liquid NH_3 and then dried, they redissolve in organic solvents. Ammonia obviously breaks down some crosslinks of the original polymer and thus makes it more soluble. These polymer solutions can be processed to form fibers, coatings, → green bodies, xerogels, or aerogels (see also Section 4.5). They are converted in these forms in BN on pyrolysis to 1200 °C by elimination of ammonia. Labeling experiments show that the nitrogen in the final BN originates both from the rings and the bridges. Thus, opening of the borazene rings is a crucial reaction during pyrolysis.

5.7.2
Other Phosphorus-Containing Polymers

Polyphosphates, as well as silicates, are not considered polymers in the context of this section. However, if the bridging oxygen atom in $[PO_3^-]_n$ is formally replaced by a O–C_6H_4–O unit, the structurally related "phoryl resins" are obtained (Figure 5.16). They are prepared by reaction of $RO(O)PCl_2$ with aromatic diols. Although

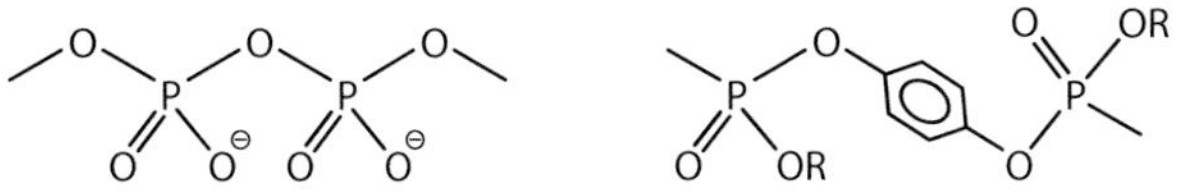

polyphosphate chain "phoryl resin"

Figure 5.16 Structural relation between polyphosphates and the phoryl resins.

they have some interesting properties, such as good transparency and → hardness, they lack long-term hydrolytic stability.

Recently, some new inorganic polymers have been developed in which some of the PR_2 units of polyphosphazenes are replaced by a CR, SR, or S(O)R unit [poly(carbophosphazenes) (**2a**), poly(thiophosphazenes) (**2b**) and poly(thionylphosphazenes) (**2c**), respectively]. They are prepared by thermal ROP of the corresponding molten heterocycles **1** [Eq. (5.49)]. Typical reaction temperatures are between 90 (for **1b**) and 165 °C (for **1c**).

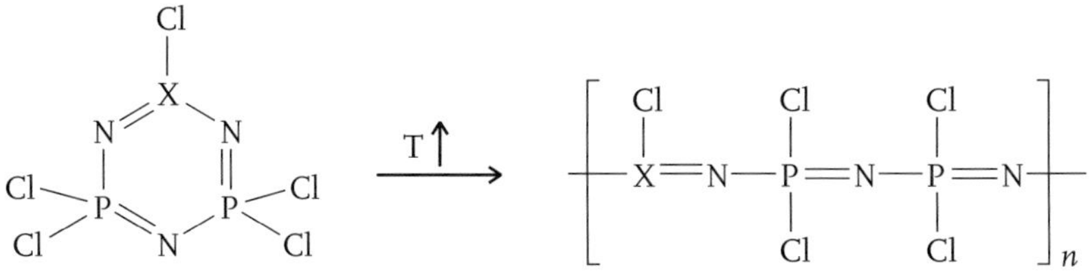

1 $\xrightarrow{T\uparrow}$ **2** (5.49)

a: X = C
b: X = S
c: X = S(O)

The polymers are very sensitive to moisture as a consequence of the hydrolytically sensitive P–Cl and/or S–Cl groups. As for the corresponding polyphosphazenes (Section 5.3), hydrolytically stable derivatives are obtained if the perchlorinated polymers **2** are treated with nucleophiles, such as aryloxides or primary amines. The most promising derivatives are currently the poly(thionylphosphazenes) (**2c**), which exhibit a much higher stability than the poly(thiophosphazenes) owing to the four-coordinate sulfur atom. In **2a** and **2b**, the C–Cl or S–Cl groups are more reactive than the P–Cl groups, which allows a regioselective substitution in every case. In the poly(thionylphosphazenes) **2c**, the regioselectivity is inverse, that is, the P–Cl group is preferentially substituted.

The glass-transition temperatures of **2** is governed by two opposing effects compared with the parent polyphosphazenes. The C=N bond in **2a** and the S=O group in **2c**, as well as the smaller size of C and S compared to P, tend to increase T_g. On the other hand, the presence of only five large substituents per repeat unit compared to six substituents for polyphosphazenes lowers T_g. When the substituents are large enough, this effect overrides the lower flexibility introduced by the replacement of a phosphorus atom by a S(O)R or C=N moiety.

5.7.3
Polymers with S–N Backbones

When gaseous S_4N_4 is passed over metallic silver at elevated temperatures, it is catalytically converted to the cyclic dimer S_2N_2. The dimer, in turn, slowly polymerizes in the solid state to $[SN]_n$ (Figure 5.17). The polymer is thermally stable to 140 °C and highly crystalline with almost planar chains. It is composed of layers of fibers, which are soft and malleable, and have a golden metallic luster. The most important property of this compound is its highly isotropic electric conductivity along the chains down to low temperatures like a metal, which led to its use as an electrode material. At −273 °C it becomes a → superconductor. $[SN]_n$ can be used for visualization of fingerprints on a wide range of media. Interaction of gaseous S_2N_2 with fingerprints induces polymerization that renders the prints visible.

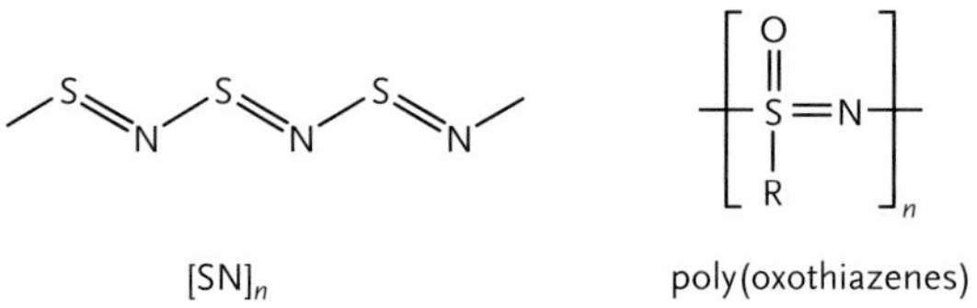

$[SN]_n$ poly(oxothiazenes)

Figure 5.17 Structural formula of poly-SN and poly(oxothiazenes).

The processing of $[SN]_n$ is hampered by its insolubility and its failure to melt without decomposition. Furthermore, it is easily oxidized. The situation could possibly be improved by having the sulfur atom in a higher oxidation state, that is, by having substituents at the sulfur atom. A series of polymers of this type, the poly(oxothiazenes) have been prepared (Figure 5.17).

Polymers with aryl or alkyl substituents at the sulfur atom can be prepared by thermal condensation of sulfonimidates, $RO(O)S(R){=}NH$ (elimination of ROH) or, at higher temperatures, of N-silylsulfonimidates, $RO(O)S(R){=}NSiMe_3$ (elimination of $ROSiMe_3$).

Poly(oxothiazenes) are highly polar. For example, $[S(O)(Me){=}N]_n$ dissolves in DMF, DMSO, or hot water. Its T_g (~60 °C) is much higher than that of the corresponding phosphazene, $[NPMe_2]_n$ (−46 °C). This shows that the backbone in poly(oxothiazenes) is less flexible, consistent with the trend already observed for poly(thionylphosphazenes) (see above).

5.8
Metal-Containing Polymers

Transition-metal compounds often have interesting physical properties. Therefore, the incorporation of transition-metal centers in the repeating polymer unit is an attractive approach to obtain new types of macromolecules with properties distinctly different from organic polymers or the inorganic polymers discussed so far.

One-dimensional metal-containing polymers (coordination polymers) are obtained when bidentate → ligands of the type X–Y–X bridge two metal centers.

This can be extended to two- or three-dimensional network polymers either by means of multidentate ligands or by coordinating more than two X groups to a given metal center.

$$\ldots X{-}ML_n{-}X{-}Y{-}X{-}ML_n{-}X{-}Y{-}X{-}\ldots$$

ML_n is a metal complex fragment containing n coligands L not integrated in the polymer backbone; X is a coordinating group that provides a stable link to the ML_n fragment; and Y is an organic or inorganic spacer.

Not any bi- or tridendate ligand can be used, because → chelation is entropically favored. This is related to the ring/chain problem previously discussed for other inorganic polymers (Section 5.1). Therefore, the groups Y should be rigid.

The definition of a metal-containing polymer is less straightforward than that of any other polymer type and we refrain from attempting such a definition here. Many compounds form one-, two-, or three-dimensional aggregates through bridging groups in the solid state but disintegrate when dissolved. For example, organotin or organolead derivatives $Me_3M{-}X$ (M = Sn, Pb; X = halide, carboxylate, etc.) assemble in the solid state through bridging X groups, but are monomers in solution. On the other hand, many coordination polymers with interesting properties only exist as solids, such as the three-dimensional metal-organic framework structures which will be discussed in Section 5.8.2.

There are three main approaches to metal-containing polymer synthesis:

- Complexation of metal centers with di- or multidentate → ligands through Lewis-acid–Lewis-base interactions. Most examples discussed in this section were prepared by this method.
- Coordination of metals to already formed polymers that contain complexing groups. This approach corresponds to crosslinking via metal coordination as was already discussed in Section 5.1.2.
- Polymerization of ligands that already complex the metal. An example is the ring-opening polymerization of bridged ferrocenes [Eq. (5.50)]. This approach is related to polycarbosilane preparation according to Eq. (5.38).

5.8.1 One-Dimensional Coordination Polymers

We restrict ourselves to two groups of polymers as representative examples. The first is based on phthalocyanines as the metal-complex moiety (Figure 5.18). There is a great number of phthalocyanines and structurally related porphyrines of various metals (M = Cr, Mn, Fe, Co, Rh, Ru) that can be stacked by appropriate groups X–Y–X. Substituents can be introduced at the aromatic rings to modify the polymer properties. Stacking can be achieved by a single oxygen atom (the resulting polymers have a –M–O–M–O– backbone). More extended X–Y–X entities have also been used, such as CN^- (giving M–CN–M–CN–linkages), 4,4′-bipyridine, pyrazine derivatives, 1,4-diisocyanobenzene, 4,4′-bipyridylacetylene, and so on. The synthesis of the mostly insoluble polymers is accomplished by reacting the

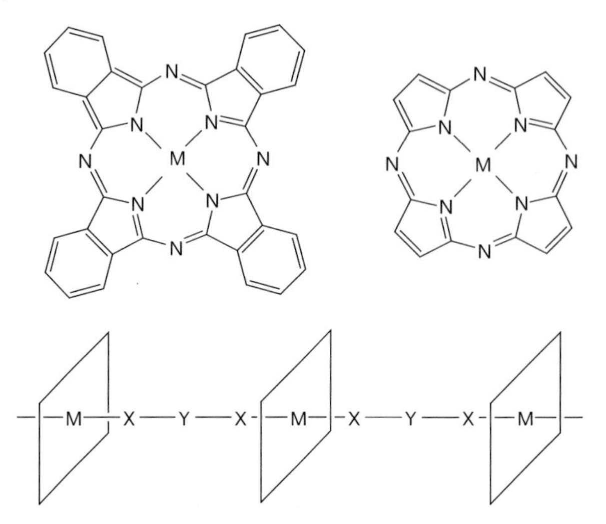

Figure 5.18 The general formula of a metal phthalocyanine (top left) and a metal porphyrine (top right). Linkage of the phthalocyanine or porphyrine units (represented by squares) through X–Y–X units (bottom).

coordinatively unsaturated or dihalogenated metal phthalocyanine or porphyrine with the bidentate → ligand.

Some of these polymers have an excellent thermal and chemical stability. They can become electrically conducting after chemical or electrochemical doping. One-dimensional conductivities up to 0.1 S cm^{-1} were obtained. Such polymers may also have interesting magnetic or electro-optical properties. An example is ferrimagnetic $[Mn(III)(tetraphenylporphyrin)]^+[TCNE]^-$ with antiparallel coupling of the electron spins in the p-orbitals of the tetracyanoethylene [TCNE, $(NC)_2C{=}C(CN)_2$] radical anion and the d-orbitals of Mn(III) (Figure 5.19).

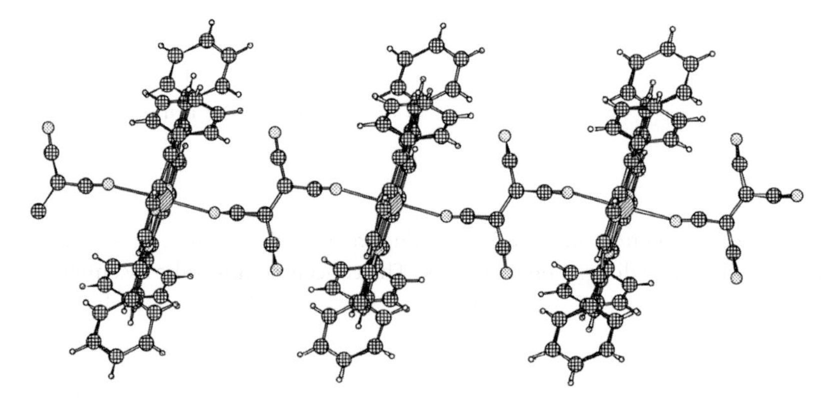

Figure 5.19 The polymeric structure of ferrimagnetic $[Mn^{III}(tetraphenylporphyrin)]^+$ $[TCNE]^-$. The planes of the porphyrin units are perpendicular to the chain axis. Two of the four CN groups of $(NC)_2C{=}C(CN)_2$ coordinate to the Mn atoms.

The second example of metal-containing polymers with potentially interesting properties are those with ferrocenyl units incorporated into the polymer backbone. The polymers exhibit high thermal stability and interesting redox properties of the ferrocenyl unit. The most promising route is ROP of ferrocenophanes [Eq. (5.50)].

(5.50)

Bridging of the cyclopentadienyl rings by a single atom ([1]ferrocenophanes) induces tilting of the rings (the rings are parallel in non-bridged ferrocene). The thus-generated strain is the driving force for ROP. For example, while high molecular weight polymers are obtained from SiR_2-bridged ferrocenophanes (tilt angle $\sim 21°$), the corresponding disilanyl-bridged [2]ferrocenophanes ($X = SiR_2SiR_2$) are not strained enough (tilt angle 4–5°) for ROP. In contrast to that, the shorter C–C distance in ethylene-bridged [2]ferrocenophanes ($X = CH_2CH_2$, tilt angle $\sim 21°$) allows the preparation of poly(ferrocenylethylenes). Although most attention has been paid to the properties of poly(ferrocenylsilanes), the corresponding polymers with $X = GeR_2$, PR, P(S)R or S were also prepared by ROP of the corresponding [1]ferrocenophanes.

Most poly(ferrocenylsilanes) are soluble in organic solvents, despite their high molecular weights. While the $SiMe_2$ derivative is thermoplastic, the $SiHex_2$ derivative is an amorphous rubber with $T_g = -26\,°C$. Electrochemical investigations showed two reversible oxidations in a 1 : 1 ratio. This proves an interaction between the ferrocenyl units along the chain. It was concluded that, initially, only every second iron atom along the chain is oxidized.

5.8.2
Metal-organic Frameworks

In three-dimensional coordination polymers, the so-called metal-organic frameworks (MOFs), metal ions or metal clusters (the *connectors*) are bridged in three dimensions with organic *linkers*. Linkers and connectors assemble through Lewis-acid–Lewis-base or ionic interactions, where the metals are the Lewis-acidic centers or cations, and the organic linkers the Lewis-basic centers or anions. The connectivity of the connectors and linkers (Figure 5.20) determines the structure of the resulting three-dimensional network. The concept to direct the assembly of ordered frameworks through the coordination geometry of the connectors has also been called reticular synthesis.

The construction principle of MOFs is related to that of zeolites (Section 6.3.1) where, in the language of MOF chemistry, the tetrahedral Si or Al atoms are the connectors and the oxygen atoms the two-coordinate linkers. In both classes of solid compounds, the three-dimensional framework defines interconnected

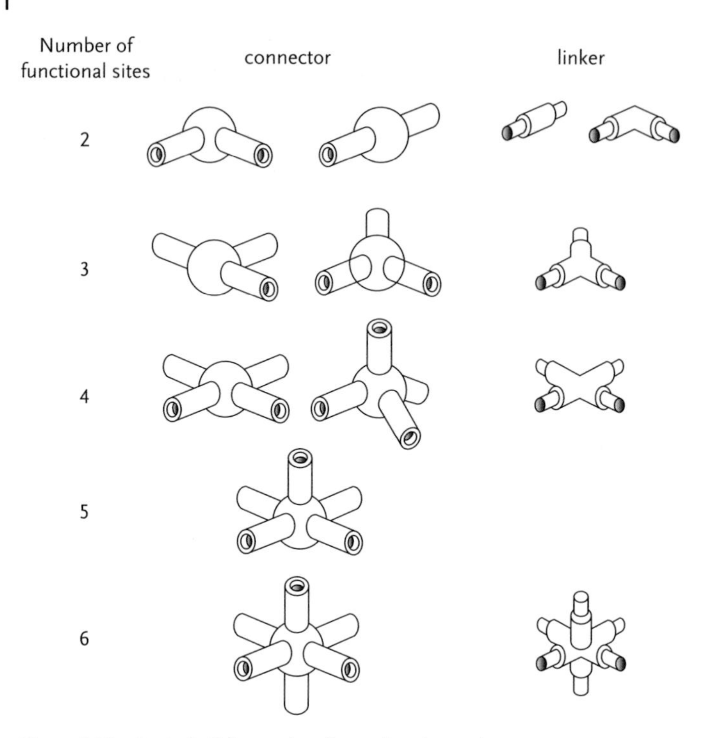

Figure 5.20 Basic building units of coordination polymers.

channels and cages. Porosity is thus an inherent part of the network structures. However, there are some important differences:

- The connectors in MOFs are mostly metal ions, with various coordination numbers and coordination geometries or → polynuclear compounds (clusters). The latter can be formed *in situ*, as Zn_4O(terephthalate)$_6$ discussed below, or can be employed as preformed entities.
- The linkers are organic or inorganic → bi- or → multidentate ligands with various linking directionalities.
- The interaction between the connectors and the linkers is based on coordinative or ionic interactions (rather than covalent bonds as in zeolites).
- The skeleton of most MOFs is neutral (compared to anionic framework structures in zeolites).
- Thermal stability of MOFs is limited to 350–400 °C due to the presence of the organic groups.

The greater variety of coordination geometries of metal ions or clusters, and the possibility to design linker → ligands with certain geometrical and chemical properties allows constructing new and unusual topologies. Topologies of MOF frameworks are determined by several factors:

- The structural characteristics of the connector ion or cluster, especially the number and spatial orientation of the coordination sites.

- The number and geometrical arrangement of the coordinating groups of the linker, as well as the rigidity of the spacer group connecting the coordinating groups.
- The nature of the donor atoms of the linkers.
- The nature of the counterions of the starting metal salt.
- Auxiliary ligands that may block coordination sites at the metal ion or cluster.
- Additional non-covalent or non-ionic interactions (hydrogen bonds, π–π stacking, hydrophobic interactions, etc.).
- The presence of organic guest molecules acting as templates, or non-coordinated counterions of the employed metal salts.

Some frameworks that can be obtained by using various connectors and just one type of linker (linear) are shown in Figure 5.21. It is easy to imagine how many combinations are possible when other linkers are being used.

One of the earliest and most prominent examples of a MOF structure is Zn_4O(terephthalate)$_6$ (terephthalate = $^-OOC–C_6H_4–COO^-$), prepared from zinc nitrate and terephthalic acid (MOF-5, Figure 5.22). A three-dimensional cubic array is obtained by combination of the $[Zn_4O]^{6+}$ cluster units with terephthalate ions as linear linkers. The crystals have a very low density (0.59 g cm^{-3}) and their surface area and pore volume are higher than in most zeolites.

When the length of the dicarboxylate is varied, the framework topology is preserved, but the cavity size can be varied in a rather wide range. This creates series of so-called "isoreticular solids." For example, the diameter of the largest (spherical) molecule that could fit inside the pore of MOF-5 without contacting the van der Waals radii of the atoms making up the walls is about 130 pm. The size increases to about 290 pm when the terephthalate ligands are replaced by $^-OOC–C_6H_4–C_6H_4–C_6H_4–COO^-$.

The six edges of the tetrahedral Zn_4O core are bridged by six terephthalate units, that is, the connector unit is octahedrally coordinated. The $[Zn_4O]^{6+}$ cluster unit can thus be regarded as a big (expanded) pseudoatom with octahedral coordination geometry. The following comparison shows that MOF structures are often analogs to common inorganic structures. In the crystal structure of ReO_3, the metal atoms are octahedrally coordinated by six oxygen atoms, and each oxygen atom bridges two rhenium atoms. This is a very common structure type for inorganic MX_3 compounds. If the bridging O^{2-} in the ReO_3 structure are replaced by bridging ^-OOC-R-COO^- ("expanded linkers"), and Re(VI) by the $[Zn_4O]^{6+}$ cluster unit ("expanded connector"), the MOF-5 structure is obtained, that is, the MOF-5 structure is a metal-organic, and highly porous, analog of the ReO_3 structure.

Different framework topologies are of course obtained when linker and /or connector have different spatial orientation of the coordinating groups or the coordination sites. In Figure 5.23 (left) the basic structural unit is shown when the edges of the tetrahedral Zn_4O core are bridged by six trigonal planar 1,3,5-benzene-tricarboxylate groups (rather than linear terephthalate units as in MOF-5). Figure 5.23 (right) shows the reverse situation, namely replacement of the tetrahedral Zn_4O core by the trigonal prismatic Fe_3O core.

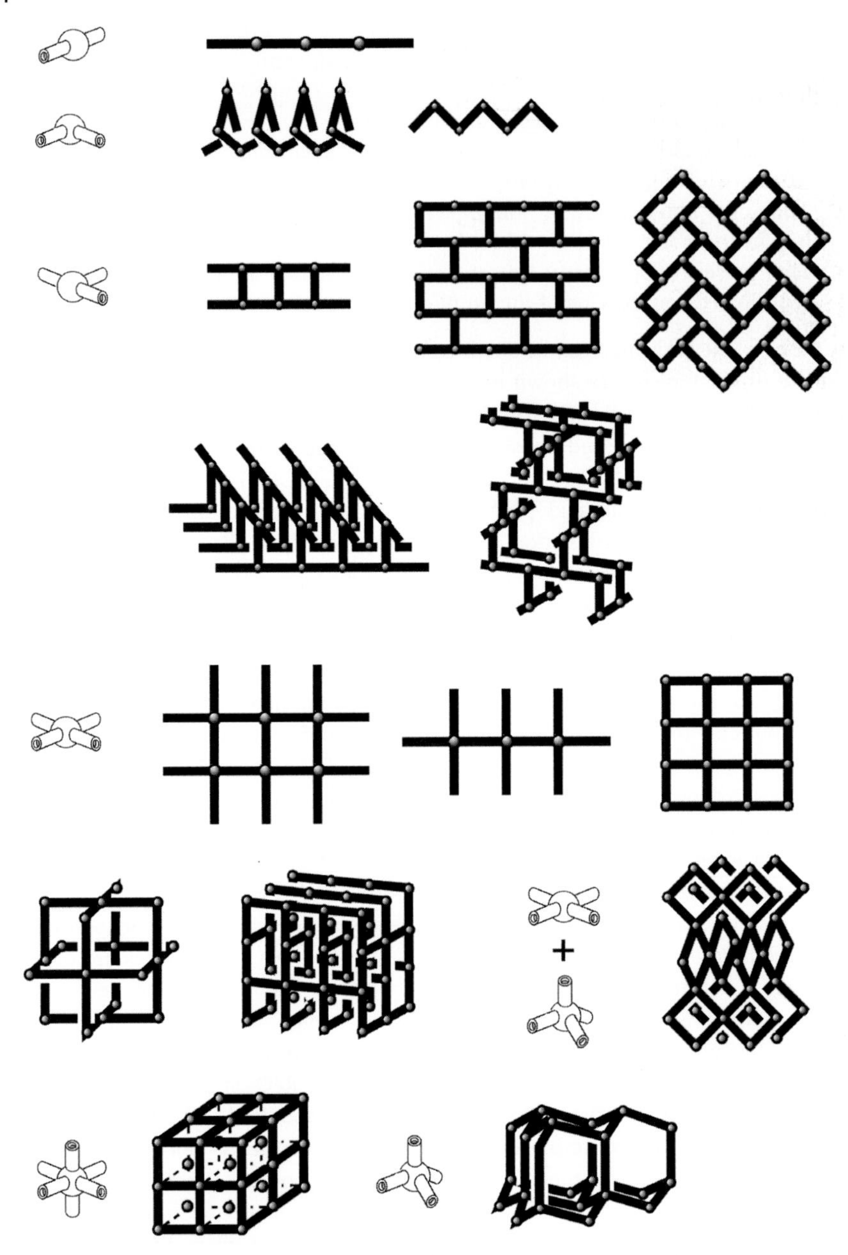

Figure 5.21 Examples for structural frameworks that can be constructed from 2 to 6 coordinated connectors and linear linkers.

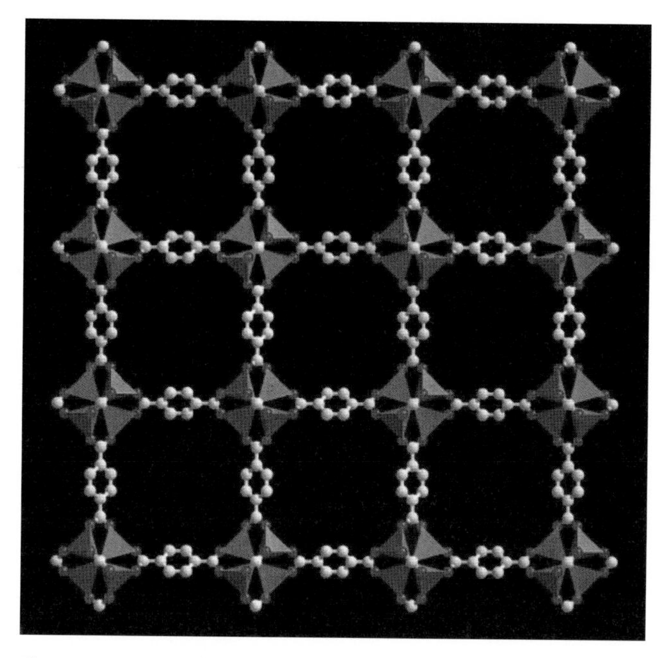

Figure 5.22 A two-dimensional section through the crystal structure of Zn_4O(terephthalate)$_6$ (MOF-5).

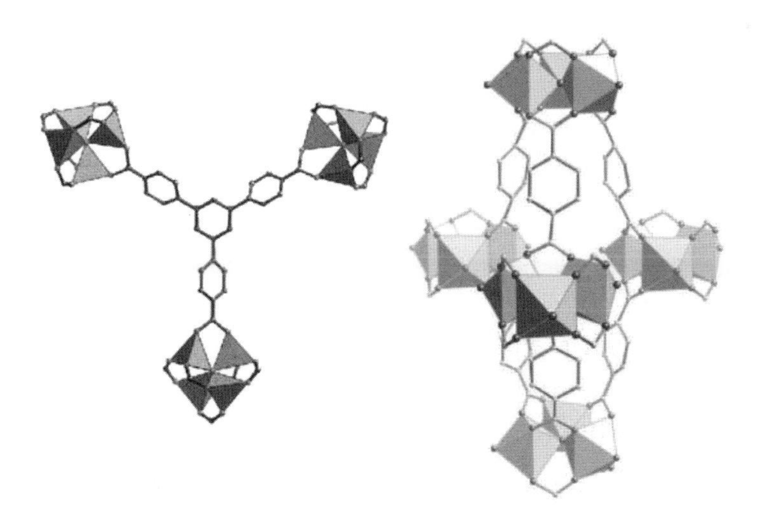

Figure 5.23 Left: combination of the Zn_4O connector with another linker, trigonal-planar 1,3,5-benzene-tricarboxylate (MOF-177). Right: combination of the linear terephthalate linker with the trigonal-prismatic $M_3^{III}O$ connector (M = Fe, Cr) (MIL-88B).

A general feature common to almost all linkers used for the formation of MOFs is rigidity. Clearly, if a →ligand is too flexible and has a number of possible conformations, several products can be formed and it is increasingly difficult to get ordered networks (crystalline compounds). Popular ligands include nitrogen or oxygen-donor atoms, such as 4,4′-bipyridine, pyrazine, benzene-1,4-dicarboxylates (terephthalate), benzene-1,3,5-tricarboxylate, bisphosphonates or related compounds, to name only a few (Figure 5.24).

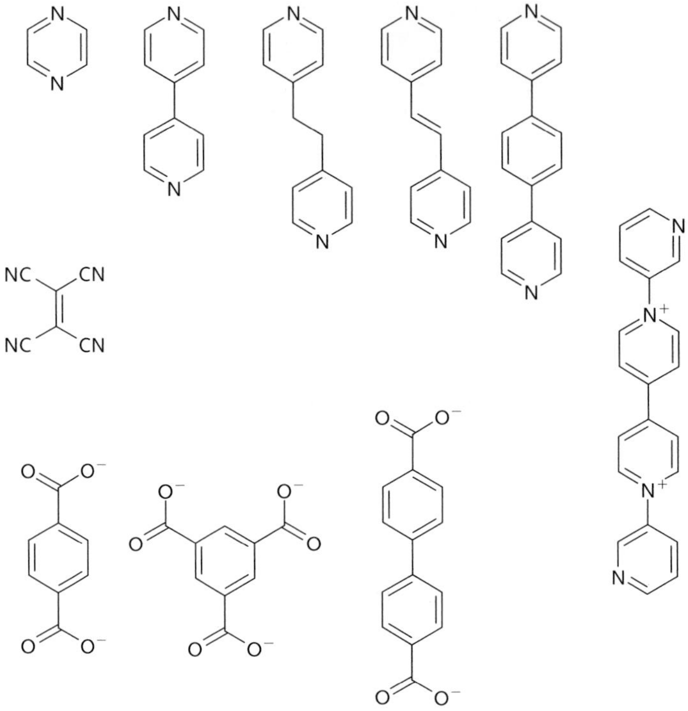

Figure 5.24 Examples of linkers used in MOFs.

Early MOF syntheses involved techniques known to grow high-quality crystals of simple inorganic salts by reducing the crystal nucleation rate, such as slow evaporation of solvents, layering of different solvents or slow diffusion of one component into the solution of another. Amines can be added when a ligand needs to be deprotonated prior to coordination (i.e., when carboxylate linkers are employed).

MOFs are now most commonly prepared by solvothermal, especially hydrothermal methods (see Section 4.4). To this end, the precursors are typically combined in dilute solutions of polar solvents, or solvent mixtures, and heated in sealed vessels. The different solubility of the inorganic and organic components is thus overcome. Solvothermal conditions facilitate equilibration of the system, that is, formation of ordered structures. This is enabled by the weaker coordinative bonds (compared with covalent bonds) that allow detachment of incoherently assembled monomers or building blocks and reattachment of the ligands to the metal centers in thermodynamically more favorable positions. Thus, the formation of crystalline MOFs is in principle self-assembly of the basic building units (connectors and linkers). There is an inverse relationship between the metal–ligand bond strength (i.e., the reversibility of the bond-formation process) and the robustness of the formed frameworks. This requires a very deliberate consideration of the synthesis protocol.

Microwave synthesis has been recently successfully applied to MOFs as an alternative to conventional heating. Since MOFs can be prepared by this method in considerably shorter reaction times, with better yields and purities, a continuous flow of products appears to be feasible.

MOFs feature amongst the largest pores and largest specific surface areas known for crystalline compounds. The open voids, cavities, and channels in these porous coordination polymers can make up a large proportion of the volume of the crystal.

MOFs were classified in three generations:

- **First generation**: the frameworks are only sustained with guest molecules and collapse irreversibly upon removal of the guests.
- **Second generation**: the frameworks are stable and robust, and show permanent porosity without any guest. An example is the compound shown in Figure 5.22.
- **Third generation**: the frameworks are flexible and dynamic, and respond to external stimuli (temperature, pressure, light, electric fields, guests, etc.) and change their channels or pores reversibly. The expanding and shrinking (breathing) frameworks can act similar to sponges; the drastic volume changes – with retention of the topology – are induced by strong host–guest interactions.

The coordination polymers of the second and third generations are promising future materials, for example as gas storage/separation or catalytic materials, or as nanoreactors. Interest in the gas-sorption properties is currently focused on fuel gases such as methane or hydrogen. The main advantage of MOFs compared with other porous materials could be the easy adjustment of the pore and channel diameters (due to the easy variation of the linkers), the potential integration of chemical functionalities in the linkers, and the fact that the pore walls are constructed of organic entities, providing a "light material."

Further Reading

1 Abd-El-Aziz, A.S., Carraher, C.E. Jr, Pittman, C.U. Jr, and Zeldin, M. (2005) *Metal-Coordination Polymers*, Wiley-Interscience, Hoboken, New Jersey.

2 Allcock, H.R. (2003) *Chemistry and Applications of Polyphosphazenes*, Wiley-Interscience, New York.

3 Archer, R.D. (2001) *Inorganic and Organometallic Polymers*, John Wiley & Sons Ltd, New York.

4 Baldus, H.-P. and Jansen, M. (1997) Novel high-performance ceramics–amorphous inorganic networks from molecular precursors. *Angew. Chem. Int. Ed. Engl.*, **36**, 328–343.

5 Baldus, H.-P., Jansen, M., and Sporn, D. (1999) Ceramic fibers for matrix composites in high-temperature engine applications. *Science*, **285**, 699–703.

6 Birot, M., Pillot, J.-P., and Dunogues, J. (1995) Comprehensive chemistry of polycarbosilanes, polysilazanes, and polycarbosilazanes as precursors for ceramics. *Chem. Rev.*, **95**, 1443–1477.

7 Bruins, P.F. (1970) *Silicone Technology*, John Wiley & Sons Ltd, New York.

8 Chandrasekhar, V. (2005) *Inorganic and Organometallic Polymers*, Springer, Berlin, Heidelberg.

9 Clarson, S.J., Fitzgerald, J.J., Owen, M. J., Smith, S.D., and van Dyke, M.E. (2003) Synthesis and properties of silicones and silicone-modified materials: a concise overview. *ACS Symp. Ser.*, 838.

10 Drake, R., MacKinnon, A.I., and Taylor, R. (1998) Recent advances in the chemistry of siloxane polymers and copolymers, in *The Chemistry of Organic Silicon Compounds* (eds. Z. Rappoport and Y. Apeloig), John Wiley & Sons Ltd, pp. 2217–2244.

11 Férey, G. (2008) Hybrid porous solids: past, present, future. *Chem. Soc. Rev.*, 191–214.

12 Gates, D.P. and Manners, I. (1997) Main-group-based rings and polymers. *J. Chem. Soc. Dalton*, 2525–2532.

13 Greil, P. (2000) Polymer derived engineering ceramics. *Adv. Eng. Mater.*, **2**, 339–348.

14 Interrante, L.V., Liu, Q., Rushkin, I., and Shen, Q. (1996) Poly (silylenemethylenes) – a novel class of organosilicon polymers. *J. Organomet. Chem.*, **521**, 1–10.

15 James, S.L. (2003) Metal–organic frameworks. *Chem. Soc. Rev.*, **32**, 276–288.

16 Janiak, C. (2003) Engineering coordination polymers towards applications. *Dalton Trans.*, 2781–2804.

17 Jones, R.G. (2003) Polysilanes: formation, bonding and structure, in *Silicon Chemistry: from the Atom to Extended Systems* (eds P. Jutzi and U. Schubert), Wiley-VCH, Weinheim, pp. 139–158.

18 Jones, R.G., Ando, W., and Chojnowski, J. (eds) (2000) *Silicon-Containing Polymers*, Kluwer, Dordrecht.

19 Kendrick, T.C., Parbhoo, B., and White, J.W. (1989) Siloxane polymers and copolymers, in *The Chemistry of Organic Silicon Compounds* (eds S. Patai and Z. Rappoport), John Wiley & Sons Ltd, pp. 1289–1361.

20 Kitagawa, S., Kitaura, R., and Noro, S. (2004) Functional porous coordination polymers. *Angew. Chem. Int. Ed. Engl.*, **43**, 2334–2375.

21 Lacave-Goffin, B., Hevesi, L., Demoustier-Champagne, S., and Devaux, J. (1999) Synthesis and properties of polysilanes: versatile new organic materials. *ACS-Models Chem.*, **136**, 215–236.

22 Laine, R.M. and Sellinger, A. (1998) Si-containing ceramic precursors, in *The Chemistry of Organic Silicon Compounds* (eds Z. Rappoport, and Y. Apeloig), John Wiley & Sons Ltd, pp. 2245–2316.

23 Livage, J., Sanchez, C., and Babonneau, F. (1998) Molecular precursor routes to inorganic solids, in *Chemistry of Advanced Materials – an Overview* (eds L. V. Interrante and M. Hampden-Smith) Wiley-VCH, New York, pp. 389–448.

24 Matyjaszewski, K., Cypryk, M., Frey, H., Hrkach, J., Kim, H.K., Moeller, M., Ruehl, K., and White, M. (1991) Synthesis and characterization of polysilanes. *J. Macromol. Sci. Chem.*, **A28**, 1151–1176.

25 Miller, R.D. (1989) Polysilanes – a new look at some old materials. *Angew. Chem. Adv. Mater.*, **101**, 1773–1780.

26 Miller, R.D. and Michl, J. (1989) Polysilane high polymers. *Chem. Rev.*, **89**, 1359–1410.

27 Narula, C.K. (1995) *Ceramic Precursor Technology*, Marcel Dekker, New York.

28 Noll, W. (1968) *Chemistry and Technology of Silicones*, Academic Press.

29 Paine, R.T. and Narula, C.K. (1990) Synthetic routes to boron nitride. *Chem. Rev.*, **90**, 73–91.

30 Richter, R., Roewer, G., Böhme, U., Busch, K., Babonneau, F., Martin, H.P., and Müller, E. (1997) Organosilicon polymers – synthesis, architecture, reactivity and applications. *Appl. Organomet. Chem.*, **11**, 71–106.

31 Riedel, R. (1996) Advanced ceramics from inorganic polymers, in *Materials Science and Technology*, Vol. **17B** (eds R.W. Cahn, P. Haasen, and E.J. Kramer), Wiley VCH, Weinheim, pp. 1–50.

32 Rochow, E.G. (1987) *Silicon and Silicones*, Springer Verlag, Berlin.

33 Rowsell, J.L.C. and Yaghi, O.M. (2004) Metal–organic frameworks: a new class of porous materials. *Microporous Mesoporous Mater.*, **73**, 3–14.

34 Seyferth, D. (1995) Preceramic polymers: past, present, and future. *Adv. Chem. Ser.*, **245**, 131–160.

35 Sheats, J.E., Carraher, C.E., Pittman, C. U., Zeldin, M., and Currell, B. (eds) (1990) *Inorganic and Metal-Containing Polymeric Materials*, Plenum, New York.

36 Stark, F.O., Falender, J.R., and Wright, A.P. (1982) Silicones, in *Comprehensive Organometallic Chemistry*, Vol. **2** (eds G. Wilkinson, F.G.A. Stone, and E.W. Abel), Pergamon Press, Oxford, pp. 305–363.

37 West, R. (1989) Polysilanes, in *The Chemistry of Organic Silicon Compounds* (eds S. Patai and Z. Rappoport), Wiley, pp. 1207–1240.

38 Whittell, G.R., Hager, M.D., Schubert, U.S., and Manners, I. (2011) Functional soft materials from metallopolymers and metallosupramolecular polymers. *Nature Mater.*, **10**, 176–188.

39 Whittell, G.R. and Manners, I. (2007) Metallopolymers: new multifunctional materials. *Adv. Mater.*, **19**, 3439–3468.

40 Yaghi, O.M., O'Keeffe, M., Ockwig, N. W., Chae, H.K., Eddaoudi, M., and Kim, J. (2003) Reticular synthesis and the design of new materials. *Nature*, **423**, 705–714.

41 Yaghi, O.M. and Li, Q. (2009) Reticular chemistry and metal–organic frameworks for clean energy. *Mater. Res. Soc. Bull.*, **24**, 582–590.

42 Zeigler, J.M. and Fearon, F.W.G. (1990) Silicon-based polymer science. *Adv. Chem. Ser.*, **224**.

6
Templated Materials

Porous and high surface area materials are of interest to many scientific communities and are used for various applications. One synthetic approach towards porous materials with a high degree of control over the structural and textural processes is based on templating strategies. Over recent years, templating procedures have been developed in a way that structure and texture on length scales between nanometers and micrometers can be deliberately tailored for many materials with unprecedented precision.

Other methods for synthesizing porous and high surface area materials were already discussed in other chapters of this book. Examples include biomorphic SiC ceramic in Section 2.1.3, pillared clays in Section 2.3.3, porous biogenic materials in Section 4.3.1 or aerogels in Section 4.5.6.

Structural design via templating can – in principle – rely on two different approaches: Molecules, supramolecular arrangements or larger objects can be added to the mixture of starting compounds and are occluded in the growing solid. Pores are then generated by removal of these templates. This process is called endotemplating. The other approach – termed exotemplating, in analogy to the term "exoskeleton" in biology – is based on the infiltration of a preformed porous scaffold with a precursor mixture. After solidification, the porous scaffold or mold is removed and a porous material remains (Figure 6.1). Either small particles or porous networks can be generated by the latter approach.

The concept of → templating was first introduced into zeolite science. However, the pore structure that is obtained from specific templates is not always an exact replication of the template structure. There are cases where the shape and size of the template correlates with the final porous structure, but in many cases the function of the template is more complicated and not very well understood. This is especially true for microporous solids, such as zeolites. Here, the notion "structure-directing agent" instead of template is more appropriate.

Since the templating strategies that are presented in this chapter will typically lead to porous materials and in most textbooks inorganic porous materials are not even mentioned, a short introduction to porous materials and porosity will be given in Section 6.1, followed by a brief overview of different types of porous and templated solids and their applications. The chosen examples are somewhat arbitrary, since it would be impossible to cover all types of inorganic templated porous materials.

Synthesis of Inorganic Materials, Third Edition. Ulrich Schubert and Nicola Hüsing.

Published 2012 by WILEY-VCH Verlag GmbH & Co. KGaA

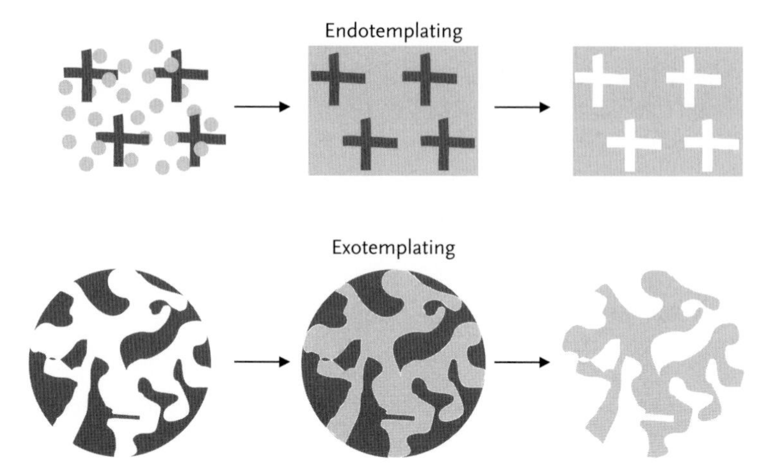

Figure 6.1 Schematic drawing of the endotemplating and exotemplating approach to synthesize porous and high surface area materials.

In this chapter, we introduce the reader to

- materials with different types of porosities regarding the size (from the nanometer range up to millimeters) and arrangement of the pores (irregular or ordered);
- materials with different chemical compositions, for example, from metals to oxides; and
- different preparative templating approaches to induce and control porosity in materials.

6.1 Introduction to Porosity and High Surface Area Materials

In nature, many materials are porous, including wood, cork, sponge, bone, and so on, or even the skeleton structure of the most simple organisms such as diatoms. Most synthetic materials are also – at least to some extent – porous. Although the evolution of natural porous structures is complicated and complex (see Section 4.3.1), the principle of minimal material and mass consumption for an optimum stability of the whole structure governs the formation process. In addition, the materials must be optimally adapted for their desired use.

Porous materials are also called cellular solids. The word "cell" is derived from the Latin *cella*: a small compartment, an enclosed space. Cellular solids are therefore clusters of cells, assembled by solid edges or faces. If the solid material is contained in the cell edges only, so that the cells are connected through open faces, the material is said to be open-celled (Figure 6.2). If the faces are solid, each cell is separated from its neighbors; the material is then called closed-celled. Everybody is familiar with organic porous materials such as polymeric foams used to make a vast range of items, from disposable coffee cups to the crash padding of an aircraft cockpit. Techniques now exist for foaming not only organic materials but also metals, ceramics, and glasses.

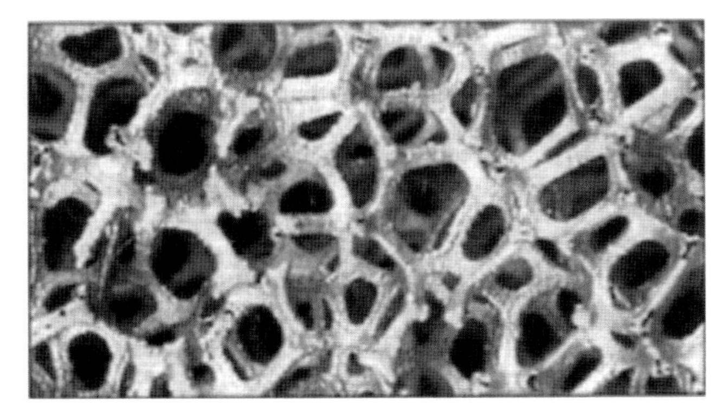

Figure 6.2 Structure of a high-porosity, open-cell metal.

These materials are used increasingly, for example, for insulation or cushioning, to absorb the kinetic energy from impacts, and for catalysis.

Many physical properties such as density, thermal conductivity, strength, and so on, depend on the porosity and the pore structure of a solid. Especially for industrial applications, a deliberate control of porosity is of great importance, for example in the design of catalysts, industrial adsorbents, membranes, structural materials, and ceramics. Furthermore, porosity is one of the factors that influence the chemical reactivity and the physical interactions of solids with gases and liquids.

A solid is called porous when it contains pores, that is, cavities, channels, or interstices, which are deeper than they are wide. There are two ways to look at porous materials: They can be characterized by describing the pores, but also by describing the cell (pore) walls. Some porous materials are based on agglomerated or aggregated powders in which the pores are formed by interparticle voids, while others are based on continuous solid networks around pores.

Not all pores are similar, and different types must be distinguished. A classification is usually done by describing their accessibility to an external fluid (Figure 6.3). From this viewpoint, pores totally isolated from their neighbors are

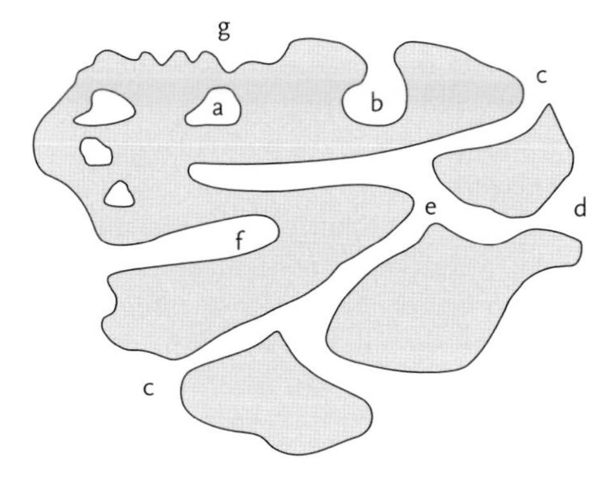

Figure 6.3 Different types of pores.

called *closed pores* (a). They influence the macroscopic properties of the solid, but they are inactive in terms of chemical reactions. On the other hand, pores that are open to the external surface of the solid are called *open pores*, such as (b), (c), (d), (e), and (f) in Figure 6.3. Some may be open only at one end, such as (b) and (f). They are described as *blind* (i.e., *dead-end* or *saccate*) pores. Others may be open at both ends (*through* pores) like (e). Pores may also be classified according to their shape: they may be *cylindrical* (either open (c) or blind (f)), *ink-bottle*-shaped (b), *funnel*-shaped (d), or *slit*-shaped. Close to, but different from porosity is the *roughness* of the external surface, represented around (g).

To describe qualitatively and quantitatively a porous solid, more information is necessary, such as the porosity, the density, the specific surface area, or the pore size and the pore size distribution of the porous solid. Table 6.1 shows the definitions of some terms.

Table 6.1 Definition of terms used to characterize porous solids.

Density	True density	Density of a material excluding pores and interparticle voids (density of the solid network)
	Apparent density	Density of a material including closed and inaccessible pores
	Bulk density	Density of the material including pores and interparticle voids (mass per total volume, with volume = solid phase + closed pores + open pores)
Pore volume	V_p	Volume of the pores
Pore size		Also called pore width (diameter): the distance of two opposite walls of the pore
Porosity		Ratio of the total pore volume V_p to the apparent volume V of the particle or powder
Surface area		The accessible (or detectable) area of solid surface per unit mass of material

For almost all these terms, the measured value depends strongly on the method used. Some methods detect only open pores, for example, methods that use adsorption of molecules into the cavities, others may have also access to closed pores, for example, spectroscopic, diffraction, and scattering techniques. Moreover, for a given method the value can vary with the size of the molecular probe, as shown in Figure 6.4 for a gas adsorption experiment.

For most applications the pore size is of major importance. However, it is not very susceptible to a precise measurement, because the pore shape is usually highly irregular and variable, leading to a variety of pore sizes, or a broad distribution. Nevertheless, three different pore size regimes are defined by IUPAC:

- *micropores*, which have diameters smaller than 2 nm;
- *mesopores*, which have diameters between 2 and 50 nm;
- *macropores*, which have diameters larger than 50 nm.

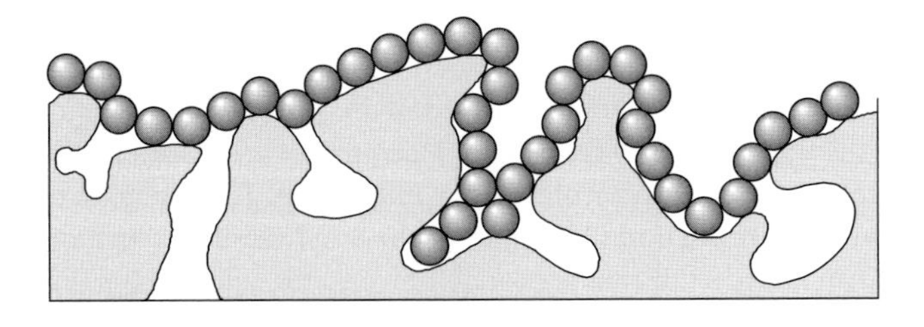

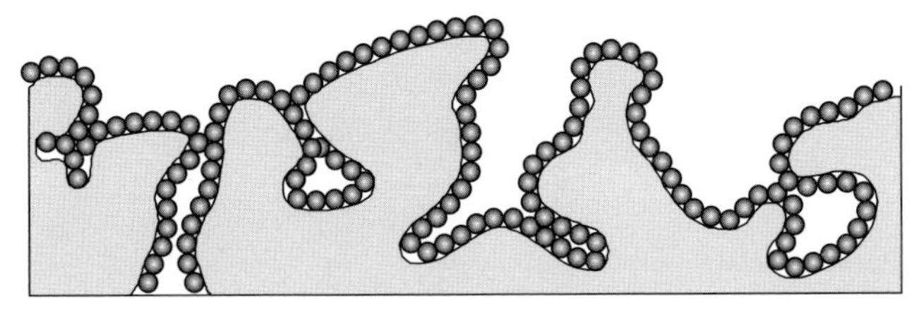

Figure 6.4 Schematic of gas adsorption on a porous material by gases of different molecular size. The surface area, when measured with large molecules, is smaller than when smaller molecules are used.

This nomenclature is associated with the transport mechanisms occurring in the different types of pores (Figure 6.5).

Macropores These are much larger than the mean free path length of a typical gaseous imbibing fluid. The dominant transport mechanisms are molecular (bulk) diffusion and, if a total pressure gradient exists, viscous flow. Overall → permeabilities are highest in macroporous solids, but permselectivities are lowest.

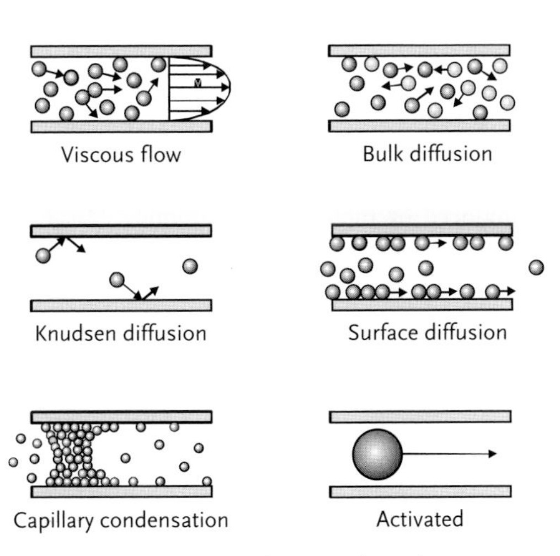

Figure 6.5 Transport mechanisms through pores.

Mesopores These are on the same order or smaller than the mean free path and Knudsen diffusion becomes important. The classical inverse-square-root dependence of flux on molecular weight applies. In addition, surface diffusion of the molecules along the pore walls may contribute. This process is sensitive to the specific adsorption features of the molecule–solid combination. If a component has a sufficiently low volatility, then multilayer adsorption and capillary transport may occur, leading to an enhanced flux of the condensable species along the walls.

Micropores These are encountered as the pore size is decreased to the size of the molecules. In this regime, much higher permselectivities are possible that depend on both molecular size and specific interactions with the solid. The so-called activated transport dominates in this class of porous solids.

6.2 Metallic Foams and Porous Metals

Porous metals and metallic foams are → composite materials in which one phase is gaseous and the other is a solid metal. The primary distinction between a porous metal and metallic foam is the relative density. Porous metals have a high bulk density (high volume fraction of solid) and independent, distributed voids, while metallic foams have a low bulk density and interconnected voids. The porosity within these metallic materials varies from 30 to 98 vol%.

As foamed plastics (the organic counterparts), these metallic materials possess a unique combination of properties due to the porosity such as impact energy absorption capacity, air and water → permeability, unusual acoustic properties, low thermal conductivity, and good electrical insulating properties. Therefore, foamed metals have a variety of different possible applications such as energy-absorbing systems, porous electrodes, sound absorbers, filters, insulation materials, heat exchangers, construction materials or for electromagnetic shielding. Some devices made from foamed aluminum are shown in Figure 6.6.

A number of different methods can be used for the production of metallic foams (Figure 6.7). Early attempts concentrated on casting techniques, similar to those used for plastics, with a gas serving as blowing agent. Casting techniques include all methods in which the base metal or → alloy is molten and then solidifies as foam. Metallic foams can also be produced by metallurgy techniques. Metallurgical processes mainly consist of preparing the structural constituents, such as powder particles, chips, fibers, wire meshes, and so on and subsequently compacting and sintering them. Not all the processes presented here are typical templating approaches, however, to give a better overview, other processes are also described briefly.

Not every process is applicable for every metal, and the materials obtained from the different techniques are quite different with respect to their structural characteristics. While casting techniques usually result in highly porous metal structures, processes based on powder metallurgy such as powder sintering give materials with a medium porosity.

Figure 6.6 Aluminum foam components for different applications.

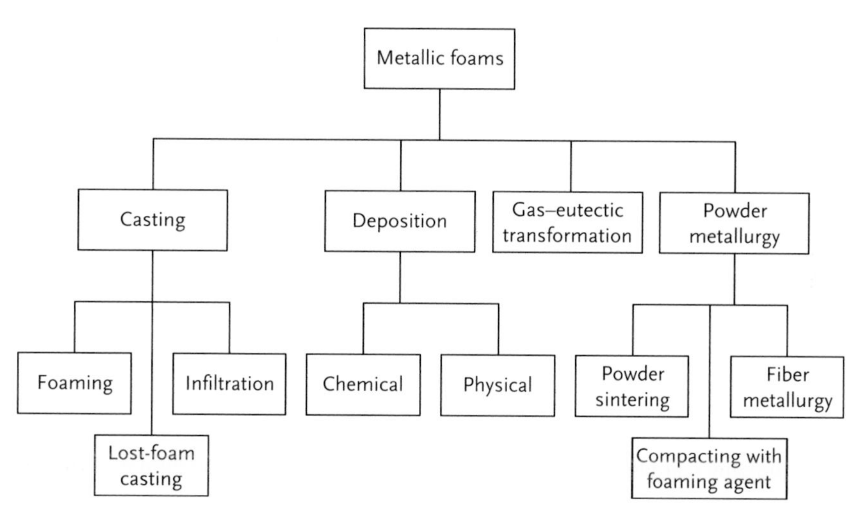

Figure 6.7 Summary of methods for the synthesis of porous metallic structures.

6.2.1 Casting Techniques

6.2.1.1 Foaming

In this method, a blowing agent, for example, a metal hydride, is added to a molten metal. The mixture is heated to decompose the blowing agent by evolution of a gas. The gas expands, causing the molten metal to foam. After foaming, the resultant body is cooled to give a solid (Figure 6.8) that looks like a continuous sponge with bubbles encapsulated.

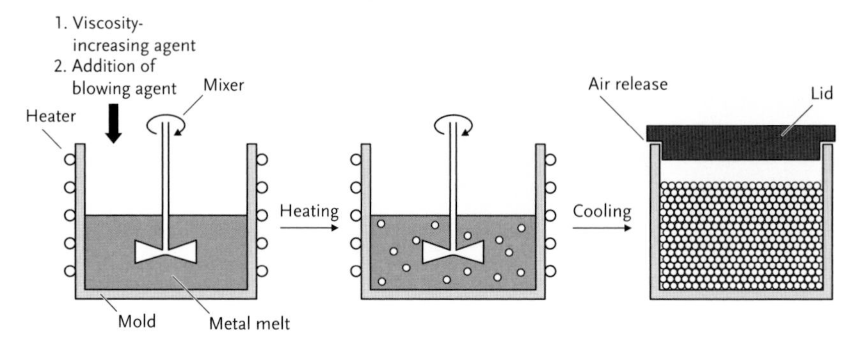

Figure 6.8 Foaming of a melt with a blowing agent.

Controlling this foam-generation process is rather difficult, and the foamed metal produced contains large bubbles that are distributed non-uniformly throughout the material. Different approaches have been used to overcome some of the problems, such as vigorous high-speed mixing to provide a good and uniform distribution of the foaming agent in the melt, or thickening of the melt, that is, increasing the viscosity of the molten metal to prevent the escape of the gas bubbles. Problems, however, persist due to the relatively short time between the introduction of the foaming agent and the generation of the foam.

Numerous variations of foaming exist; in some cases foam is generated by introducing gas directly into the melt during solidification.

The porosities typically reached by foaming techniques are very high (60–97%), and closed-cell materials are obtained. The walls separating the cells (or pores) are usually poreless.

6.2.1.2 **Lost-Foam Casting**

Connected pores in organic (plastic) foam are filled with a castable inorganic material, for example, gypsum, which is then hardened. Upon heating, combustion of the organic foam leads to formation of a sponge-like solid. This solid is used as a mold (template) for the metal that solidifies in its pores. The mold material (template) is then removed and a metallic foam having the same sponge form as the original plastic is obtained. The process is used primarily for making cellular metals with low melting points such as copper, aluminum, lead, zinc, tin and their alloys (Figure 6.9). The porosity of the final metallic material depends primarily on the porosity of the plastic foam used as → template; porosities up to 95% can be reached.

6.2.1.3 **Infiltration**

This method produces an interconnected cellular structure of sponge metal with porosities of about 70% by casting metal around densely packed granular

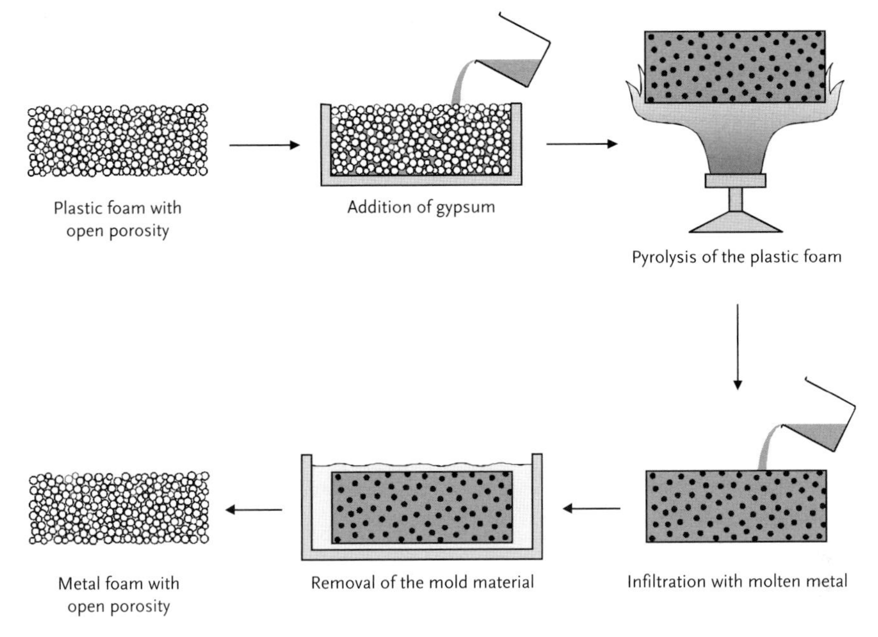

Figure 6.9 Principle of lost-foam casting.

→ templates introduced into the casting mold. These templates can be soluble (but heat-resistant) solid compounds, such as sodium chloride, which is later leached out to leave a porous metal. Other inorganic materials such as expanded clay, glass spheres, hollow corundum spheres, and so on can also be used as space-filling material that is removed in a second step (Figure 6.10). The advantage of using hollow spheres as the template is that it does not have to be removed.

As an alternative to casting molten metal around granules, these → templates can also be incorporated into metal melts. In this process, the metal is molten in a crucible, and to this vigorously stirred melt granules are introduced and thus uniformly dispersed. During the mixing, the metal is already allowed to cool until the mixture is sufficiently viscous to prevent segregation.

6.2.2
Gas–Eutectic Transformation

This process is usually applied for metal–hydrogen systems and is based on the saturation of the molten metal by the gas. Making the material consists of two steps:

1. Charging the melt with hydrogen to reach the → eutectic composition.
2. Solidification of the melt. During the solidifying process the saturated melt decomposes into a solid and a gas phase.

Foaming is not observed, because the gas is evolved as the melt freezes. The process is in many ways similar to conventional → eutectic solidification, the

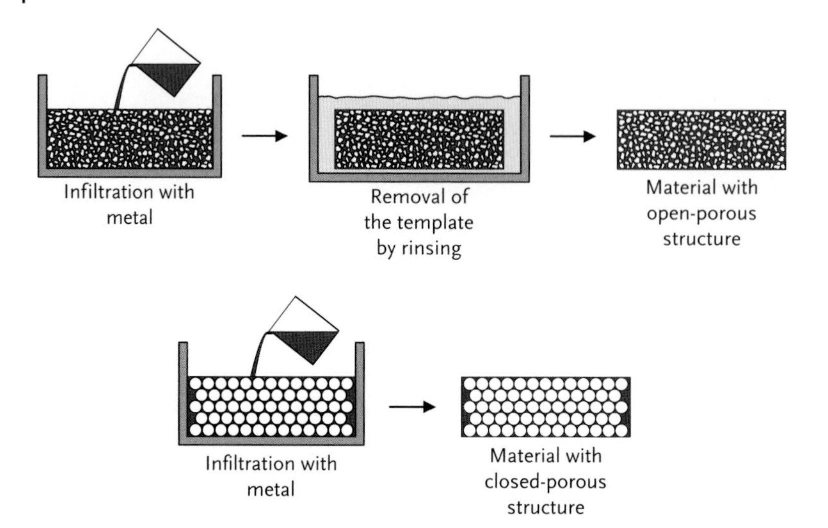

Figure 6.10 Process of casting metals around templates. Upper row: two-step process in which the template is removed after infiltration of the metal; lower row: one-step process using hollow spheres.

distinction being that the liquid decomposes in a solid and a gas rather than into two solids. The main process variables that govern the porosity and the size, shape, and orientation of the pores are the hydrogen level in the melt, gas pressure over the melt during solidification, direction and rate of heat removal, and chemical composition of the melt. Porosities of about 70% and pore sizes in the range of 10 μm to 10 mm are typically obtained. Materials produced by this method are called "gasars," which is an abbreviation of the Russian term for "gas-reinforced."

6.2.3 Powder Metallurgy

6.2.3.1 Powder Sintering

Powder sintering is one of the simplest methods of making porous metals. In the easiest case, the metal powder is filled in a mold and then sintered. This process is called loose-powder sintering. According to the sintering mechanisms (see Section 2.1.5), contacts between powder particles are established and grow during the time that the powder particles are heated in contact with each other. Application of pressure is not necessary. The materials produced are porous metals rather than foams, with porosities ranging from 40 to 60 vol%.

To increase the porosity, pore-forming or spacing agents are frequently added to the blend. Such agents decompose or evaporate during sintering, or are removed by sublimation or dissolution. The porosity can increase up to 90% and a foam can be obtained. In general, the structure of a sintered porous metal depends on the shape of the particles or other constituents to be sintered, the degree of compacting before sintering, and the sintering conditions.

The maximum porosity is usually achieved by sintering of hollow spherical particles. First, the hollow spheres are close-packed, followed by a pre-sintering step of this structure. The metal powder is filled into the interstices of this material and the process is finalized by the actual sintering process (Figure 6.11).

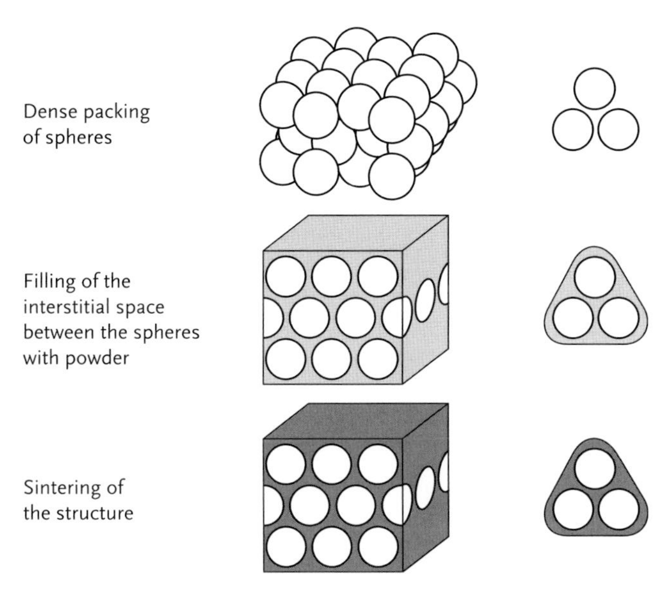

Figure 6.11 Principle of sintering with hollow spheres.

Besides the above-discussed processes, alternative techniques can be used for the production of metallic foams such as sintering of a slurry-saturated sponge, or foaming from a slurry.

6.2.3.2 **Compacting with Foaming Agents**

The metal powder is mixed with a foaming agent and subsequently compacted. As a result, a product is obtained in which the foaming agent is homogeneously distributed within a dense, virtually non-porous metallic matrix. This foamable material can be processed into sheets, rods, profiles, and so on using conventional techniques. The foamed metal parts are obtained by heating the material to temperatures above the melting point of the matrix metal. After expansion of the material to the desired degree, the foaming process is terminated by cooling below the melting temperature. The density of metal foams can be controlled by adjusting the amount of the foaming agent and several other foaming parameters. If metal hydrides are used as foaming agent, a content of less than 1% is sufficient in most cases. The materials obtained by this process are usually foams with porosities in the range of 60–90 vol%.

6.2.4
Metal Deposition

Different deposition techniques can be used: CVD (see Section 3.2), electrochemical deposition, and PVD. Deposition techniques are often used for making filters, catalysts, and structural parts. During electrochemical deposition, the metal is deposited on porous organic substrates, typically polyurethane; the substrate is pre-treated to make it more rigid and conductive. Typical porosities are around 90%.

PVD methods are carried out in a vacuum chamber containing a cold porous substrate and a vapor source of the base metal or → alloy. Metal atoms condense on the substrate, forming a continuous three-dimensional grid of preset thickness. The substrate is then removed by a thermal, chemical or other method, leaving a porous metal the macrostructure of which is a replica of the substrate. The process is suited for making metals with porosities up to 95%.

Another very remarkable process is the production of porous silicon by anodization. A silicon wafer is immersed in a solution of hydrofluoric acid, ethanol, and water, and subjected to an electric current for a brief time. The anodizing process gives an interconnected network of pores with a cell size of 10 nm. The density of the obtained porous silicon is as low as one-tenth of that of the metallic silicon, yet the material remains crystalline.

6.3
Soft Templates/Endotemplating

In this section, we will introduce the reader to endotemplating approaches towards ordered porous materials. In general, three steps are involved in forming porous materials by templating: synthesis, drying, and template removal.

Ordered porosity means that the pores are arranged in a regular fashion usually with a quite narrow pore size distribution. Figure 6.12 shows representative materials in different pore size regimes with the corresponding pore size distribution. As indicated there, materials in almost all pore size regimes exist.

One of the major differences between the materials, besides the average diameter of the pores, is their size distribution. For example, pillared clays (see Section 2.3.3) and porous glasses have broad pore size distributions, while zeolites and M41S molecular sieves show a narrow size distribution.

Probably the best known *microporous* materials with an ordered porosity synthesized by templating approaches are zeolites. The templating species are molecules, however, the mode of action is in many cases not fully understood. Zeolites are characterized by an extremely narrow pore size distribution because the pores are intrinsic features of the crystal lattice. Similar materials can be formed by coordination chemistry of for example, transition metals and organic linkers; the so-called metal-organic frameworks (see also Section 5.8.2).

In the second part of this section the reader will be introduced to a method by which structured *mesoporous* materials with narrow pore size distributions, designated as M41S-materials, are prepared.

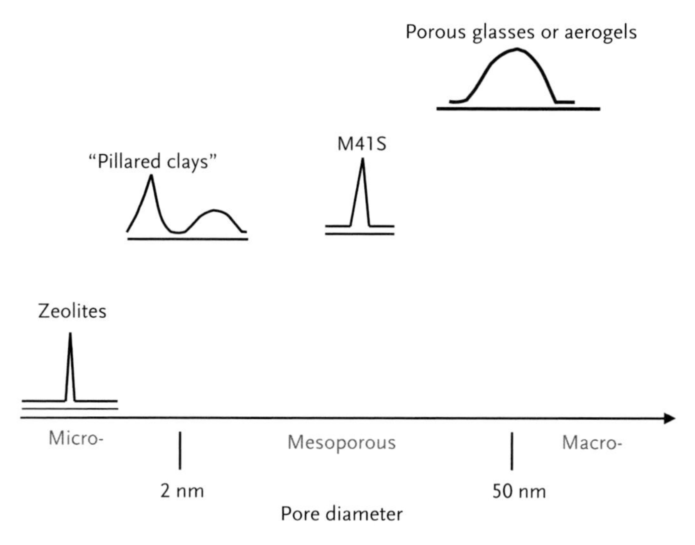

Figure 6.12 Representative porous materials with different pore sizes and the corresponding pore size distributions.

6.3.1
Templating with Molecules

Molecular or ionic compounds usually form crystal structures in which the molecules or ions are packed as closely as possible. In contrast, network-forming compounds (see also Section 4.1.1) can also crystallize in open framework structures in which the available space is not optimally occupied. In this section, the focus is on zeolites as an example of crystalline materials with network structures where the porosity is a result of the special three-dimensional, periodic arrangement of the network. Since the pores are part of the crystal structure, their narrow size distribution is not surpassed by any other material. Some other compounds, such as aluminophosphates or clathrasils (aluminum-free microporous silica) have similar structures.

Microporous inorganic solids are finding wide use as catalysts, molecular sieves and sorption media. The control of pore geometry and diameter is a key feature for many applications, especially those that rely on size and/or shape selectivity and a ready access to the pores. The porous nature of these crystalline ceramic oxides also allows their use as molecular-sized reaction vessels, as specific hosts, and as shape-selective conversion media.

Zeolites At present, about 40 naturally occurring zeolite minerals are known, and many more have been synthesized in the laboratory. Zeolites are crystalline aluminosilicates. The name zeolite, from the Greek *zeo* = to boil and *lithos* = stone, was coined to describe the behavior of the mineral stilbite, which loses water on heating and thus seems to boil.

The framework forming the regular one-, two-, or three-dimensional channel systems and cavities of molecular size, is constructed from fully linked tetrahedra. The most common zeolites are structurally based on corner-sharing silicon and aluminum TO_4 tetrahedra. If the network is built from $[SiO_4]$ tetrahedra only, the overall charge of the framework is zero, as already discussed for glass in Section 4.1.1. Isomorphic substitution of Si(IV) by the similarly sized Al(III) results in a net negative charge of the framework, which is equal to the number of the constituent Al atoms ($[AlO_4]^{5-}$ tetrahedra). This charge is balanced by exchangeable cations (M) present in the pores, in which also adsorbed water is located to remain electroneutrality.

The formula for a zeolite thus is:

$$M_{x/n}(Al_xSi_yO_{2(x+y)})\cdot zH_2O$$

The Si/Al ratio is adjustable over a wide range from 1 to infinity. The exchangeable cation M^{n+} is generally a group I or II ion, although other metal, non-metal, and organic cations may also be used. Many elements which form tetrahedral TO_4 structural units, can take part in the construction of zeolites such as $[PO_4]$, $[BeO_4]$, $[GaO_4]$, $[GeO_4]$, and $[ZnO_4]$.

Synthesis The diversity of structures in zeolites is enormous and, obviously, the synthesis conditions are crucial to which structure is obtained. The synthesis of zeolites is in principle based on two previously discussed synthetic approaches: The sol–gel process (see Section 4.5) and hydrothermal processing (see Section 4.4). Zeolites are synthesized from aqueous solutions of sodium silicates and aluminates, which contain either alkali metal hydroxides or organic bases to achieve a high pH. A gel is formed from the silicate and aluminate ions by condensation reactions, as has been discussed in Section 4.5. If the silica content of the zeolite is low, the product can often just be crystallized at 70–100 °C. Silica-rich zeolites, on the other hand, are mostly obtained by hydrothermal treatment of the gels. In this case, the gel is put in an autoclave for several days; the zeolites are typically formed between 100 and 350 °C. Variable parameters that determine the type of zeolite obtained include the composition of the starting solution, the pH, the temperature, and time program, and some history-related factors such as the aging conditions, the rate of stirring, and even the order of mixing.

A complicated interplay of all the different parameters governs the final zeolite structure. For example, if the Si/Al ratio of a given zeolite is to be changed, this is not possible by just changing the Si/Al ratio in the precursor solution. Instead, the whole set of parameters must be adjusted. Even today the synthesis of zeolites is mostly based on trial-and-error approaches.

Table 6.2 provides some rules of thumb on how different components of the reaction mixture influence the final zeolite structure.

Table 6.2 Influence of different components of the reaction mixture on the zeolite structure.

Reaction mixture composition	Primary influence
SiO_2/Al_2O_3 ratio	Framework composition
H_2O/SiO_2 ratio	Rate; crystallization mechanism
OH^-/SiO_2 ratio	Silicate molecular weight
Inorganic cations/SiO_2 ratio	Structure; cation distribution
Organic additives/SiO_2 ratio	Structure; framework Al content

Besides acting as counterions to balance the zeolite framework charge, the metal or ammonium cations present in the reaction mixture additionally appear to have a structure-determining effect. The latter effect is also attributed to uncharged organic additives such as amines. With the extension of additives to molecules such as amines and quaternary ammonium ions, the number of zeolite structures has been greatly extended mainly because variation of the silica/alumina ratio over a wider range is possible. Typical examples of organic additives, for example, diamines, pyridine, tetraalkylammonium ions, and many more, are shown in Figure 6.13.

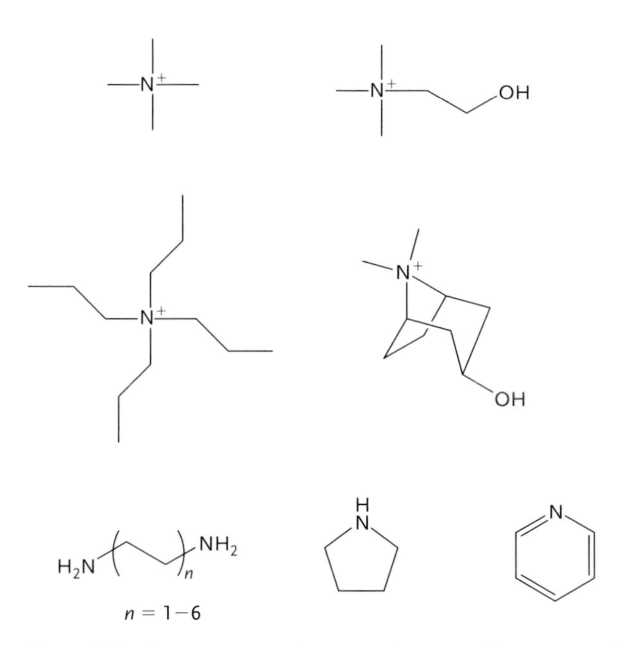

Figure 6.13 Some representative ammonium or amino compounds used as organic additives in zeolite syntheses.

A templating theory has been postulated for the role of the cations and the organic additives in stabilizing the formation of structural subunits in the reaction mixture. A chemical species is considered to be a template or structure-directing agent, if crystallization of a specific zeolite structure is induced that could not be formed in the absence of the reagent.

The notion that the zeolite structure grows around the → template is probably too simple. For example, it has been shown repeatedly that, in the presence of a particular organic species, crystallization of various zeolite structures can occur. Conversely, one structure can be directed by a number of different amines.

The role of the "template" is more diverse and complex than initially perceived. It can:

- behave as a structure-directing or templating agent;
- act as a gel modifier, particularly influencing the Si/Al ratio;
- act as a void filler; and
- influence physically and chemically the formation and aging of the gels, and the crystallization process.

Structure There are a large number of complex zeolite structures known, which are characterized by the building units, size and arrangement of the pores, and their interconnection.

The framework structure of zeolites can be constructed by assembling structural units with increasing complexity starting from the tetrahedral building blocks; the pore structure is a result of the particular network structure. However, some general rules must be obeyed:

- The basic building units are tetrahedral.
- All tetrahedra share all their corners with neighboring tetrahedra.
- The arrangement of the building blocks must be periodic (crystallinity).

Another important feature of the traditional zeolite structures is that there are only very few examples known having Al–O–Al linkages. This has the consequence, that with only Si–O–Al and Si–O–Si bonds forming the network, the maximum substitution rate of silicon by aluminum is 50%.

The three-dimensional arrangement of the TO_4 tetrahedra – also called basic or primary building units (BBU/PBU) – can occur in numerous ways. The wealth of zeolite structures is additionally due to the flexibility of the T–O–T bond angle, which can assume values in the extensive range of 120°–180°.

Sometimes it is useful to combine BBUs to construct larger composite or secondary building units (CBU/SBU) that characterize the topology of the network. These CBUs can be single rings or single chains, but also more complex structures such as branched chains or even polyhedral structures. For clarity it should be noted that in the graphic representation of CBUs each node represents the center of a tetrahedron, that is, the position of the silicon or aluminum atoms. This is shown for one of the CBUs – a six-membered ring structure – in Figure 6.14, as an example. Several more CBUs are shown in Figure 6.15.

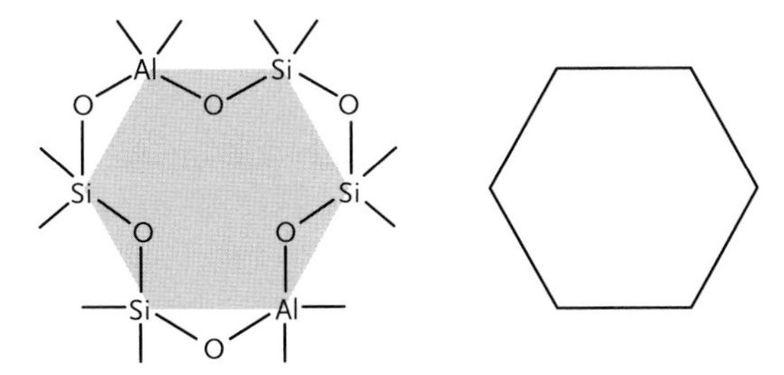

Figure 6.14 Six-membered ring structure of [SiO_4] and [AlO_4] tetrahedra (left) that can be found in zeolites, and its schematic representation as a CBU (right).

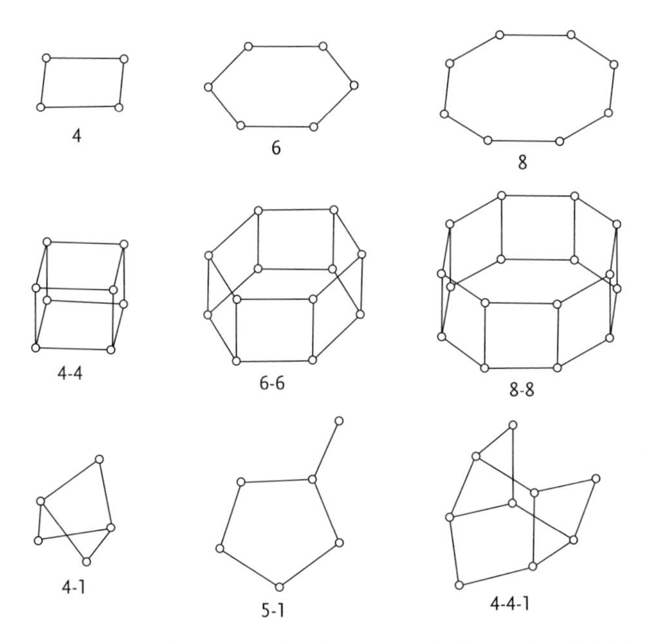

Figure 6.15 Selected examples of composite building units (CBUs) found in zeolite structures.

The CBUs provide a convenient method of topologically describing and relating different zeolites. Figure 6.16 shows an example in which the network structure (ZSM-5) can be imagined to be built only from 5-1 CBUs. The pores with openings formed from 10 [TO_4] tetrahedra are a consequence of the particular arrangement of the CBUs.

For most zeolite applications the pore size is one of the key features. The pore or channel openings range from 0.3 to 1 nm, depending on the structure of the

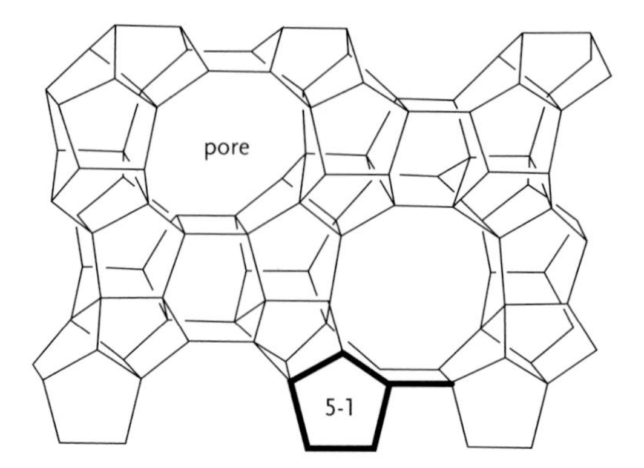

Figure 6.16 Network structure of ZSM-5.

zeolite. Figure 6.17 shows cross sections through pores of different zeolites. For ZSM-5 the pores are built from 10 tetrahedra, but for other zeolites larger or smaller sizes can be achieved by different arrangements of the CBUs. These pores can also differ in the shape of the opening.

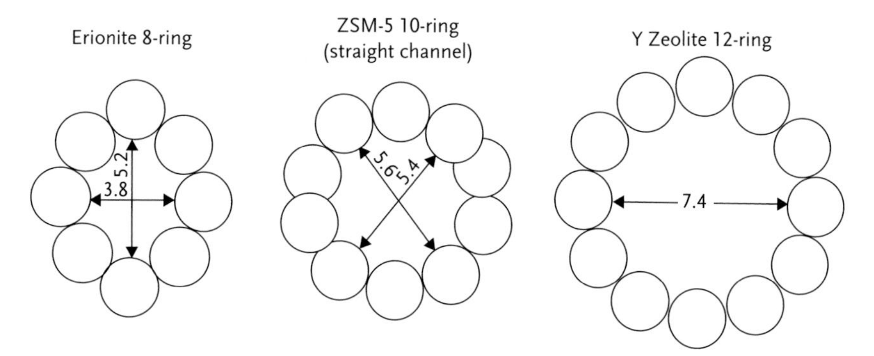

Figure 6.17 Examples of three types of pore openings in zeolites: Erionite (left) contains an eight-ring pore opening, ZSM-5 (center) a 10-ring system, and type-Y zeolite (right) a 12-ring pore system (the diameters are given in Å).

The pore system in zeolites can be described in many different ways which are schematically displayed in Figure 6.18 for zeolite A.

In Figure 6.18a the sodalith cages, zeolite A is built-up from, are presented. A *cage* is defined as a polyhedral pore, whose windows are too narrow to be penetrated by guest species larger than H_2O. In contrast, the polyhedral pore shown in Figure 6.19b is called a *cavity*, since at least one face is large enough to be entered by guest species. However, in a cavity, the pore is not infinitely extended as it is in

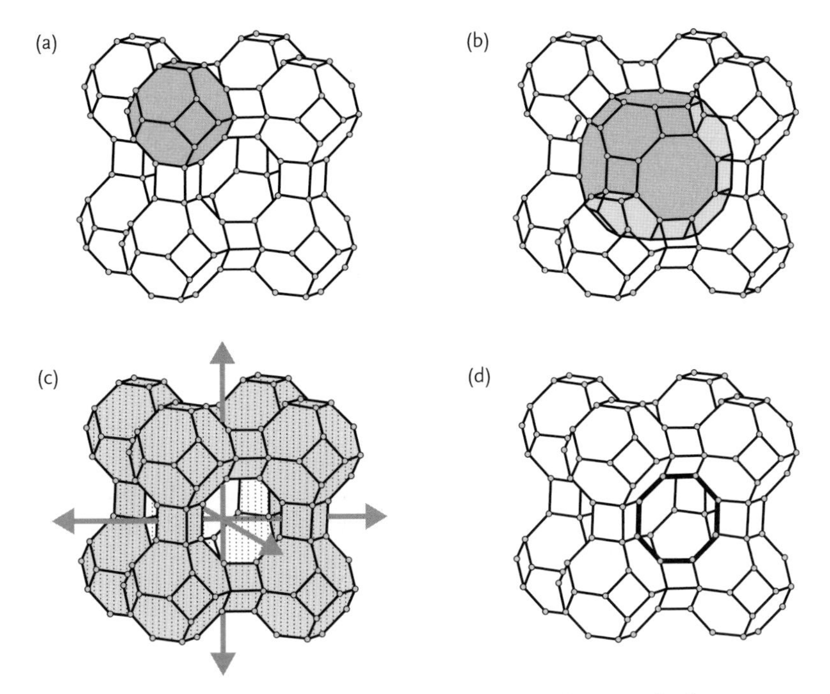

Figure 6.18 Features of the pores in zeolite A: (a) the sodalite cage, (b) the a-cavity, (c) the three-dimensional channel system, and (d) the eight-membered ring defining the effective channel width.

a *channel*, which is shown in Figure 6.18c. A channel is at least in one dimension infinitely extenuated and large enough to allow guest species to diffuse along its length. Channels can also intersect to form two- or three-dimensional channel systems. The *effective width* of the channel (Figure 6.18d) is its most important feature, since it determines the size of guest molecules that can access the zeolite structure.

In passing, it should be pointed out that the combination of the CBUs is not as simple as playing with Lego. In the representation shown in Figure 6.18, the primary units were reduced to one point (the corners of the CBUs). In reality, the three-dimensional shape must also be considered. For example, when we look at the tetrahedra that comprise the simplest CBU, the four-membered ring, it is immediately obvious that there is more than one way of joining together the four-ring building units (Figure 6.19).

Last, but not least, one must also consider the connectivity of the pores, which can be viewed as one-, two-, or three-dimensional tubes or channels. For example, analcime consists of a system of non-intersecting, one-dimensional channels, while mordenite is a two-dimensional, intersecting channel system. Three-dimensional channel systems can be found in ZSM-5 (Figure 6.20).

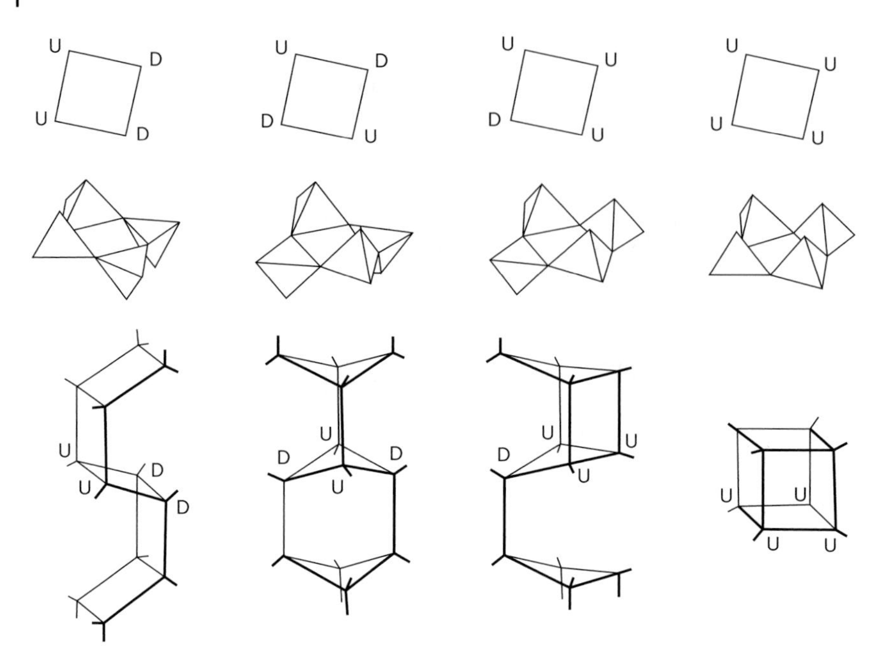

Figure 6.19 Possible linkages for the four-membered ring CBU. Upper row: schematic representation of the CBU; U = up and D = down. Center row: the tetrahedral view. Bottom row: chain sequences formed from the differently oriented CBUs.

6.3.2
Templating with Supramolecular Arrangements of Molecules

In 1992, a new concept for the synthesis of porous materials was introduced by scientists of Mobil Research and Development Corporation. In contrast to zeolites – which are synthesized with single, solvated organic molecules or metal ions as the template – supramolecular arrangements of molecules were used. Larger pore sizes are accessible with this approach.

Typically, amphiphilic surfactant molecules (or → lyotropic liquid crystals) are used as structure-directing agents.

The pores in the obtained materials are about an order of magnitude larger than those in zeolites, and can be tailored within the 2–10 nm range. The materials exhibit large specific surface areas, large pore volumes and, owing to the regularity of the supramolecular template, a → monomodal, narrow distribution of pore sizes. In contrast to zeolites, the pore walls are not crystalline, but amorphous. Mesoporous silica has been prepared as transparent hard spheres, glass sheets, hollow spheres, fibers, thin films, or monoliths.

Owing to the larger pores and high surface areas, these new materials, denoted as M41S phases, are of great interest for applications in size- and shape-selective processes, for example, as catalysts, catalyst supports, adsorbents, or as host structures for nanometer-sized guest compounds.

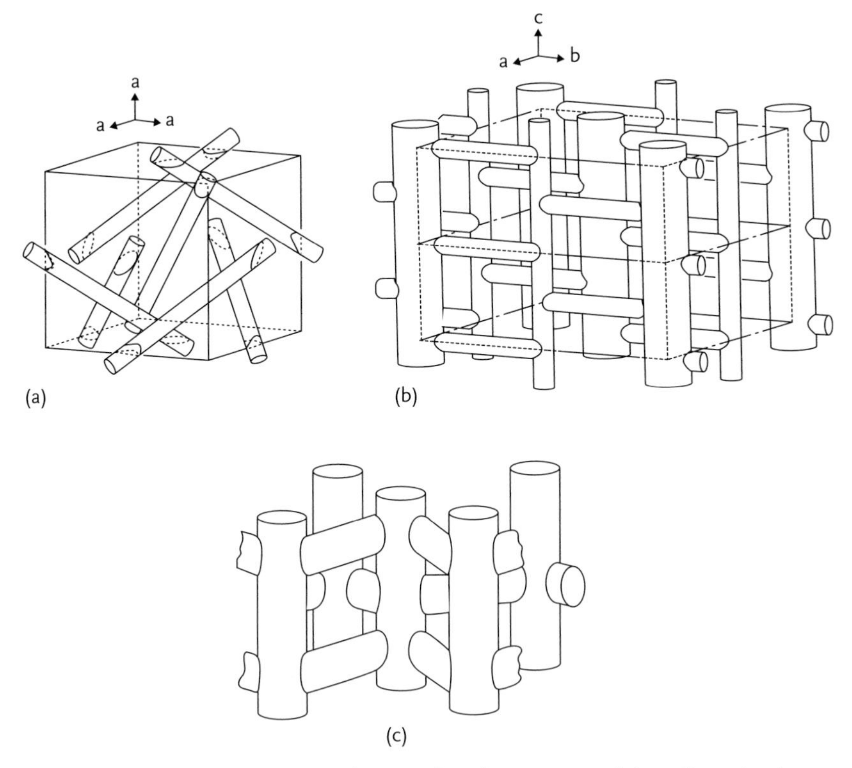

Figure 6.20 Channel representation for examples of one-, two-, and three-dimensional systems; (a) analcime; (b) mordenite; (c) ZSM-5.

Three important members of M41S materials that are distinguished by their different pore structure are shown in Figure 6.21:

- **MCM-41**: one-dimensional hexagonally ordered cylindrical channel structure.
- **MCM-48**: three-dimensional bicontinuous cubic channel structure.
- **MCM-50**: two-dimensional system of lamellar silica sheets interleaved by surfactant bilayers. (MCM = Mobil composition of matter.)

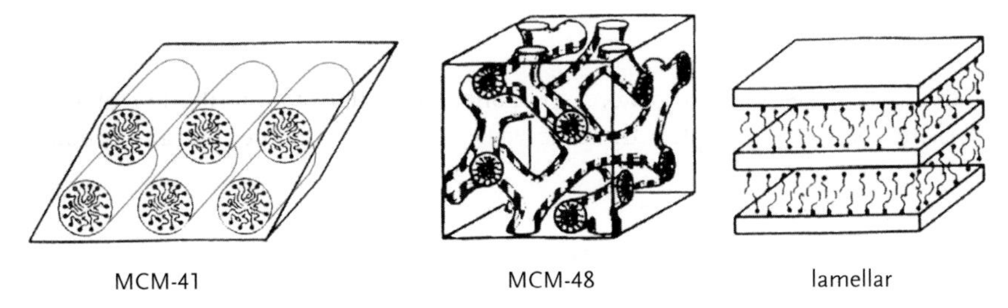

Figure 6.21 Different structural types of the M41S family: hexagonal MCM-41; bicontinuous cubic MCM-48; lamellar MCM-50.

Similar to the zeolite synthesis, four reagents are required for the M41S synthesis: water, an →amphiphilic molecule (Figure 6.22), a soluble inorganic precursor, and a catalyst. Depending on the inorganic precursor, the synthesis conditions vary over a wide range: from seconds to days and from 180 to −14 °C. In the following sections the different steps of the synthesis of these mesostructured materials are described, starting with the formation of a supramolecular arrangement of the → template molecules, followed by the actual templating step, and finally the removal of the template.

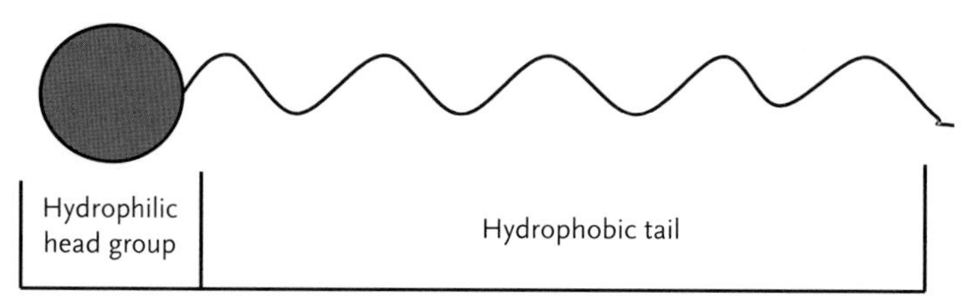

Figure 6.22 Amphiphilic surfactant molecule.

Supramolecular Arrangement of the Template Molecule As a result of their amphiphilic nature, surfactants can associate into supramolecular arrays. For example, cetyltrimethylammonium bromide (CTAB, $CH_3(CH_2)_{15}N(CH_3)_3{}^+Br^-$) will form spherical → micelles. This arrangement minimizes the unfavorable interaction of the hydrocarbon tails with water, but introduces a competing unfavorable interaction – the repulsion of the charged head groups. The balance between these competing factors determines the relative stability of the micelles.

The extent of micellation, the shape of the micelles, and the aggregation of micelles into → liquid crystals depends on the surfactant concentration. A schematic phase diagram for a cationic surfactant is shown in Figure 6.23.

At very low concentrations the surfactant is present as individual molecules dissolved in solution. When the surfactant concentration is gradually increased the following successive stages can be observed:

1. The surfactant molecules form small spherical aggregates (→ micelles) above the critical micelle concentration (cmc).
2. These spherical micelles can coalesce to form elongated cylindrical micelles when the amount of solvent available between the micelles decreases.
3. Formation of → liquid-crystalline (LC) phases occurs at slightly higher concentrations, starting with a hexagonal close-packed LC arrangement, followed by the bicontinuous cubic phase and finally resulting in lamellar structures (Figure 6.23).
4. At very high concentration, inverse phases can exist; here water is solubilized in the interior of the → micelle, and the hydrophilic head groups point inward.

Surfactants with a wide variety of sizes, shapes, functionalities, and charges exist, and can be used as → templates for the synthesis of mesoporous solids.

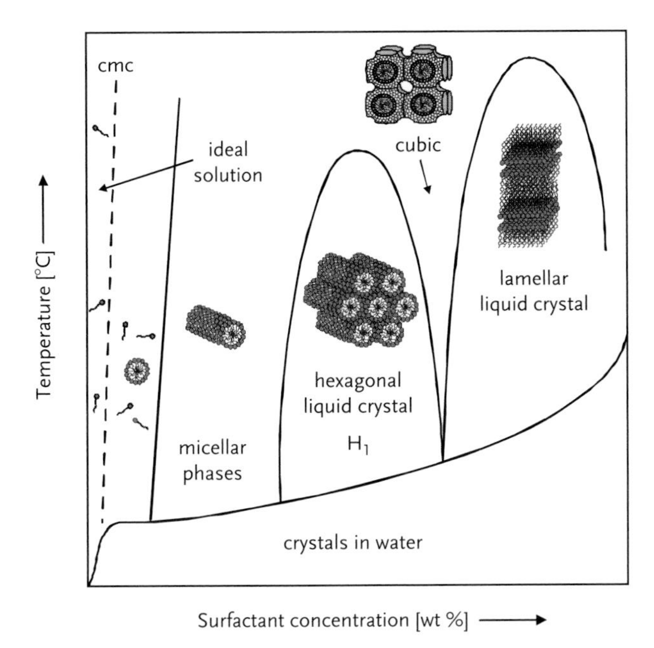

Figure 6.23 Schematic phase diagram for a surfactant in water.

Templating The original MCM-41 synthesis was carried out under hydrothermal conditions by using CTAB as the structure-directing agent and tetraethoxysilane as the silica source.

Two mechanistic pathways leading to the formation of these mesoporous materials are discussed (Figure 6.24).

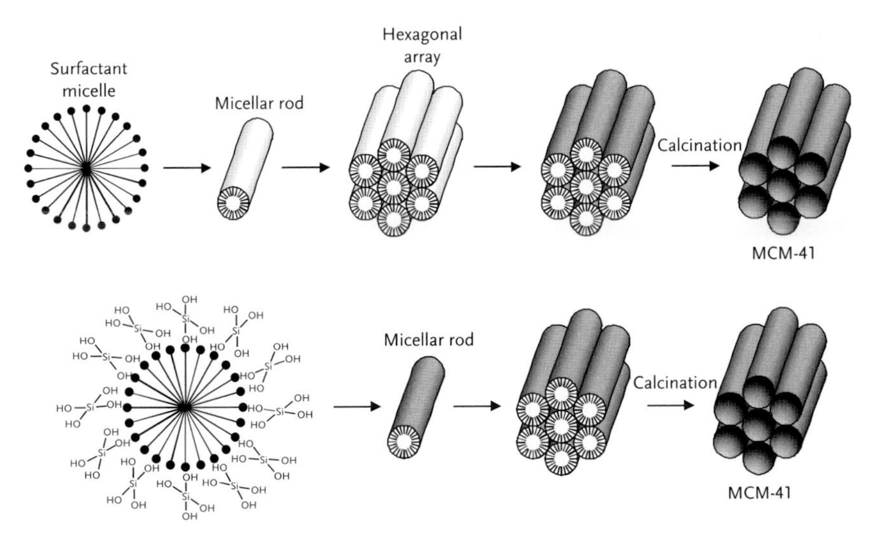

Figure 6.24 Possible mechanistic pathways for the formation of MCM-41; the gray shading represents the silica.

1. The surfactant molecules organize first, independent of the inorganic silicate → polycondensation. In this case, which is also called a liquid-crystal-initiated pathway, the siliceous framework polycondenses around the preformed surfactant aggregates.
2. The second mechanism assumes that anionic siliceous species associate first with the surfactant. This assembly then organizes in supramolecular arrays. This pathway is also called a silicate-anion-initiated pathway.

The two pathways discussed are only limiting cases. Experiments show that the formation of mesostructured materials is almost independent of which aggregation stage of the surfactant the inorganic species is added. This implies that the inorganic source does not simply petrify the LC array, but instead cooperatively coassembles at some stage with the surfactant to form LC phases.

The M41S phases described so far are generally synthesized via an electrostatic templating procedure based on surfactants with cationic head groups, and anionic inorganic building units (S^+I^-; S represents the surfactant, I the inorganic species). A huge variety of this type of surfactants exist that can be utilized as → templates. Based on this charge interaction, other combinations of ion pairs, including charge-reversed (S^-I^+) and counterion-mediated ($S^+X^-I^+$ or $S^-M^+I^+$, with X^- being halides and M^+ being alkali metal ions; Figure 6.25) synthesis pathways are possible. Besides these Coulombic forces, other interactions at the inorganic/organic interface can be used for templating such as covalent bonding of the → template to the inorganic network-forming species (S-I) or neutral templating routes based on hydrogen bonding (S^0I^0) (Figure 6.25).

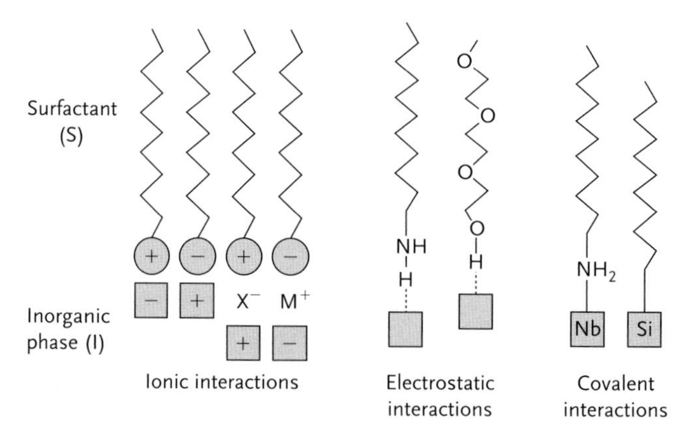

Figure 6.25 Schematic presentation of possible interactions at the inorganic – organic interface (S = surfactant, I = inorganic phase, M = metal, X = halide).

By carefully choosing the template molecule, the pore diameter can be adjusted over a wide range. An increase in pore diameter can be achieved by the addition of an auxiliary organic molecule such as mesitylene, which is trapped in the → micelle by increasing the volume of the hydrophobic core (Figure 6.26).

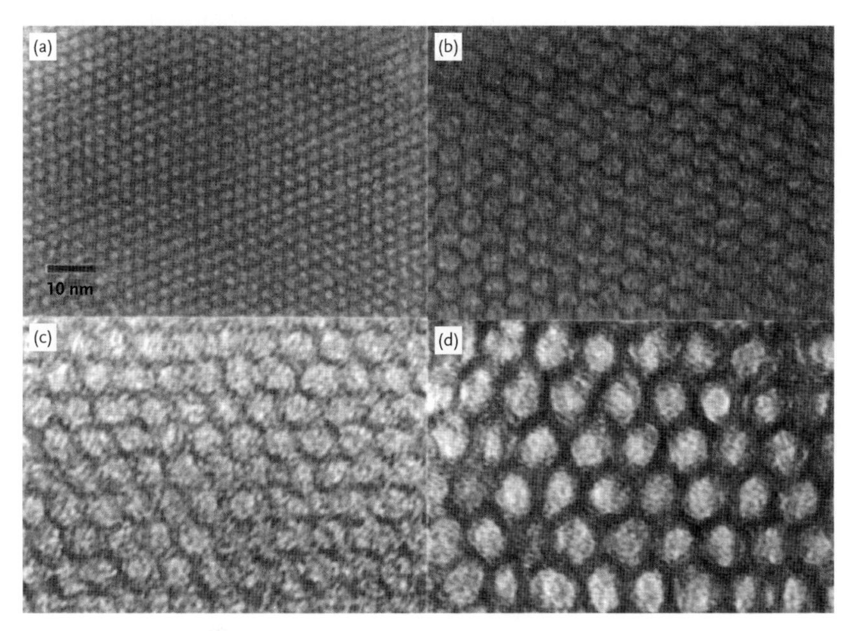

Figure 6.26 Transmission electron micrographs of MCM-41 materials having 2, 4, 6, and 10 nm hexagonally arranged pore channels (increasing pore size from (a) to (d); the black areas represent the silica walls, the lighter areas the pores).

Removal of the Template The removal of the → template is the last step in the synthesis of mesoporous structured materials. This can generally be performed in different ways:

- by solvent extraction,
- by → calcination,
- by oxygen plasma treatment, or
- by supercritical drying (see also Section 4.5.6).

The first two methods are most often applied. Extraction is performed by several washing steps. Often, organic solvents with dissolved acids, that is, ethanol with HCl, are used to remove the template from electrostatically templated materials. Covalently bonded templates cannot be removed by this procedure. Calcination is typically performed between 400 and 600 °C in various atmospheres such as nitrogen or air.

While framework structures such as the hexagonal or cubic phase are typically stable upon removal of the template with only a small degree of shrinkage, lamellar structures collapse during the removal of the → template.

The examples discussed show that a deliberate design of pores is possible in a wide pore size range, even with narrow pore size distributions. The application of more complex template systems could result in a huge extension of this approach also in the macroporous regime. In some cases a clear distinction between endotemplating and exotemplating – as discussed below – for such ordered mesoporous materials in which a preformed liquid crystal phase is applied, is not possible.

The underlying principle discussed above for silica materials may also be applied to the formation of other mesoporous materials with different chemical composition.

6.4
Hard Templates/Exotemplating

Exotemplates are rigid or compacted structures, thus hard templates, with a pore system in which another solid can form. The exotemplating approach can also be termed "confined space synthesis" (see also Section 7.3.3) or "compartment solidification." Depending on the connectivity of the pore system of the hard template or mold, differently dispersed solids are formed, ranging from small particles to fully continuous solid phases (Figure 6.27). If the pore system of the exotemplate is continuous and high loadings are achieved, a continuous porous solid with voids corresponding to the template is obtained ("nanocasting"). If the pores of the mold

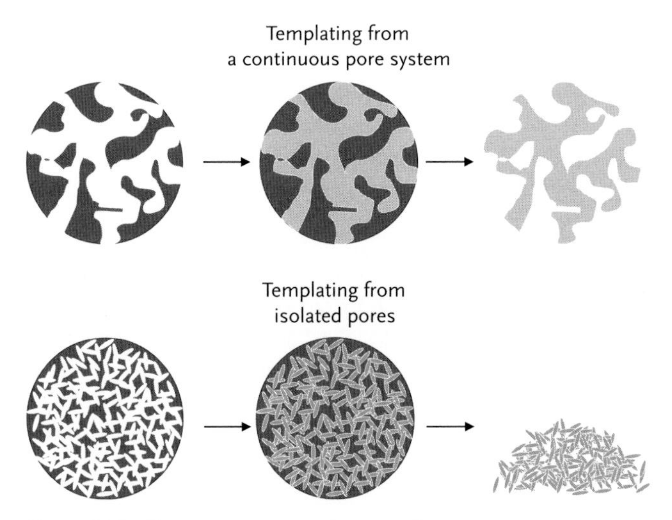

Figure 6.27 The two prototypes of exotemplating.

are not three-dimensionally connected and/or the filling of the pore system is not sufficiently high, small particles are obtained with high surface areas, but predominantly textural porosity. The nanocasting approach is discussed below in more detail.

6.4.1 Nanocasting

The application of casting techniques in the formation of porous metals was already discussed in Section 6.2.1. By filling the void of the mold with a precursor, or the material to be cast, followed by processing and removal of the mold, a porous replica structure is obtained. If this process is scaled down to the nanometer scale, "nanocasting" is the most suitable word to describe the procedure. In principle, nanocasting is not limited to hard templates, but for hard templates the relation between mold and replica structure is often well defined.

The formation of nanostructured or nanoporous materials through the nanocasting approach involves three different processing steps:

1. formation of the mold /template;
2. casting with and processing of the precursor solution; and
3. removal of the template.

Typical mold/template structures include, for instance, zeolites, ordered mesoporous silica, ordered mesoporous carbon or even larger structures such as alumina membranes, and colloidal spheres. In the following, two different examples of nanocasting procedures will be discussed.

6.4.1.1 Ordered Silica as Template/Mold

Mesoporous silica with a regular arrangement of the pores (see also Section 6.3.2) is a very interesting template/mold material due to its controllable morphology and structure. The pores can be deliberately tailored with respect to size and connectivity, and the material can be prepared in various shapes from spherical to rod-like particles or even as monoliths or coatings. One requirement for an optimal mold material is a simple and highly efficient removal step by either chemical processes (e.g., leaching, combustion) or physical methods (e.g., thermal treatment). Silica can easily be removed via solution-based processing such as leaching with hydrofluoric acids or strong bases without any thermal treatment.

Typically, the precursors for the desired casted material are infiltrated into the template/mold structure. They must either be gaseous, liquid, or highly soluble at moderate conditions to achieve the required high loadings for the infiltration. In addition, conversion to the desired target material should be simple and result in as little volume shrinkage as possible without any reaction with the mold material.

Porous carbons are of interest for a variety of applications, for example, sorption, catalysis, energy storage, and so on, but mesoporous carbons with a regular arrangement of the pores are difficult to obtain through solution-based processes. Nanocasting offers a simple and efficient route towards such materials. In brief, the syntheses can be described as follows (Figure 6.28): mesoporous silica with a specific pore structure as the template is impregnated with a carbon-containing precursor solution to give a polymerizable mixture. A variety of suitable precursors can be applied, such as sucrose, furfuryl alcohol, phenolic resins, mesophase pitch, polydivinylbenzene, acrylonitrile, and so on. After polymerization and subsequent carbonization of the precursor in an inert atmosphere, a carbon/silica composite material is obtained, from which silica is removed by leaching with hydrofluoric acid or under basic conditions. The mentioned precursors do not only decompose upon heat treatment, but give a high carbon yield, thus ensuring that the pores are well filled and a negative replica structure of the silica mesopore system is obtained.

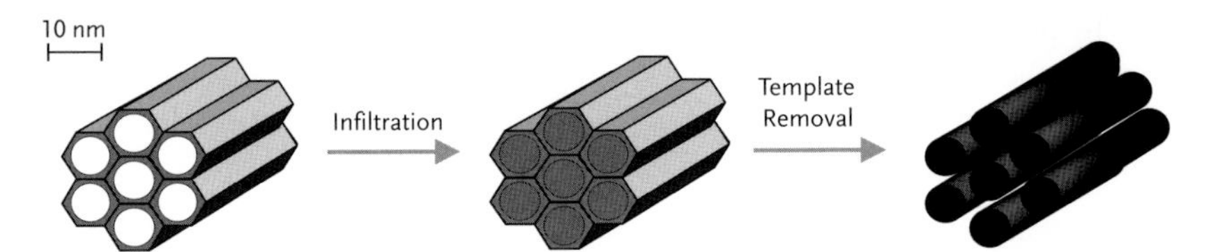

Figure 6.28 Schematic illustration of the nanocasting pathway applying silica with periodically arranged pores as mold/template.

6.4.1.2 **Colloidal Crystals as Templates/Molds**

Materials with a regular arrangement of the pores can not only be prepared in the micro- and mesoporous regimes, but also in the macroporous one. Here, the structures are typically not templated by molecules or supramolecular arrays of molecules, but by the packing of a solid material itself. A similar approach has already been described in the previous section for the preparation of metal foams (see Section 6.2.1).

Solid templating from sphere packings, also called colloidal crystal templating, is one of the most often used approaches for the preparation of ordered macroporous solids. The general concept and the different production steps are displayed in Figure 6.29 and are rather simple: First, a colloidal crystal is formed of close-packed, uniformly sized spheres (these three-dimensional colloidal crystals resemble naturally occurring opals), the interstitial spaces of which are filled in a

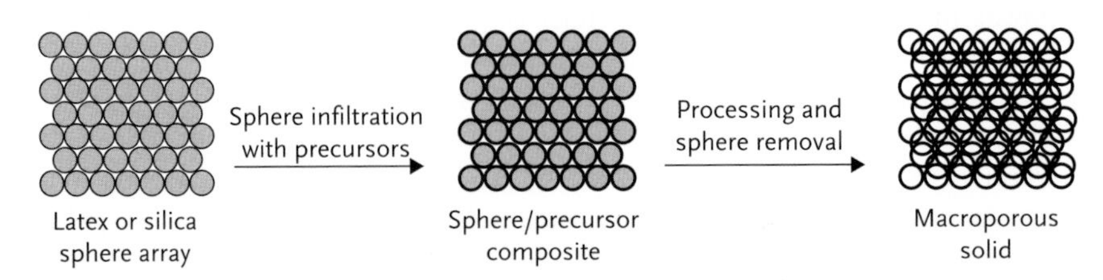

Figure 6.29 General synthesis scheme for periodic macroporous solids.

second step with a fluid precursor capable of solidification, and finally the template is removed to obtain a porous inverse replica of the template array, also called inverse opal.

For the first step, which is the choice and optimal packing of spheres to give a colloidal crystal, typically in either a face-centered cubic (fcc) or hexagonal dense packing (hdp), several requirements must be fulfilled:

- The template must be removable.
- It must be compatible with the process conditions.
- Wettability with the network-forming precursor solution must be given.
- The spheres should have a narrow particle-size distribution to achieve an optimal packing.

These requirements reduce the choice of templates to almost only two different classes:

- Silica spheres produced by the Stöber process that can be prepared from less than 50 nm to 2 μm in diameter (see also Section 4.5.2; Figure 4.53).
- Organic polymer spheres produced by → emulsion polymerization that are accessible in the range of a few 100 nm to several μm, such as polystyrene (PS) or poly(methylmethacrylate) (PMMA) spheres.

The spheres can be arranged in close-packed structures by several methods including centrifugation, gravity sedimentation, or → electrophoresis.

The colloidal crystals formed from the above-mentioned spheres are typically very stable and can be used for impregnation. The key point in the synthesis is the filling of the interstices in the sphere packing with a fluid, and the conversion of the fluid into a solid in a second step. Low-viscosity fluids are desirable to maximize penetration and achieve homogeneous and complete filling of the void space. These fluids may be pure liquids, solutions, or dispersions, for example, metal alkoxides or sols (e.g., by sol–gel chemistry; see also Section 4.5), molten metals, and so on, but also vapors (see Section 3.2.7 on chemical vapor infiltration).

Porous structures are finally produced by removal of the templates. For organic polymers this is often carried out by calcination, during which the infiltrated precursor is simultaneously converted into a solid. Another option is extraction of the colloidal crystal by appropriate solvents. Silica sphere templates can be removed by dissolution in aqueous hydrofluoric acid.

According to the list of possible infiltration techniques, the range of macroporous materials prepared by this method includes not only oxides such as silica (Figure 6.30), titania, and various other oxides (iron oxides, tungsten oxide, Co_2TiO_4), but also metals (Ni, Co, Fe, Au), metal → chalcogenides (CdS, CdSe), various allotropes of carbon, polymers (polyurethanes), and hybrid silicates (e.g., $O_{1.5}Si\text{-}CH_2CH_2\text{-}SiO_{1.5}$). Thus, insulators, semiconductors and metallic compositions can be prepared.

A target application for materials prepared by colloidal crystal templating, besides the typical applications for porous materials in areas such as sorption, catalysis, and so on, is for photonic crystals (where pores are not the key feature,

Figure 6.30 Scanning electron micrographs of the different processing steps of a periodic macroporous silica material prepared from a colloidal crystal polystyrene template.

but instead refractive-index differences). Photonic crystals are materials that exhibit periodic modulations in their dielectric constant.

6.5
Templating Towards Multiscale Porosity

For many applications, chemists in materials science seek to copy the properties of natural materials. That is especially true with regard to structural features of the material, for example, hierarchical organization of the network from the nanometer up to the micrometer scale (as found in almost all biominerals, e.g., cortical or trabecular bone (see also Section 4.3.1 on biominerals), sea sponges, etc.), and with respect to multifunctionality, for example, hydrophobicity, catalytic functions, and so on. Materials comprising several levels of pore sizes (micro–meso, meso–macro, micro–meso–macro, meso–meso–macro, or other combinations) are termed materials with multiscale porosity or a hierarchical organization of the pores. Manmade hierarchically porous materials have still to achieve the level of sophistication and the tailored architecture of their biological counterparts.

Simultaneous control over size from Ångstrom to micrometers, shape, and spatial distribution of the pores potentially enables the fabrication of hierarchical porous structures exhibiting novel properties and multiple functions. They are desired for a broad variety of applications, including chromatography, sensing, controlled release, scaffolds for biomedical applications and catalysis, due to the multiple benefits that arise from each of the pore size regimes. For example, uniform mesopores (2–50 nm) provide size- or shape selectivity and high specific surface areas, and macropores (>50 nm) facilitate mass transport through the material and to the active sites. The applicability of a material, however, not only depends on its pore size and size distribution, but on all characteristics of the porosity, such as total proportion of pores, ratio of closed to open pores, tortuosity and interconnectivity, gradients, and so on and very importantly, its chemical composition.

Different preparative routes have been developed for the formation of porous materials (mostly powders) with hierarchical organization of the pores, including, powder processing (e.g., sintering, see Section 2.1.5), foaming (Section 6.2.1), hard templating (e.g., nanocasting of preformed gels), impregnation of biological templates or colloidal crystals, nanotectonics (e.g., impregnation with preformed porous particles) as well as multiple templating routes and phase-separation processes. The latter two strategies will be discussed in the following.

6.5.1 Dual or Multiple Templating

Maximum control of the porous structure can be obtained by combining different templating strategies. Larger templating objects such as colloids, microspheres (latex spheres), macromolecules such as bacterial threads and emulsions are exploited to create macroporosity on different length scales from submicrometer to several hundred micrometers. Supramolecular arrays of molecules such as lyotropic phases of amphiphilic surfactants or block-copolymers can be used as structure-directing agents for a deliberate design of pore systems in the mesoscopic regime (up to $\sim$30 nm), and molecular templating can be applied to give micropores (see Section 6.3.1).

In principle, a combination of the above mentioned templating strategies can be used for the design of multiscale porosity with the intrinsic difficulty in the preservation of the existing levels of organization while introducing another. This has recently been demonstrated for bimodal and even trimodal pore size distributions.

One example based on the use of colloidal crystals (see Section 6.4.1) as hard templates in combination with soft templating methods is the generation of smaller (micro-, meso-, or small macro-) pores in the walls of porous materials. In principle, porosity within the walls may be present even without a secondary template, resulting from textural mesoporosity between network-forming nanoparticles. A better control of the wall architecture is possible, when a secondary or even a tertiary template is employed, including ionic and non-ionic surfactants, block-copolymers, ionic liquids, structure-directing agents for zeolites, or colloids. Depending on the choice of the soft template, for example, surfactants (see Figure 6.31), different pore structures may be present (as discussed in Section 6.3.2). In this case, however, confinement effects and interaction between the various templates may also play a decisive role in the formation of the mesopores, resulting in different structures and orientations. The different orientation of the mesopores can be seen in Figure 6.31 for the two surfactants applied: While the application of the alkyloligoether $C_{16}EO_{10}$ resulted in two-dimensional hexagonal channels that are oriented perpendicular to the polymer spheres, the ethylene-oxide–propylene-oxide-based polymer ($EO_{20}PO_{70}EO_{20}$) gave pores that run parallel to the latex spheres.

In another approach, differently sized hard templates are employed, such as a combination of 465-nm diameter polystyrene spheres and 84-nm diameter PMMA

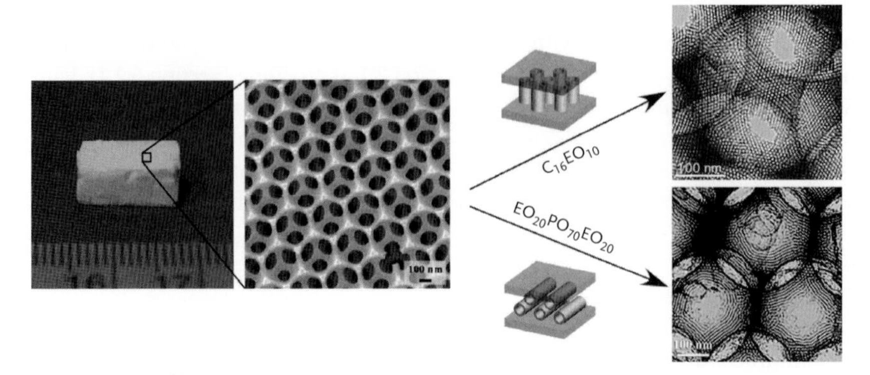

Figure 6.31 A monolithic inverse opal material. The SEM image shows the macropore structure templated from latex spheres (left), and the TEM images (right) two different mesopore orientations depending on the choice of the surfactant.

spheres in the presence of 6-nm silica particles. An inverse opal structure is obtained after calcination, in which the silica walls around the larger macropores comprise ordered voids templated by the PMMA spheres (Figure 6.32).

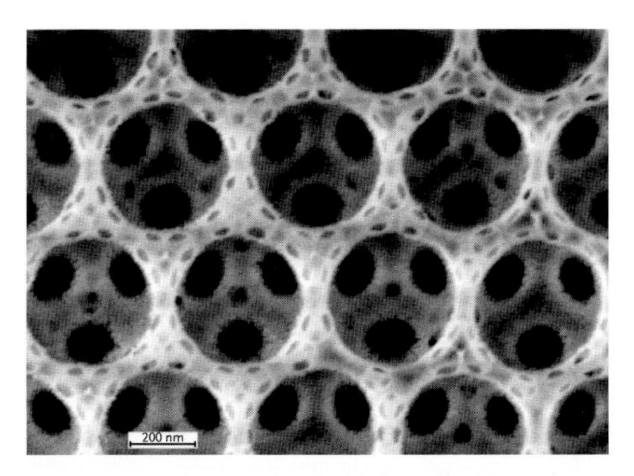

Figure 6.32 SEM image of silica with hierarchical pore structure prepared from multiple hard templates.

6.5.2
Phase-Separation Strategies

Another well-established route to produce materials with multiscale porosity is based on phase-separation processes during sol–gel processing. As discussed in Section 4.5, formation of a gel from metal alkoxides is based on hydrolysis and condensation reactions. Depending on the pH value the structure of the condensed species can be varied from polymeric to colloidal. Assuming that the condensation reactions are performed in an acid medium, differently sized polymeric particles are formed and with reaction time, the average molecular mass of the condensing species is increasing. From the thermodynamics of a solution containing polymerizing species we know that the mutual solubility among the constituents becomes lower as the average molecular mass of the polymerizing species increases. This is mainly due to the loss of entropy of mixing among the constituents, which leads to the increase of the free energy of mixing. When the sign of free energy of mixing of the system becomes positive, a thermodynamic driving force for phase separation is generated. For classical sol–gel mixtures this phase separation does not occur and macroscopically homogeneous materials with textural micro- or mesoporosity are typically obtained.

Addition of poor solvents for the oligomers formed during sol–gel processing, such as various water-soluble polymers, or cationic or non-ionic surfactants, can induce phase separation in the course of the sol–gel reaction. In the presence of these additives, inorganic condensation occurs concurrently with → spinodal decomposition, resulting in the formation of a bicontinuous gel, constituted by two interconnected phases on the micrometer scale. One phase is formed by the inorganic condensed species, while the other phase is a solvent-rich phase including the water-soluble polymer or surfactant. This spinodal decomposition, with comparable volume fractions of the conjugate phases, typically exhibits a bicontinuous structure, where both the gel phase and the solvent phase are continuous and highly interconnected.

Among the many reaction parameters of a sol–gel reaction (see Section 4.5), those that strongly influence the mutual solubility of the constituents and/or the hydrolysis and polycondensation reaction kinetics, play an important role in determining the final size of the phase-separated domains. A bicontinuous structure, in which both separated phases are interconnected, is obtained only when the onset of phase separation occurs at the point of the sol–gel transition. If the timing of the freezing of the structure by gel formation is not set correctly, other structures can be obtained such as fragmented materials with a dispersed phase in a continuous phase or even non-porous materials (Figure 6.33).

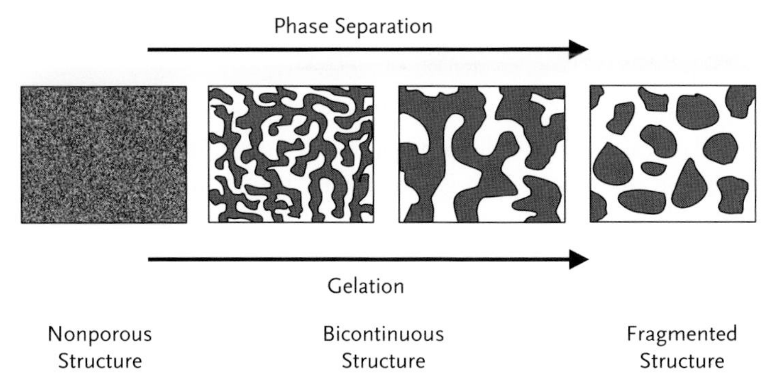

Figure 6.33 Schematic illustration of coarsening of phase-separated domains. Bicontinuous structures are obtained by inducing the phase separation parallel to gelation.

A variety of different oxides (silica, titania, alumina) with hierarchical macro–mesoporous network structures are accessible through this approach (Figure 6.34). By variation of the sol–gel conditions, for example, by the application of highly reactive and polar precursors or by the addition of a secondary soft template, even periodically arranged mesopores are accessible.

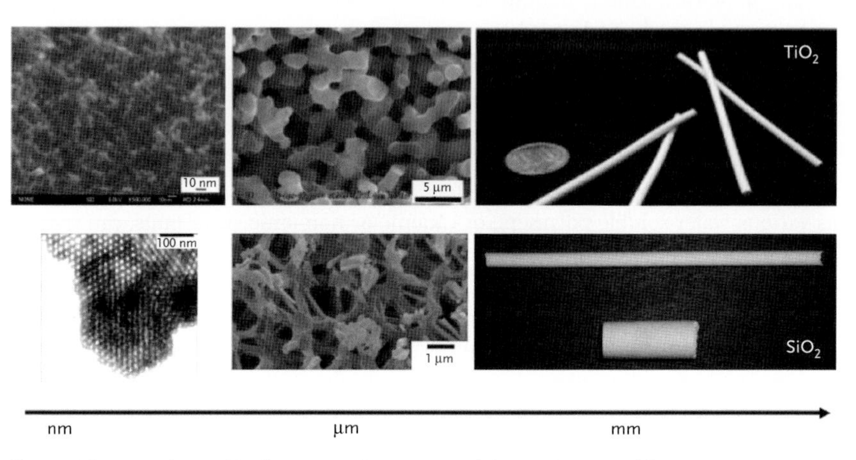

Figure 6.34 Top: hierarchically organized titania monoliths (top) prepared from titanium alkoxides in the presence of formamide. Bottom: silica monoliths prepared from tetrakis(2-hydroxyethyl)orthosilicate in the presence of a non-ionic block-copolymer template (Pluronic P123).

6.6 Incorporation of Functional Groups into Porous Materials

We have learned that porosity is an important and integral part of the architectural design of zeolites and related meso- and macroporous materials such as aerogels, M41S or inverse-opal materials. Even though the materials can be very different with respect to their pore sizes and wall structures (amorphous vs. crystalline), chemical composition, and synthesis routes, one of the key challenges to utilize them in the various applications is to induce (multi)functionality, for example, via the incorporation of guest species in the pores, onto the walls, or as an integral part of the pore walls.

As mentioned before, porous systems offer a large variety of materials compositions, from the classical alumina–silica and pure silica systems, metal-organic frameworks, inorganic phosphates, inorganic–organic hybrid systems, to pure carbon-based materials, to mention only a few. Not only because of this variety, different options arise for the functionalization of molecular sieve systems with guest species, some of which are summarized schematically in Figure 6.35.

Inclusion of molecules or guest entities, such as dyes or biomolecules, by adsorption into a preformed porous matrix from solution is possible, as well as the *in-situ* incorporation of non-covalently bonded functionalities during the synthesis

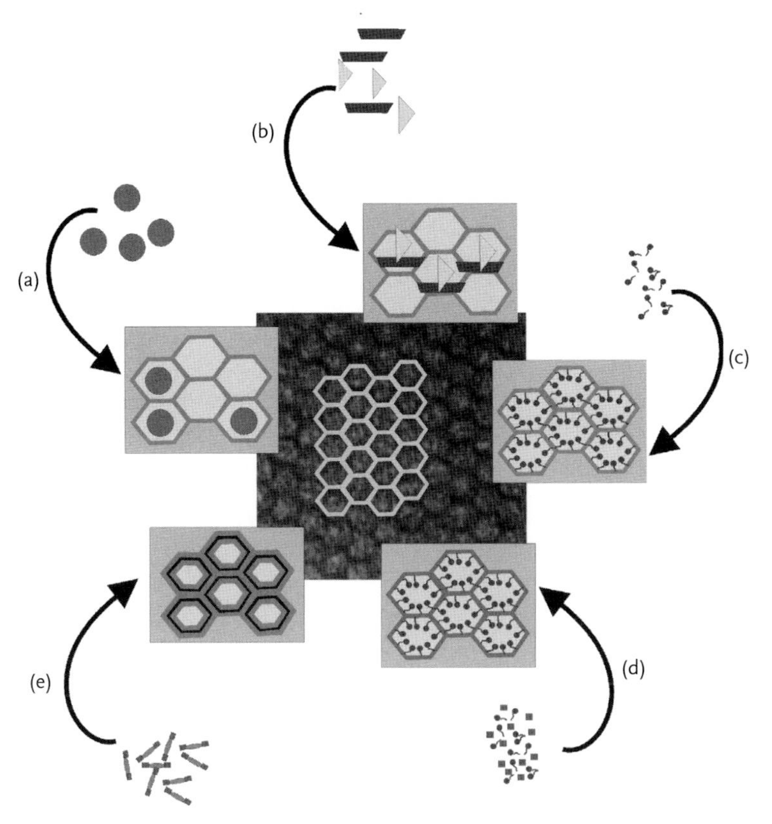

Figure 6.35 Inclusion of guest species in a molecular sieve, which is represented schematically by the MCM-41 hexagonal channel structure: (a) inclusion by treatment of a preformed matrix via the gas or liquid phase, (b) ship-in-the-bottle synthesis, (c) grafting of functional entities by covalent bonds in a post-treatment step, (d) cocondensation reactions of network forming and functionalizing molecules, (e) bridged organofunctional silanes in the synthesis of PMOs.

of the host matrix (Figure 6.35a). One of the main problems for the latter approach (with respect to ordered porous arrays) arises from the fact that the active entity should then also act as a template molecule, because otherwise the periodic pore structure is often disturbed. An alternative approach is the synthesis of, for example, dye molecules within a preformed porous host matrix in a step-wise manner (Figure 6.35b). This method, also termed *"ship-in-a-bottle synthesis,"* is especially attractive, since it allows the incorporation of large molecules, for example, dyes, for which it would not be possible to diffuse into the matrix due to size restrictions. As an additional advantage, leaching of the active molecule is thus prevented.

A different approach is based on condensation reactions of functional entities and the host matrix (Figure 6.35c/d/e). Hereby, the guest molecule is covalently attached to the porous host matrix. This procedure is called (i) *grafting* if done in a two-step synthesis by reaction of the preformed matrix with condensable molecules, or (ii) true co*condensation* if done in a single step by mixing two different condensable precursor molecules.

PMOs (periodic mesoporous organosilicas) can be prepared by cocondensation reactions of bifunctional, bridged organosilanes either with or without a tetrafunctional silane in the presence of surfactant molecules to form a hybrid inorganic–organic material. In these, the organic function is an integral part of the pore walls and PMOs represent another unique hybrid organofunctional material.

Further Reading

1 Antonietti, M. and Göltner, C. (1997) Mesoporous materials by templating of liquid crystalline phases. *Adv. Mater.*, **9**, 431–436.

2 Baumeister, J. (1992) Verfahren zur Herstellung von Metallschäumen. *Techn. Mit.*, **92**, 94–99.

3 Behrens, P. and Stucky, G.D. (1996) Novel materials based on zeolites. *Comprehen. Supramolecular Chem.*, **7**, 721–772.

4 Beck, J.S. and Vartuli, J.C. (1996) Recent advances in the synthesis, characterization and applications of mesoporous molecular sieves. *Curr. Opin. Solid State Mater. Sci.*, **1**, 76–87.

5 Davies, G.J. and Zhen, S. (1983) Metallic foams: their production, properties and applications. *J. Mater. Sci.*, **18**, 1899–1911.

6 Degischer, H.-P. and Kriszt, B. (eds) (2002) *Handbook of Cellular Metals*, Wiley-VCH, Weinheim.

7 Gibson, L.J. and Ashby, M.F. (1997) *Cellular Solids – Structure and Properties*, Cambridge University Press, Cambridge.

8 Guizard, C.G., Julbe, A.J., and Ayral, A. (1999) Design of nanosized structures in sol–gel derived porous solids. Applications in catalyst and inorganic membrane preparation. *J. Mater. Chem.*, **9**, 55–65.

9 Hüsing, N. and Schubert, U. (1998) Aerogels–airy materials: chemistry, structure, and properties. *Angew. Chem. Int. Ed. Engl.*, **37**, 22–45.

10 James, S.L. (2003) Metalorganic frameworks. *Chem. Soc. Rev.*, **32**, 276–288.

11 Janiak, C. (2003) Engineering coordination polymers towards applications. *Dalton Trans.*, 2781–2804.

12 Kitagawa, S., Kitaura, R., and Noro, S. (2004) Functional porous coordination polymers. *Angew. Chem. Int. Ed. Engl.*, **43**, 2334–2375.

13 Lu, A.-H. and Schüth, F. (2006) Nanocasting: a versatile strategy for creating nanostructured porous materials. *Adv. Mater.*, **18**, 1793–1805.

14 McCusker, L.B., Liebau, F., and Engelhardt, G. (2001) Nomenclature of structural and compositional characteristics of ordered microporous and mesoporous materials with inorganic hosts. *Pure Appl. Chem.*, **73**, 381–394.

15 Nakanishi, K. (2006) Sol–gel process of oxides accompanied by phase separation. *Bull. Chem. Soc. Jpn.*, **79**, 673–691.

16 Occelli, M.L. and Kessler, H. (eds) (1997) *Synthesis of Porous Materials – Zeolites, Clays, and Nanostructures*, Marcel Dekker, New York.

17 Raman, N.K., Anderson, M.T., and Brinker, C.J. (1996) Template-based approaches to the preparation of amorphous, nanoporous silicas. *Chem. Mater.*, **8**, 1682–1701.

18 Roland, E. and Kleinschmit, P. (2000) Zeolites. *Ullmanns Encyclopedia*, **A28**, 475–504.

19 Rouquerol, J., Avnir, D., Fairbridge, C.W., Everett, D.H., Haynes, J.H., Pernicone, N., Ramsay, J.D.F., Sing, K.S.W., and Unger, K.K. (1994) Recommendations for the

characterization of porous solids. *Pure Appl. Chem.*, **66**, 1739–1758.

20 Schaefer, D.W. (1994) Engineered porous materials. *Mater. Res. Soc. Bull.*, **14** (4), 14–17.

21 Schüth, F. (2003) Endo- and exotemplating to create high-surface-area inorganic materials. *Angew. Chem. Int. Ed.*, **42**, 3604–3622.

22 Schüth, F., Sing, K.S.W., and Weitkamp, J. (eds) (2002) *Handbook of Porous Solids*, Wiley-VCH, Weinheim.

23 Slowing, I.I., Vivero-Escoto, J.L., Trewyn, B.G., and Lin, V.S.-Y. (2010) Mesoporous silica nanoparticles: structural design and applications. *J. Mater. Chem.*, **20**, 7924–7937.

24 Szostak, R. (1989) *Molecular Sieves*, Thomson Science, London.

25 Tenev, P.T., Butruille, J.-R., and Pinnavaia, T.J. (1998) Nanoporous materials, in *Chemistry of Advanced Materials – an Overview* (eds L.V. Interrante and M. Hampden-Smith), Wiley-VCH, New York, pp. 329–388.

7
Nanostructured Materials

Many properties of solid matter are determined by the "infinite" three-dimensional arrangement of its building blocks rather than the properties of the building blocks themselves. How small can the dimensions of a distinct material in one to three dimensions become, yet still have the same properties as the bulk? For example, a macroscopic piece of metal is electrically conducting, but a single metal atom or a small metal cluster is not. Thus, the question is, at which size will a piece of metal lose its electric conductivity, magnetism, and so on when it is cut into smaller and smaller pieces? Or, conversely, at which size will a metal atom cluster become electrically conducting when it gradually becomes larger?

It has been shown that the properties of matter can dramatically change at a particle size somewhere between 1 and 100 nm. The term "nanostructured materials" ("nanomaterials") thus refers to materials having a characteristic length scale in the lower nanometer range that influences their physical or chemical properties. Nanostructured materials can be crystalline or amorphous, and can be metals, ceramics, semiconductors, or polymers. They can be classified according to their dimensionality and/or the nature of the nanosized feature (crystallites, pores, etc.). Nanoparticles are classified as 0D (zero-dimensional; no dimension outside the nm range), nanotubes and nanowires as 1D (one dimension outside the nm range), and ultrathin layers as 2D nanomaterials.

Nanoparticles (Section 7.3) are three-dimensional objects with all three dimensions being in the nanometer range. The particles can be spherical or may have special shapes, such as oblong or plate-like. Nanoparticles are the best investigated type of nanomaterials, and much attention has been paid to their synthesis, modification and properties. → Semiconductor nanoparticles are also called quantum dots.

Nanocomposites (Section 7.3.5) are → composite materials in which nano-objects are embedded in a polymeric or ceramic host phase. Supported nanoparticles of metals or metal oxides are widely used as heterogeneous catalysts. A high dispersion of the active component maximizes the contact area of the catalyst with reactant and support. The fracture toughness of ceramics can be considerably enhanced by dispersing nanoparticles or nanocrystalline fibers as a second phase.

In *nanocrystalline materials* the dimension of the → grains is in the nanometer range. Nanocrystalline materials can be considered to consist of two structural components: the small crystallites with different crystallographic orientations; and

Synthesis of Inorganic Materials, Third Edition. Ulrich Schubert and Nicola Hüsing.

Published 2012 by WILEY-VCH Verlag GmbH & Co. KGaA

a network of intercrystalline regions (the "interfacial components"), consisting of grain boundaries and triple junctions.

Mesoporous materials contain pores with diameters in the nanometer range. These are discussed in Section 6.3.2.

In *nanotubes and nanowires* (Section 7.4) two dimensions are in the nanometer range. Carbon nanotubes belong to the best-investigated nanomaterials and are close to being used for technical applications.

Nanometer-scale layers (Section 7.5) play a very important role in biological systems as membranes (see Sections 4.3.1). Artificial inorganic (multi-)layers are important in microelectronic devices and have highly interesting optical, electric and magnetic properties. The most prominent two-dimensional nanomaterial is probably graphene, single sheets of the graphite structure (Section 7.5.1).

7.1
The Origin of Nanoeffects

Nanomaterials have the potential of revolutionizing materials design for many applications. Their novel or unusual physical and chemical properties are mainly due to finite-size and surface/interface effects and differ distinctly from those of both the bulk materials and isolated atoms, molecules or small (molecular) clusters of the same chemical composition.

Finite-size effects Confining electrons to small geometries gives rise to "particle in a box" energy levels. In the nanometer regime, neither quantum chemistry nor solid-state physics hold. The former describes the properties of atoms and molecules, and the latter the properties of solids.

Linear combination of n atomic orbitals (AO) results in n molecular orbitals (MO). With an increasing number of atoms the energy difference between adjacent MOs becomes smaller, and eventually the MOs merge to energy bands. Figure 7.1a shows an arbitrarily drawn MO scheme of a molecular cluster with a small number of atoms (with discrete bonds between the atoms) and Figure 7.1c the band structure (i.e., a quasicontinuous density of electronic states) for a bulk material of the same composition. The electronic structure of a nanoparticle is in between (Figure 7.1b). When the size of a particle is scaled down to the lower nanometer range, electronic bands are gradually converted to molecular orbitals, although the discrete energy levels of molecular compounds are not yet reached in nanostructures. Note that the Fermi level (E_F, the energy of the highest occupied orbital [in quantum chemistry terminology] or, in solid-state physics terminology, the energy up to which the band is filled with electrons) is not changed.

Surface and interface effects In nanomaterials, a high percentage of the atoms are surface or interface atoms. For example, the ratio between surface and bulk atoms in a cubic AgBr crystal of $1 \times 1 \times 1$ cm is $1 : 5 \times 10^6$, but it is $1 : 4$ for a 10 nm crystal. About 50% of the atoms in a spherical 5-nm particle are surface atoms. As

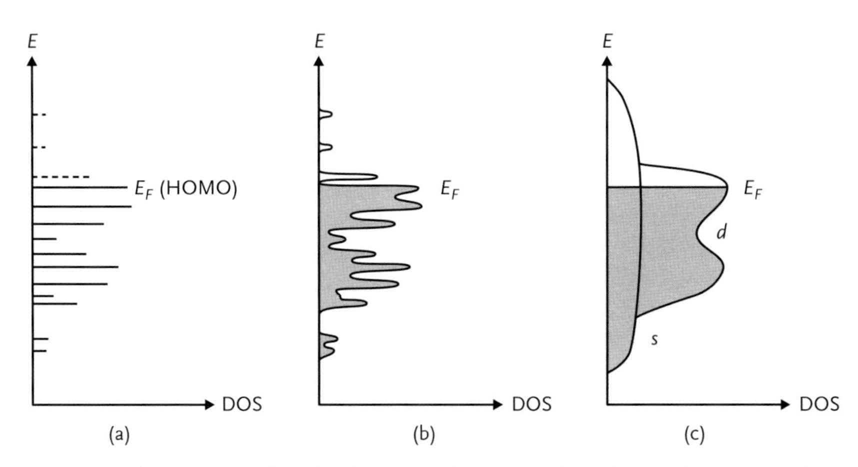

Figure 7.1 Development of the band structure from a hypothetical molecular compound (a) via a nanosized particle (b) to the fully developed band structure (c) consisting of s and d band. E_F = Fermi energy, DOS = density of states.

has already been pointed out in Section 2.1, the surface of solid matter increases rapidly if the particle size is decreased (see also Figure 2.5). Nanoparticles can therefore be considered surface matter in macroscopic quantities.

Surface atoms are different from bulk atoms because they are coordinatively unsaturated and have "dangling bonds." Therefore, their electronic contribution to the behavior of the particle is different from that of the inner atoms that are fully coordinated. For this reason a high amount of energy is stored in the surface. The high surface energy renders nanostructures inherently unstable and would lead to easy aggregation or particle growth (see Section 4.5.1) if the surface is not protected. Methods for attaching organic groups to the surface of nanostructures and options for using these surface groups to influence the properties of the nanostructures will be discussed in Section 7.3.4.

This influence of the surface layer of atoms has been studied for well-defined giant clusters. These clusters close the gaps between the "normal" metal clusters with lower nuclearity and metal → colloids. Probably the best-investigated example is $Au_{55}(PPh_3)_{12}Cl_6$ (Figure 7.2), which is a two-shell cluster with a diameter of 1.4 nm (without → ligands). Other examples of giant metal clusters are $Ni_{147}(PPr_3)_{12}Cl_{24}$ and $Pd_{561}(phen)_{60}(OAc)_{180}$ (phen = *o*-phenanthroline). The Mössbauer spectra of $Au_{55}(PPh_3)_{12}Cl_6$ show contributions from four different types of gold atoms: 13 inner atoms forming the nucleus of the cluster, 24 uncoordinated surface atoms, 12 atoms coordinated to PPh_3 ligands, and six chlorine-bonded atoms. Only the signal of the Au_{13} core is close to that of bulk gold. These and other investigations have shown that the electronic and structural disturbance of the structure of a cluster or a nanoparticle, respectively, by the coordinatively unsaturated surface atoms only affects the outer 1–2 atom layers.

The → grain size of "normal" ceramic powders is typically >10 μm. As the grain size is reduced ("nanocrystalline," "nanoscale," or "nanophase" powders), the

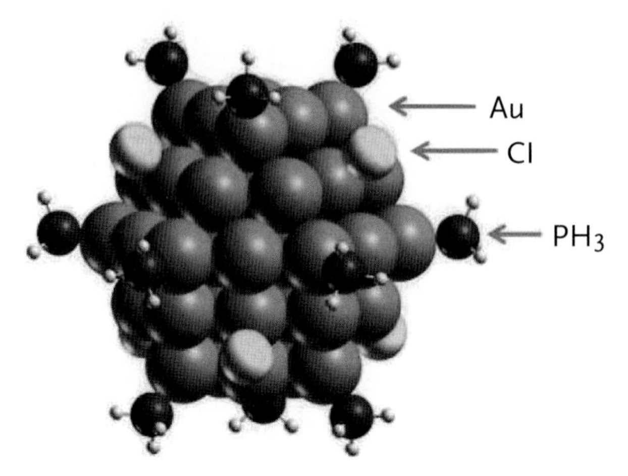

Figure 7.2 Structure of the cluster $Au_{55}(PPh_3)_{12}Cl_6$ (PPh_3 groups replaced by PH_3).

fraction of the grain-boundary volume becomes increasingly higher. For example, for a grain size of 10 nm, 15–25% of all atoms reside in a region within 0.5–1.0 nm of a grain boundary. Grain boundaries are thought to play an important role in the deformation of nanocrystalline materials. They act, *inter alia*, as sources and sinks for dislocations.

7.2 Properties of Nanomaterials

Property changes of nanostructured materials (relative to bulk materials or molecules) will be discussed in this section. Although the emphasis is on nanoparticles, the general principles are the same for other types of nanomaterials. Distinguishing between properties brought about by surface effects and quantum effects, respectively, is somewhat artificial, because coordination unsaturation of the surface atoms also has consequences for the electronic structure. Quantum effects mainly play a role in metal and semiconductor particles. Property changes due to surface effects are *additionally* observed.

7.2.1 Properties Originating from Surface Effects

7.2.1.1 Chemical Properties

Oxide powders have acidic or basic properties due to their surface OH groups (see Section 4.5). Many catalytic reactions make use of this property (e.g., dehydration of alcohols, cracking, isomerization of alkanes and alkenes, esterification and alkylation reactions). Nanoparticles are better catalysts because of their higher proportion of surface atoms. The relative larger surface also results in better adsorption properties.

The investigation of chemical phenomena related to nanosized metal structures such as metal → colloids or highly dispersed metals is not particularly new. What has changed is the perspective. As chemical reactions are governed by electron affinities/ionization potentials and orbital energies, there must be a relationship between the electronic structure of nanoparticles and their chemical properties. For example, in heterogeneous catalysis with supported metals one finds size-sensitive ("demanding") reactions, where the reaction rates depend on the metal particle size, and size-independent ("facile") reactions. In "demanding" reactions, there appears to be a connection between the ionization potential of the metal particle and the reaction rates.

Metallic nanoparticles may not only be more reactive than bulk metals, the example of gold nanoparticles shows that they may also exhibit a *different* chemical behavior. While bulk gold, as a noble metal, is chemically inert, gold nanoparticles with a size <3 nm are catalytically active. This is a consequence of the different electronic structure.

7.2.1.2 Photocatalysis

Photocatalytic reactions occur on the surface of certain semiconductors, such as TiO_2, ZnO, SnO_2, and so on, once electrons are excited from the valence band into the conduction band by illumination. Apart from oxidation of organic compounds (e.g., for water purification or "self-cleaning" windows), hydrogen production by water splitting would be a particularly interesting application for photocatalysts. Nanoparticles are especially suitable photocatalysts because of their high surface areas. This allows a better utilization of the incoming light and provides a higher contact area for the substrate molecules.

The charge carriers, that is, an electron (e^-) in the conduction band (CB) and a hole (h^+) in the valence band (VB), are generated upon bandgap irradiation and must migrate to the surface to induce chemical reactions with surface-adsorbed species (Figure 7.3). The big advantage of semiconductor nanoparticles is that the photogenerated charge carriers can reach the surface before they recombine.

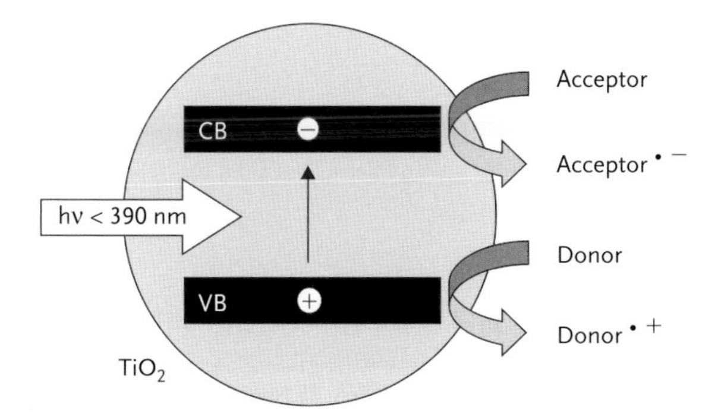

Figure 7.3 Exciton (= electron–hole pair) generation upon irradiation (hν), and redox reactions at the surface of a TiO_2 nanoparticle.

Recombination is a fast process, and therefore the distance the exciton has to travel to reach the surface must be short for high quantum yields. After trapping the electron and hole by surface groups, such as metal–OH groups, an oxidation and a reduction reaction must proceed simultaneously. Otherwise, the particle would be charged. In an aqueous environment, the trapped hole can react with adsorbed water to produce $OH^{\bullet}$ radicals [Eq. (7.1)]; the latter are capable of oxidizing most organic pollutants. The trapped electron can react with oxygen to give a hyperoxide (O_2^{-}) radical anion, which eventually can also be converted to $OH^{\bullet}$ radicals [Eq. (7.2)] or other reactive species.

$$h^+ + H_2O_{ads} \rightleftharpoons H^+ + OH^{\bullet}_{ads} \tag{7.1}$$

$$e^- + O_{2ads} \rightleftharpoons O_2^{\bullet -} \xrightleftharpoons{+H_2O} OH^{\bullet}_{ads} \tag{7.2}$$

7.2.1.3 **Melting Points**

The Lindemann criterion states that a crystal melts when the *mean* thermal displacement of the atoms (δ) becomes larger than a certain fraction of the interatomic distance. Surface atoms have smaller coordination numbers and thus a two- to fourfold higher displacement (δ_s) than the atoms in the bulk (δ_v). When a particle becomes smaller, the portion of surface atoms becomes higher, and thus the mean δ increases. The melting point is therefore only a constant as long as the proportion of surface atoms can be neglected. This is not the case for nanoparticles. Figure 7.4 shows the melting point of gold particles as an example.

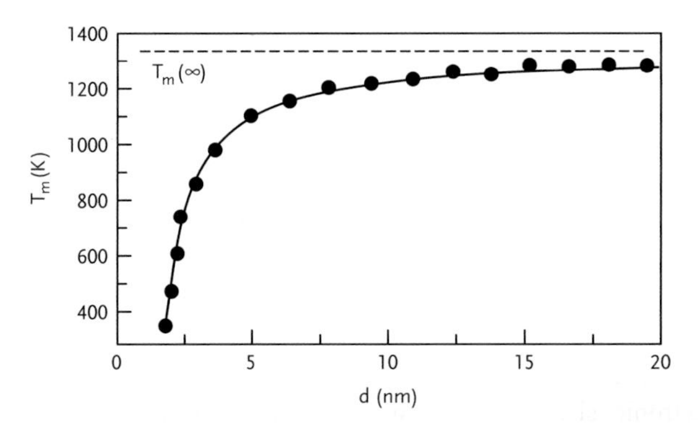

Figure 7.4 Size dependence of the melting point of gold particles. The solid line represents the theoretically predicted curve. T_m is the melting point of bulk gold.

7.2.1.4 **Phase Transitions**

Differences in surface energy may result in different relative stabilities of crystallographic phases. The Gibbs free energy (G) of a nanoparticle is $G_{bulk} + G_{surface}$. The contribution of $G_{surface}$ is large, as discussed before. If two polymorphs of a compound have different surface energies, the relative stabilities of phases may reverse.

For example, the thermodynamically most stable phase of titania (TiO_2) at ambient conditions is rutile. At the isoelectric point (see Section 4.5.1) of titania (about pH 6), the anatase phase is more stable, when the particle size decreases below 15 nm. The particle size at which phase transition occurs depends on the groups covering the particle surface. This introduces the possibility of inducing phase transitions by changing the surface chemistry. Protonated or deprotonated surfaces (i.e., below or above the isoelectric point) significantly influence both the shape of the nanocrystals and the anatase-to-rutile transition size.

7.2.1.5 **Diffusion and Sinterability**

Nanopowders can sometimes be densified and sintered under substantially milder conditions (lower temperature and pressure, shorter time) than those required for conventional ceramic materials. As has been discussed in Section 2.1.4, mass transport occurs from convex particle surfaces to concave particle necks or pores. The better sinterability is due to the higher curvature of the particles (higher surface energy, see Section 4.1.2), short diffusion paths and a facilitated diffusion in the interphase. The easier diffusion also allows → alloy phases to be synthesized at temperatures much lower than those usually required in other systems.

→ Hardness and strength of sintered nanophase ceramics are considerably higher than those observed in conventional materials. For example, tensile and compressive strengths in nearly all nanoscale material systems studied have shown anomalously high values. One reason for this is that the strength of crystalline materials typically increases with decreasing grain size (Hall–Pech relationship). Hardness and → yield strength, however, do not increase indefinitely, but instead exhibit a maximum. The reason why hardness and yield strength start to increase again at extremely small grain sizes could be due to fundamentally different mechanisms of hardening/softening compared to coarser-grained materials.

7.2.2
Properties Originating from Quantum Confinement

7.2.2.1 **Optical Properties**

→ *Semiconductor* nanoparticles of about 1–20 nm diameter (quantum dots) possess short-range structures that are essentially the same as the bulk semiconductors, but have optical properties that are dramatically different from the bulk. This is due to the special electronic situation in semiconductor nanoparticles. The bandgap between the conduction band and the valence band widens as the particle size decreases (Figure 7.1). As a consequence, the corresponding absorption band progressively blueshifts and becomes sharper (Figure 7.5). The optical properties of semiconductor nanoparticles can be perturbed significantly by adsorption of solvent molecules or organic → ligands.

The wavelength of emitted photons consequently shows the same dependency on the particle size. It is also influenced by the chemical composition of the particle. For example, CdTe nanoparticles emit photons of less energy compared to CdSe particles of the same size (see also remarks on "semiconductor alloys" in

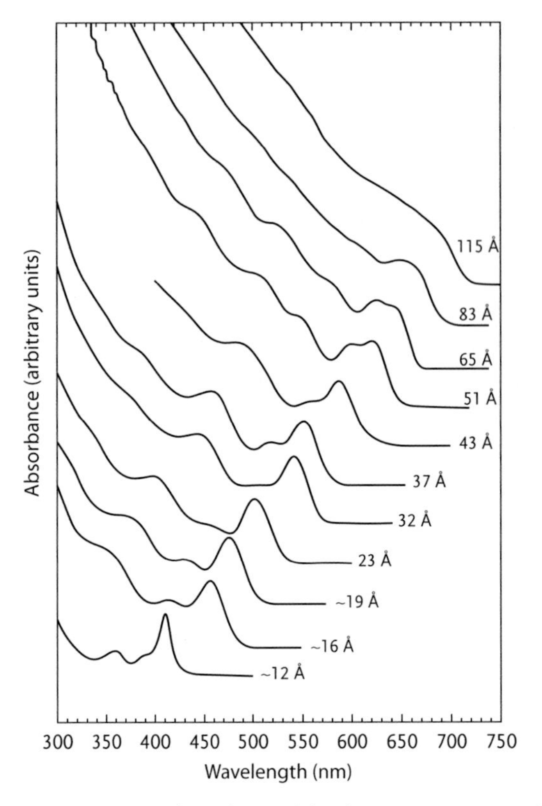

Figure 7.5 Size dependence of the absorption spectrum of CdSe nanoparticles dispersed in hexane.

Section 3.2.6). Emission of quantum dots can thus be tuned over a broad wavelength range by changing size and composition. An attractive application of semiconductor nanoparticles is therefore their use as diagnostic markers.

When semiconductor nanoparticles are doped into polymers or glasses, the resultant nanocomposite materials may be luminescent. This is a surface-related phenomenon, originating from interaction of the polymer with surface atoms.

The unique optical properties of very small *metal particles* were discovered in the nineteenth century by Michael Faraday (1791–1867) when he studied the various colors of gold → colloids. These colors can range from ruby red through purple to blue, depending on the colloid particle size. This color is due to an intense band in the absorption spectra of small metal particles that is not observed in bulk metal.

In a three-dimensional piece of bulk metal, electrons spread in waves of different wavelength (λ), depending on the energy of the electron. A delocalization of the electron in the conduction band is only possible if the diameter (d) of the particle is a multiple of λ. For most metals, the electron mean free path is in the range 5–60 nm. Delocalization is no longer possible if $d \approx \lambda$. This corresponds to a "particle in a box" situation, and the electrons are located between the atoms as in molecules.

Gustav Mie developed in 1908 the theory of plasmon resonance that explains the visible absorption bands exhibited by small particles. Interaction of light with free electrons in a metal can give rise to collective oscillations. Since the electromagnetic wave only has a certain penetration depth (<50 nm for Au or Ag), only electrons at the surface are the most significant. Their collective oscillations are termed surface plasmons. Surface plasmons represent a charge density wave that can travel across the surface. Peaks appear in the extinction spectra when surface plasmons are excited by light under resonance conditions. For Ag and Au nanoparticles, these peaks are in the visible range.

The color change of gold particles from gold to blue, purple and ruby red upon decreasing the particle size can be explained by Mie's theory. As the particle size varies, the resonance conditions change. When the particle becomes smaller, the band broadens and is blueshifted (shown in Figure 7.6 for Ag particles). The subsequent changes upon further reduction of the particle size, reddish-brown to orange to colorless are due to quantum size effects, as has been discussed for semiconductor nanoparticles.

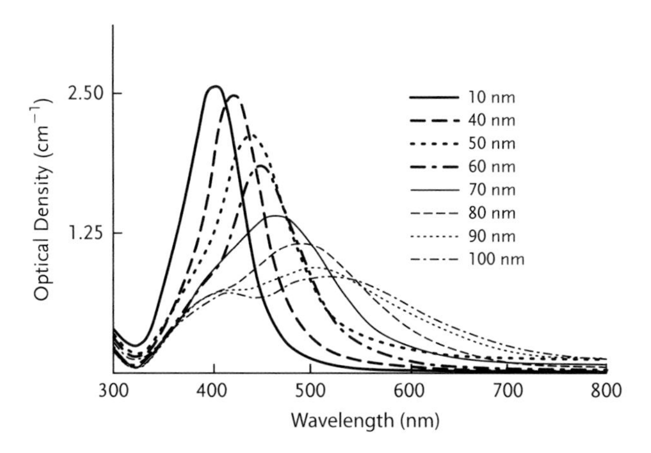

Figure 7.6 Absorption spectra of silver nanoparticles of varying diameters.

The wavelength and width of the absorption band depend on the type, size and geometry of the metal particle, and on the difference between the dielectric function of the metal and its surroundings. For example, the absorption maximum for a 5-nm gold particle is 520 nm in water and 565 nm in Al_2O_3. When a spherical particle is elongated, the surface plasmon band is split in a longitudinal band (along the long axis) and a transverse band (along the short axis).

In insulators and semiconductors, the wavelength of the plasmon resonance is much closer to the bulk absorption band, and therefore their colors exhibit a less pronounced dependency on the particle size than that of metals. As outlined before, the color changes of semiconductor particles are caused by changes in the bandgap between the conduction and valence band.

7.2.2.2 **Electrical Properties**

Bulk metals exhibit a linear dependence between current (I) and voltage (V) (Ohm's law), which corresponds to a collective movement of the electrons through the metal lattice. Reducing the size of the conducting object changes the mechanism of electrical conductivity, when the geometric dimensions reach the mean free path length of the electrons.

The unusual electron-transfer properties of metal nanoparticles can be understood considering the electrostatic energy (E_{el}) required to incorporate an electron into a nanoparticle (or to remove an electron). This quantity is given by $E_{el} = e^2/2C$, where C is the effective capacitance of the particle (which depends on the particle size) and e the charge of the electron. When E_{el} becomes larger than the thermal energy ($k_B T$) of the charge carriers, no charge can be added to or removed from the nanoparticle, that is, the particle is no longer electrically conducting. In order for this "Coulomb blockade" to be observable, the temperature thus has to be low enough. Otherwise the distinct steps are smeared.

An electron can be either incorporated or removed by single-electron transfer (SET) steps when the Coulomb energy of the particle is compensated by the external voltage. Repetition of this process gives rise to the Coulomb staircase shown in Figure 7.7. This charging/discharging of nanoparticles by SET steps is similar to that of molecular compounds or clusters.

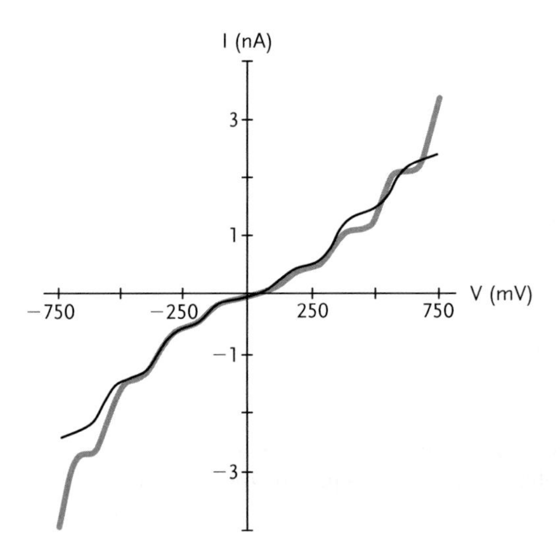

Figure 7.7 Coulomb staircase for 3.3-nm Pd particles (gray line) and the theoretical fit (black line) at $T = 300$ K.

7.2.2.3 **Magnetic Properties**

The spins of neighboring atoms in → ferromagnetic materials, for example, Fe, Co, Ni, Gd, or CrO_2, undergo a parallel alignment within magnetic domains below the Curie temperature (T_c). This results in an increase of the magnetic susceptibility by a factor of 10^7–10^{10}. The total magnetic moment in the *absence* of an external magnetic field is zero because the spins between the domains align in antiparallel fashion (minimization of the total magnetic moment).

The magnetic domains are separated from each other by Bloch walls, where the orientation of the spins changes gradually (Figure 7.8). Bloch walls have a thickness of several 10 nm. Their moving changes the direction of magnetization within a grain and gives rise to the familiar hysteresis curve of ferromagnetic materials.

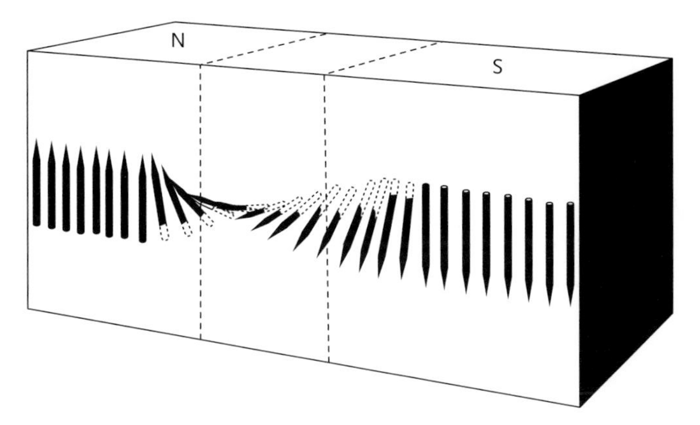

Figure 7.8 Spin reorientation from 0° to 180° between two magnetic domains through the Bloch wall.

When the size of a magnetic particle is in the same range as the size of the magnetic domains, the particle can only accommodate one domain. There is a critical size for each material below which the particles are single-domain, typical values (for spherical particles) are: Fe 14 nm, Co 70 nm, and Fe_3O_4 128 nm. The coercivity (H_c), that is, the field required to reverse the magnetization, of single-domain particles is much higher than that of the corresponding bulk material, because the spin exchange can only occur through rotation of the spins instead of shifting the Bloch walls.

Further reducing the size of single-domain particles (typically to 10–20 nm) results in the phenomenon of *superparamagnetism*, which is only present in nanoparticles. The particles behave similar to → paramagnets, but rather than having a small moment of a few Bohr magnetons, the moment is the sum of the moment of all atoms in the particle. Another useful property of superparamagnetic particles is that H_c=0. Thus, when the magnetic field is removed, thermal energy allows them to freely rotate their spins, and no external energy is needed to demagnetize the system. Below a critical temperature (T_B, the blocking temperature), superparamagnetic behavior disappears, and ferromagnetism sets in (although the particles are still single domain). There is sufficient thermal energy below T_B to allow the spins to easily realign. Superparamagnetism can also be suppressed by shape effects, for example, in needles.

The dependency of the coercivity of small particles on particle size is summarized in Figure 7.9 (see also the discussion on the size of magnetite nanocrystals in magnetospirillum bacteria [Figure 4.23] in Section 4.3.1). Large particles have many magnetic domains. The spins between the domains can rather easily be reoriented by shifting the Bloch walls, such that H_c is low. Single domain particles have the highest H_c at a certain size (D_s), below which superparamagnetic behavior sets in ($H_c = 0$ for particles smaller than D_{sp}).

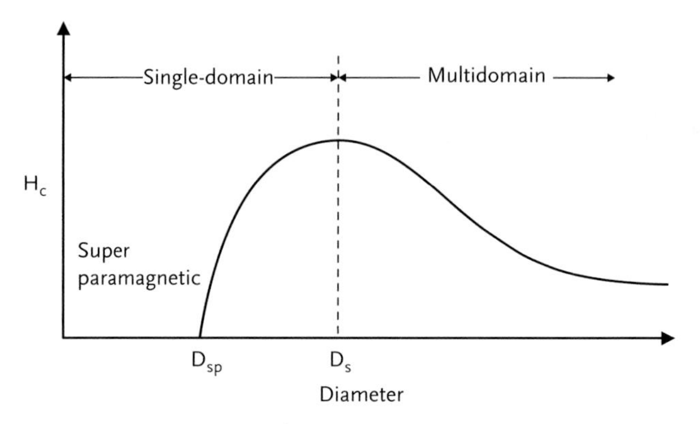

Figure 7.9 Dependence of the coercivity of small magnetic particles on particle size.

The phenomenon of *giant magnetoresistance* (GMR, a dramatic drop in the electrical resistance of certain materials as a magnetic field is applied) has been observed in certain multilayers composed of alternating → ferromagnetic and non-magnetic nanometer-thick layers such as Fe/Cr and Co/Cu. The origin of GMR is the reorientation of the magnetic moments of the ferromagnetic layers relative to each other (parallel or antiparallel). In the absence of a magnetic field, coupling between adjacent ferromagnetic layers becomes antiferromagnetic. Application of a magnetic field results in a magnetic coupling (parallel orientation) and thus a strong decrease of the resistance. Electrons can much easier move through the second ferromagnetic layer, if magnetization is parallel to the first layer. Applications of GMR materials are for high-density magnetic data storage materials, sensors, or solid-state compasses.

7.3 Syntheses of Nanoparticles

One of the most advanced applications of nanoparticles are *ferrofluids*. Ferrofluids are stable → colloidal dispersions of → ferromagnetic nanoparticles, such as magnetite (Fe_3O_4), Fe or Co, dispersed in a liquid, usually a hydrocarbon or water (up to 10 vol% particles). The particles show superparamagnetic behavior (see Section 7.2) because of their small size (typically 5–10 nm). As discussed before, the particles must be stabilized electrostatically (especially for dispersions in water) or by adsorption or bonding of an organic compound (steric barrier – see Section 4.5.1).

In the absence of an external magnetic field, a ferrofluid behaves like a normal liquid. When it is subjected to a sufficiently strong magnetic field, however, the surface of the liquid forms regular three-dimensional patterns (Figure 7.10). This is caused by the oriented and repulsive magnetic field of the individual particles, which, however, are held together by the force of the surface tension. Even with strong magnetic fields, the nanoparticles are not separated from the liquid. This allows "shaping" of a liquid!

Figure 7.10 Ferrofluid on glass, with a magnet underneath. The hedgehog-like structures are formed when the ferrofluid is subjected to high magnetic fields.

An important application of ferrofluids is to form liquid seals around the spinning shafts in hard disks. A small amount of ferrofluid is placed in the gap between the magnet and the shaft and held in place by its attraction to the magnet. The fluid forms a barrier that prevents debris from entering the interior of the hard drive. Ferrofluids are also used in loudspeakers to remove heat from the voice coil, and to passively damp the movement of the cone. They are held by strong magnets in what would normally be an air gap around the voice coil.

Emerging applications for ferrofluids are in medicine. They are used as contrast agents for magnetic resonance imaging. Dispersions of iron oxide nanoparticles can be used for cancer diagnostics. A potential cancer treatment is called magnetic hyperthermia. Ferrofluids placed in an alternating magnetic field release heat. The nanoparticles are concentrated in tumor cells and are able to destroy cancer cells when a magnetic field is applied.

There are two fundamental approaches for preparing nanoparticles, and nanostructures in general. In the bottom-up approach, materials and devices are built from molecular components, while in the top-down approach nano-objects are constructed from larger entities without control at the atomic or molecular level.

In principle, any method capable of producing solids can be used for the preparation of nanoparticles. If a phase transformation is involved (e.g., gas to solid, or precipitation from solutions), then steps must be taken to increase the → nucleation rate and decrease the growth rate during formation of the product phase to obtain nanoparticles (see Section 4.2). Additionally, the aggregation and coalescence of the particles must be prevented once they have been formed. Otherwise, the properties originating from quantum confinement and the high proportion of surface atoms would be destroyed. It should be noted at this point that the general physical principles of → colloidal dispersions discussed in Section 4.5.1 also apply to nanoparticles (sol particles are nanoparticles).

The simplest approach to stabilize nanoparticles is to find an agent that can bind to the particle surface and thereby prevents the uncontrolled growth. In many cases this may be the solvent itself acting as the stabilizer. As outlined in Section 4.5.1, polymers can be used that attach to the particle surface (steric stabilization). Methods have been developed by which protecting groups can be covalently bound to the surface of the nanoparticles (see Section 7.3.4).

7.3.1
Mechanical Attrition

The → grain diameter of powder particles can be reduced to the nanometer scale (2–20 nm) by high-energy ball milling. The basic process of mechanical attrition was already discussed in Section 2.1.2.

During mechanical attrition the powder particles are subjected to severe mechanical deformation from collisions with the hard balls. The deformation is localized at the early stage in shear bands extending throughout the entire particle and consisting of an array of dislocations with high density. Nanometer-sized → grains are nucleated within these shear bands. For longer duration of ball milling this results in an extremely fine-grained microstructure with randomly oriented grains separated by high-angle grain boundaries.

A related process to obtain nanocrystalline metals is called severe plastic deformation. This is achieved by, for example, torsion, multiple extrusion or multiple forging. Due to extensive and repeated deformation processing, the grain boundaries are highly dislocated.

Mechanical attrition and related processes do not result in individual nanoparticles, but instead in nanocrystalline materials (nanosized grains).

7.3.2
Formation of Nanoparticles from Vapors

Methods for the gas-phase growth of small, unprotected metal clusters will not be discussed here (such as supersonic beam expansion, laser vaporization, or laser photolysis of organic compounds). Rather, we will restrict ourselves to approaches leading to the production of nanoparticles on a preparative scale.

The synthesis of nanoparticles via gas-phase reactions involves the same stages as discussed in Section 3.3 for the gas-to-particle conversion route of → aerosol processes, that is, evaporation (producing a supersaturated vapor), → nucleation, particle growth, and transport of the particles.

The aerosol methods discussed in Section 3.3 result in materials formed from highly aggregated nanoparticles. Because there are hardly any possibilities to prevent the primary particles from aggregation, these methods are not considered synthesis routes for nanoparticles in the context of this chapter.

7.3.2.1 **Gas Condensation Method**

The most common apparatus is shown in Figure 7.11. This comprises an ultra-high vacuum chamber, equipped with a liquid-nitrogen-cooled finger and scraper assembly, and an *in-situ* compaction unit to consolidate the powders collected in the chamber. The vacuum chamber is first pumped to a vacuum $<10^{-5}$ Pa, and then filled with a few hundred Pa of a high-purity inert gas, typically He.

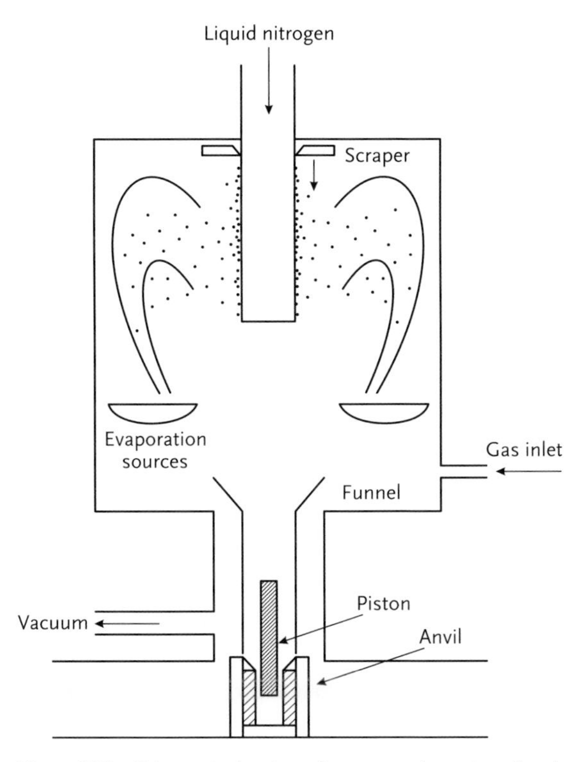

Figure 7.11 Schematic drawing of a gas condensation chamber for the synthesis of nanoparticles.

A solid compound with a high vapor pressure – normally a metal – is evaporated by resistive heating from a high-temperature crucible (or by laser or electron beam evaporation, etc.). An evaporation temperature corresponding to a vapor pressure of about 10 Pa for the precursor is chosen. The formation of small clusters of fairly homogeneous size by → homogeneous nucleation takes place near the vaporization source. Farther from the source, the clusters grow – mainly by cluster –cluster condensation – to give nanoparticles with a broader size distribution. A convective flow of the inert gas between the warm region near the vapor source and the cold surface carries the nanoparticles to the cooled finger, where they are collected. The convection may be combined with a forced flow of the inert gas.

The reason for using a relatively high pressure of an inert gas is that the frequent collisions with the gas atoms decrease the diffusion rate of the atoms away from the source region. The collisions also cool the atoms. If the diffusion is not limited sufficiently, then supersaturation is not achieved, and atoms or small clusters are deposited on the collecting surface. Thus, the main process parameters controlling the characteristics of the particles produced are the inert gas pressure, the evaporation rate, and the gas composition. The particle size can be decreased by decreasing either the gas pressure in the chamber or the rate of evaporation of the metal.

The particles are scraped from the cold finger and funneled into a piston and anvil device where they are compacted. This approach is a semibatch process with the production of gram quantities per run, and is mainly used for preparing nanoparticles of the elements. Nanocrystalline → alloys can be produced by using two or more evaporation sources. Oxide and other ceramic materials can be produced by mixing or replacing the inert gas with a reactive gas, such as O_2 or ammonia.

Metal oxide nanoparticles such as TiO_2 and Al_2O_3 have also been produced in a two-step process. First, metal nanoparticles were formed by evaporation of the metal, followed by controlled oxidation with pure oxygen or air. Magnesium oxide having an ultrafine grain size (5-nm diameter) was produced by the sublimation of MgO in a vacuum chamber. The method of postreacting small metal particles with a gaseous phase has also been applied to generate metal hydride or nitride nanoparticles, for example.

7.3.2.2 Solvated Metal Atom Dispersion Method

This is another method of growing clusters from atoms. Contrary to the gas condensation methods, the particles are not formed in the gas phase but in a cold matrix. The atoms, however, are transported from the source to the matrix through the gas phase. The metal is vaporized and codeposited with a large excess of an organic solvent at liquid-nitrogen temperature, −196°C (Figure 7.12). Metal atoms or small metal clusters (dimers, trimers, and so on) are thus trapped in a solid (frozen) organic matrix and prevented from recombination with the bulk metal. Ultrafine, highly reactive particles are formed upon warming to room temperature.

This method has been used to prepare a wide array of nanomaterials, such as highly dispersed, active metals for organic syntheses or catalysis, bimetallic particles, or metal particles in polymers.

Generally, slow warming of the matrix using a large molar excess of solvent to metal or the use of more polar (more strongly coordinating) solvents results in smaller particles.

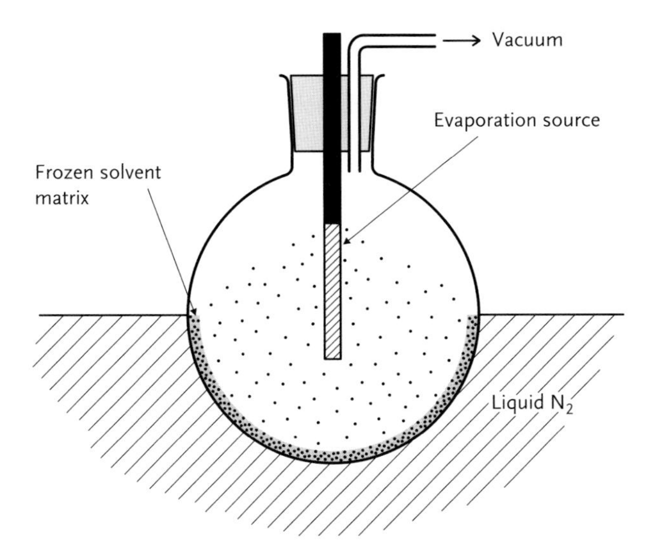

Figure 7.12 Schematic drawing of the reaction vessel for the solvated metal atom dispersion method.

The particle formation can be understood as follows:

- Upon slight warming and matrix softening, solvated atoms or small clusters become labile and begin to oligomerize.
- This aggregation is reversible within a narrow range of solvent viscosity and temperature. It competes with the reaction between the metal atoms and the solvent.
- As the clusters or particles grow, they become less mobile. Eventually, coordination of the solvent to the atoms at the cluster/particle surface becomes more favorable and stops further cluster growth.

7.3.3 Formation of Nanoparticles from Solution

Nanoparticles can be prepared from solution by the same methods as discussed in Chapter 4.2, provided that particle growth and aggregation is carefully controlled. Methods to generate nanoparticles mainly include direct reaction of ions, redox reactions (for metal nanoparticles), decomposition reactions and hydrolysis reactions (although sol–gel processes usually do not result in non-aggregated nanoparticles). Therefore, only a few specific examples will be discussed in the following.

Redox reactions If a metal salt is reduced in the presence of some stabilizer, such as alkanethiols, amines, phosphanes, surfactants, or polymers (see also Section 7.3.4) to prevent aggregation of the particles, metal nanoparticles will form. Dialysis can be used to remove any remaining ions. Relatively narrow size distributions can be achieved with this technique, and the mean particle size can be controlled. The kind of reducing agents depends on the metal. Often used reductants are citrate, alcohols, H_2, formaldehyde, sodium borohydride ($NaBH_4$) or CO. For example, → colloidal dispersions of gold nanoparticles are easily prepared by reduction of $HAuCl_4$ in diluted aqueous solution with citric acid or trisodium citrate [Eq. (7.3)]. Citrate (see Figure 7.17) acts both as reductant and as stabilizer for the nanoparticles.

$$6Au^{3+} + C_6H_5O_7{}^{3-} + 15OH^- \longrightarrow 6Au + 6CO_2 + 10H_2O \tag{7.3}$$

Decomposition reactions Metal colloids have been prepared by the thermolysis of transition-metal carbonyls in an inert atmosphere. For example, thermal decomposition of $Fe(CO)_5$ (by which the carbonyl → ligands are cleaved from the metal) in an organic surfactant solution produces ferrofluids of iron nanoparticles.

Nanoparticles of Ni, Co, Fe, and Cu have been prepared by thermolysis of their formiates ($M(OOCH)_x \rightarrow M + CO_2 + H_2$ or $M + CO + CO_2 + H_2O$) or oxalates ($\rightarrow M + CO_2$). A special case is the photochemical decomposition of silver halide nanocrystals in elemental silver and halides. This is the basic chemical process in photographic films.

Non-metallic nanoparticles can also be prepared by decomposition reactions of suitable precursors. CdSe nanoparticles have been prepared by pyrolysis of $Cd(SePh)_2$ in 4-ethylpyridine as the stabilizing base, for example.

Two synthesis protocols have found wide application for preparing monodisperse nanoparticles from solutions: the two-phase method, where nucleation and growth of the particles only occurs at the interface of two non-miscible solvents, and a single-phase method (hot-injection method), where nucleation and growth are separated.

7.3.3.1 Two-Phase Method

The diffusion-controlled growth rate (dr/dt) of particles depends on the particle size [Eq. (7.4)].

$$dr/dt = DV_m(S_0 - S_p)/r \tag{7.4}$$

where D is the diffusion coefficient, r the particle radius, V_m the molar volume of the material, and S_p and S_0 the solubility of the nanoparticles and the corresponding bulk solid (flat platelet), respectively. According to the Ostwald–Freundlich relation [Eq. (7.5)], the solubility of small particles is higher than that of the bulk solid.

$$S_p = S_0 \exp(2\gamma V_M/rRT) \tag{7.5}$$

(γ = specific surface energy, R = gas constant). Equation (7.5) means that smaller particles grow faster than larger ones. Therefore, smaller particles can catch up with bigger particles with increasing reaction time. → Monodisperse nanoparticles can thus be obtained if both the rates of nucleation and growth are sufficiently low. Reaction probability can be decreased by dissolving the two precursors for binary nanoparticles in two immiscible liquid phases. Reaction in such a two-phase system is only possible at the interface, where the two precursors can get in contact.

Brust and Schiffrin, in the 1990s, were the first to use the two-phase approach to prepare monodisperse metal nanoparticles, especially of gold, silver, copper, platinum, palladium, and so on. Several variations of this method have been developed since then. The reactions are typically carried out in toluene-water, in the presence of tetraoctylammonium bromide as phase-transfer catalyst. In a typical experiment, an aqueous solution of $HAuCl_4$ is mixed with a solution of tetraoctylammonium bromide in toluene. The two-phase mixture is vigorously stirred until all the $HAuCl_4$ is transferred into the organic layer. Dodecanethiol is then dissolved in the organic phase. Upon addition of an aqueous $NaBH_4$ solution, Au(III) is reduced to metallic gold nanoparticles at the toluene/water interface. The gold nanoparticles are capped by the (hydrophobic) thiol and thus rendered dispersible in toluene. The alkanethiol-protected gold nanoparticles can be precipitated and then redissolved.

A similar reaction is shown in Figure 7.13 for the preparation of CdS nanoparticles. Cadmium salts of fatty acids are dissolved in the organic phase, and a sulfur compound slowly releasing sulfide ions (e.g., thiourea) in the aqueous phase. CdS nanoparticles are formed at the toluene/water interface and capped by the long-chain carboxylate → ligands. Once nuclei and nanoparticles form, they will enter the organic phase. They can only continue growing if they go back to the interface.

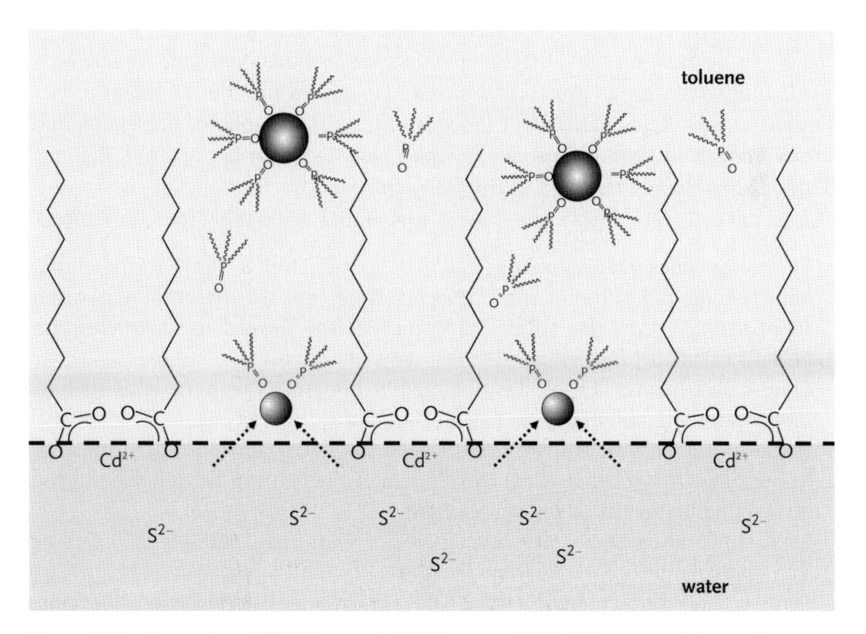

Figure 7.13 Formation of TOPO-capped (TOPO = $octyl_3PO$) CdS nanoparticles at the toluene/water interface.

Metal oxide nanoparticles can also be prepared by the two-phase approach. Metal alkoxides or metal salts of fatty acids are dissolved in the organic phase. An activator, such as tert-butylamine, is dissolved in the aqueous phase. Hydrolysis occurs at the organic/water interface; the formed nanoparticles are capped by, for example, long-chain carboxylates and are transferred in the organic phase.

7.3.3.2 **Hot-Injection Method**

The general idea of this method is to produce small particles in organic solvents by injecting solutions of precursor molecules into a hot liquid in the presence of stabilizing ligands. The injection leads to the instantaneous supersaturation and formation of nuclei. Temperature is lowered to prevent formation of new nuclei and allow the existing nuclei growing (similar to the controlled double-jet precipitation process in Section 4.2).

For example, CdSe nanoparticles of variable size were prepared by injecting a solution of $CdMe_2$ and tri-*n*-octylphosphine selenide ($octyl_3PSe$) in tri-*n*-octylphosphine into hot tri-*n*-octylphosphine oxide ($octyl_3PO$, TOPO). The CdSe particles are stabilized by coordination of TOPO to the surface Cd atoms. Subsequent size-selective precipitation by addition of acetone results in samples with narrow size distributions. The nanoparticle size can be controlled by the temperature of the reaction solution and the precursor-to-surfactant ratio. Higher temperatures and larger precursor-to-surfactant ratios produce larger particles. Other CdE or ZnE nanoparticles (E=S, Se, Te) can be prepared by the same method. The reactions are in general similar to that in CVD processes (see Section 3.2).

Preparation of nanoparticles of the same composition in aqueous solutions (such as CdS from Cd salts and a sulfide source) has clear advantages in terms of ease of synthesis using standard reagents. Size dispersions, however, are relatively broad, and postsynthesis procedures have to be applied in order to obtain monodisperse nanoparticles. Because of the very efficient separation of nucleation and particle growth in the hot injection method (see Section 4.2) narrower size distributions are already obtained during the synthesis.

Since the discovery of this route, some preparative adaptations were developed.

- The solvent does not necessarily need to be coordinating by itself. High-boiling, non-coordinating solvents such as octadecene can also be employed when coordinating agents (for surface protection of the formed nanoparticles) are present in small proportions.
- The surface properties of the obtained nanoparticles can be modified by postsynthesis ligand exchange (see Section 7.3.4).
- Hazardous organometallic precursors such as $CdMe_2$ can be replaced by other compounds, for example, CdO or $Cd(acetate)_2$, which can be dissolved in the corresponding solvent/stabilizer mixture.
- Core–shell nanoparticles can be prepared, for example, CdSe with a CdS or ZnS shell.
- Other semiconductor nanoparticles were similarly prepared by other reactions, for example, InP particles from InX_3 (X = Cl or oxalate) and $P(SiMe_3)_3$ (see solid-state metathesis in Section 2.1.4).

7.3.3.3 Syntheses in Confined Spaces

The general idea behind this approach is to confine the growth of a particle by carrying out reactions in nanosized reactors ("ship-in-a-bottle" approach). Such reactors can be pores or channels in solids, small liquid droplets, or nanosized domains in polymers.

The cage framework of zeolites This has been used to confine the growth of semiconductor nanoparticles. For example, Cd^{2+} ions were introduced into the zeolite cages by ion exchange. The dry Cd-exchanged zeolite was then treated with H_2S gas, and very small CdS clusters were formed in the zeolite framework. Other porous solid materials can be similarly used, such as membranes, or even carbon nanotubes (see Section 7.4.2, endohedral functionalization).

Reversed micelles (water-in-oil microemulsions) When a small amount of water is added to a solution of surfactants in hydrocarbon solvents, the polar heads of the surfactant molecules gather together and thus disperse very small water droplets (Figure 7.14). The water droplets of the → emulsion can act as nanosized reactors. Any reaction leading to the formation of solids in aqueous systems can be carried out within these droplets. The size of the resulting particles is governed by the size of the droplets, which is controlled by the water-to-surfactant ratio. From a practical point of view, two methods are used to get the reactants in contact:

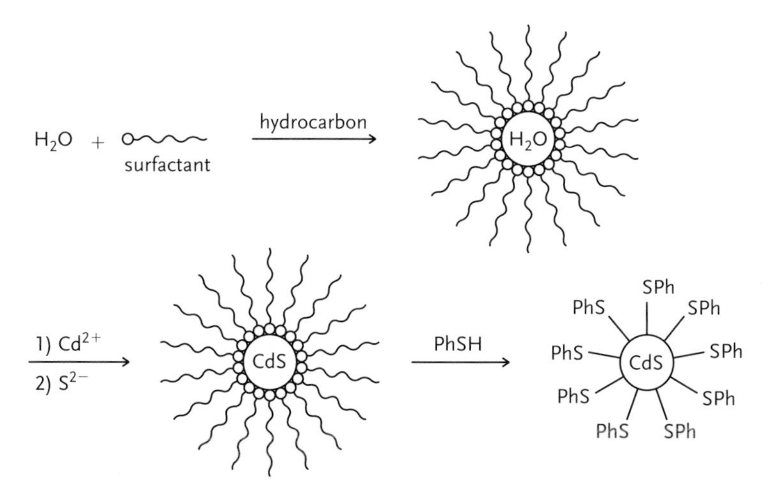

Figure 7.14 Microemulsion method of preparing surface-capped CdS nanoparticles.

- One component is diffused into the microemulsion of another. For example, when Cd salts are dissolved in the water droplets and a sulfide source is allowed to diffuse into this droplet, CdS is precipitated within the water droplet (Figure 7.14). Another example is sol–gel chemistry (see Section 4.5) within the droplets by letting $Si(OR)_4$ diffuse into the water droplet in which the catalyst is dissolved.

- Mixing two emulsions with each other. For example, emulsion 1: the aqueous solution of a hydrolysable metal salt + emulsion 2: an aqueous ammonia solution (to shift the pH, see Eq. (4.10)), which optionally may contain surface-stabilizing molecules.

Domains in microphase-separated block-copolymers When two polymers of different polarity are mixed (e.g., hydrophobic polyisoprene and hydrophilic poly (ethylene oxide)), phase separation occurs because of the unfavorable mixing entropy of two long-chain macromolecules. Macroscopic phases can no longer form in block-copolymers of the same polymers, because the dissimilar blocks are connected with each other. Only self-organization of the dissimilar blocks in nm-sized domains is possible. This minimizes the free surface energy. Depending on the composition, chain lengths and volume fraction of the two blocks, several nanostructures can be obtained, ranging from spheres to cylinders to lamellae (Figure 7.15).

Figure 7.15 Nanostructures in microphase-separated block-copolymers (A–B) depending on the block volume fraction (f). Lamellar structures (center) are obtained for $f = 0.5$. Increasing the fraction of A results in gyroid, cylindrical or spherical structures of A in B (center to right). Decreasing the fraction of A (center to left) results in inverse structures.

When reactions are carried out in such domains, or when the block-copolymers are used as templates, the structures of the obtained solids reproduce the domain structure. Figure 7.16 shows organically modified aluminosilicate nanoparticles, as an example, arranged in the hydrophilic poly(ethylene oxide) blocks of poly(isoprene-block-ethylene oxide). By varying the amount of the (hydrophilic) nanoparticles in the hybrid material, the phase diagram of a typical block-copolymer can be mapped out. Mesostructured aluminosilicate materials are obtained when the organic components are burnt out.

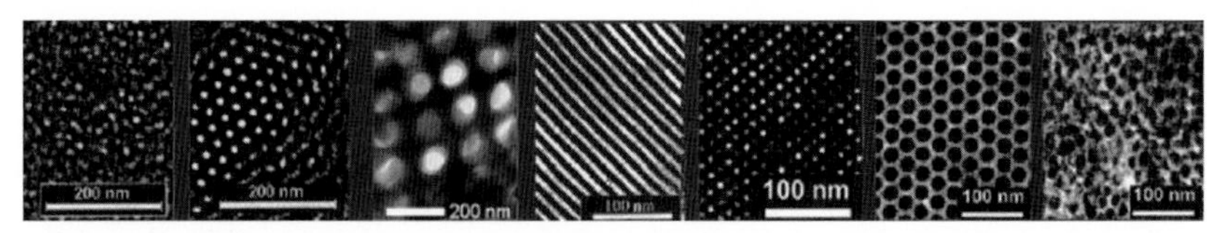

Figure 7.16 Mesostructures consisting of hydrophobic polyisoprene and hydrophilic poly (ethylene oxide) + aluminosilicate. From left to right: body-centered cubic spheres, hexagonally packed cylinders, cubic bicontinuous structure, lamellae, gyroid, inverse hexagonal cylinders, and inverse body-centered cubic spheres.

7.3.4 Surface Modification of Nanoparticles

The methods for attaching organic groups to the surface atoms of nanoparticles discussed in this section can be applied to other nanostructures too. The organic groups are called → "ligands" because the underlying chemistry is similar to that of coordination compounds. The reader is also referred to Section 6.6 where the modification of the inner surface of porous materials by organic groups has been treated. "Steric stabilization" of nanoparticles by adsorption of a polymer shell has already been discussed in Section 4.5. A combination of steric stabilization and surface protection by organic ligands is the use of polymers with coordinating groups. The modification by a layer of another inorganic compound is not included in this section, because the underlying preparative methods are the same as discussed throughout this book.

First, for what reasons is the grafting of organic ligands onto the surface of nanoparticles necessary or desirable?

- stabilization of the nanoparticles against agglomeration;
- control of the particle size and shape;
- compatibility with another phase;
- functionalization of nanoparticles by ligands with functional organic groups;
- influence on physical properties.

These options are synergetic. For example, ligands used for functionalization also protect the nanoparticles against agglomeration and may be useful for dispersing nanoparticles in another phase.

The first two issues (stabilization and shape control) were already discussed in Sections 4.5 and 4.2.

Compatibility with another phase A frequent problem, especially for biological applications, is to render nanoparticles water soluble. This can be achieved by appropriate ligands. For example, a frequently used ligand for both metal and metal oxide nanoparticles is citrate (Figure 7.17). This ligand can be coordinated through one or two carboxylate functionalities, leaving at least one carboxylic acid and the OH group exposed to the solvent and rendering the particle hydrophilic. Another example is the use of modified inorganic nanoparticles as fillers in organic polymers. Modification by suitable organic groups can avoid compatibility problems and improve the particle dispersion.

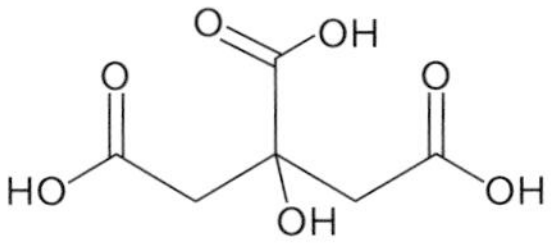

O OH
O O
HO OH
OH

Figure 7.17 Chemical formula of citric acid.

Functionalization The use of ligands with functional organic groups can add an additional quality to the nanoparticles. The organic functionalities may implement complementary physical properties or enable deliberate interactions of the nanoparticles with molecules, other nanoparticles, surfaces, and so on.

Influence on physical properties The electronic structure of the passivating ligands contributes to the overall electronic structure of the nanoparticles and may thus influence physical properties. In the case of optical properties, surface states may be blocked (see Section 7.2) and hence emission yields are affected. Surface coordination may also decrease spin–orbital couplings and surface anisotropy of magnetic nanoparticles and thus change the magnetic properties.

The most suitable type of ligands depends on the kind of nanoparticles. → Bidentate ligands often have advantages because they are more strongly bonded.

Metal nanoparticles, mainly noble-metal particles, are preferentially modified by thiols (RSH), disulfides (RSSR), amines (NR_3), or phosphanes (PR_3). Disulfides are split into two SR groups upon reaction with the particles. Stability of the metal –ligand bond decreases in the order thiols > phosphanes > amines. The interaction of thiols with metal nanoparticles is an interesting issue, which was not yet fully resolved in each case. The thiol can either retain the sulfur-bonded hydrogen atom (coordination as intact HSR) or lose it (to form thiolate anions or radicals).

The majority of *semiconductor nanoparticles* are stabilized by thiolate or related ligands. Several giant clusters were isolated and structurally characterized, such as $Cd_{32}S_{14}(SPh)_{36}(DMF)_4$, with an 82-atom core of the cubic-phase CdS capped by 36 phenyl groups and four solvent molecules (DMF), or $Ag_{172}Se_{40}(SeBu)_{92}(dppp)_4$ (dppp = bis(diphenylphosphino)propane). As pointed out in Section 7.1, such clusters are structural models for nanoparticles. Other ligands used for the stabilization of semiconductor nanoparticles include carboxylates, phosphanes or phosphine oxides (see TOPO in Section 7.3.3), or amines.

Metal oxide nanoparticles are mainly stabilized by oxygen-based ligands, such as carboxylates ($RCOO^-$), phosphinates (RPO_3^{2-}), or → organosilicates ($RSiO_3^{3-}$). Each of them can be coordinated as monodentate or → polydendate ligands. When modifying oxide nanoparticles, the surface charge has to be taken into account (see Eqs. (4.15) and (4.16)).

The most important methods for surface modification of nanoparticles by organic groups are:

1. **Synthesis of the nanoparticles in the presence of a stabilizing organic ligand.** Examples have already been mentioned, such as the synthesis of CdSe nanoparticles in the presence of TOPO, or the synthesis of gold nanoparticles in the presence of alkanethiols or citrate. The size of the obtained nanoparticles is determined by two energy terms. The first is the volume free energy that, at a certain particle size, is balanced by the surface energy term. The latter depends on the bond strength between the surface atoms and the attached ligands. Interligand interactions may also play a role.

2. **Ligand exchange**. Ligands used for stabilization of the particles during their synthesis can later be exchanged for other ligands that provide better stabilization, allow functionalization, and so on. Coordination of ligands to metal atoms in general is a dynamic process, in which ligands coordinate and decoordinate. Ligand exchange is a facile process if the incoming ligand is more strongly bonded. For example, TOPO layers on semiconductor nanoparticles are neither sufficiently stable nor sufficiently robust for many applications. Replacement of TOPO by mercaptoacetic acid ($HSCH_2COOH$) both stabilizes the particles and renders them water soluble. The COOH groups can also be used for bioconjugation.
3. **Postsynthetic modification of the ligand shell**. Functional groups of the organic ligands can be modified by standard organic reactions provided that the inorganic particle is not degraded by this reaction. For example, $-NH_2$-terminated ligands can be reacted with Cl(O)C〰X or related compounds (X = another functional group) to result in –NH–CO〰X groups. Another possibility of modifying the ligand shell is the use of → amphiphilic molecules the hydrophobic portion of which intercalates hydrophobic stabilizing ligands at the particle surface (Figure 7.18).

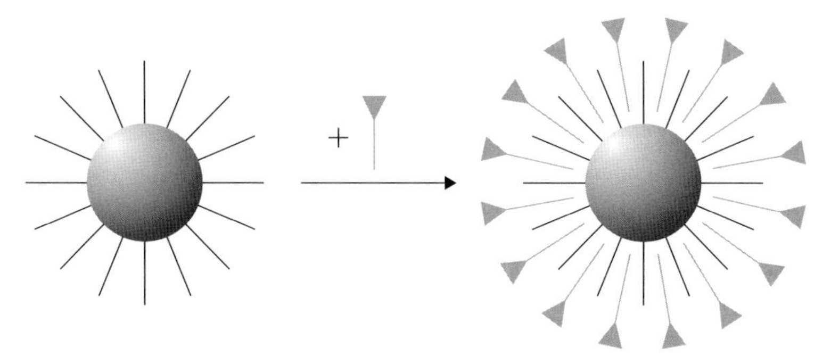

Figure 7.18 Ligands with long hydrophobic tails (dark bars) can be intercalated by the hydrophobic part of amphiphilic molecules (gray). The gray triangles represent functional groups attached to the hydrophobic part of the intercalates.

7.3.5
Nanocomposites

In → composite materials, a guest phase is dispersed in a host (matrix) phase. The term nanocomposite refers to nanoscale guest phases. Examples include metallic or semiconductor nanoparticles dispersed in polymeric, glassy or ceramic hosts, carbon nanotubes or single sheets of layered solids (see Section 2.3) in organic polymers, and so on.

The unknowing use of gold and silver nanoparticles for coloring glass and ceramics dates back to the fourth and fifth century BC in Egypt and China. Today metal nanoparticles ("highly dispersed metals") on alumina, silica, carbon, aluminosilicates, or other inorganic supports are mainly used as industrial catalysts for various applications. The automobile exhaust catalyst is a well-known example. "Dispersion" is defined as the percentage of metal atoms on the surface of a particle – this means, the smaller the particles the higher is the dispersion. The supports for catalytically active particles need to be highly porous and to have a high specific surface area. Apart from supporting the catalytically active species, they also act as a heat sink even in highly exothermic reactions (to allow for isothermal reaction conditions) and are sometimes directly or indirectly involved in the catalytic reaction.

There are many other applications for nanoparticle-containing films. Silver nanoparticles, for example, have antibacterial properties and are, for this purpose, embedded in or coated on various materials, ranging from fabrics (e.g., odor-free socks) to paints. As pointed out in Section 7.2, metal and semiconductor nanoparticles have interesting optical properties. Incorporating the particles in transparent host materials prevents the particles from agglomeration and thus allows utilization of these properties. Thin films containing metal nanoparticles are also of interest for the electronics industry, for sensors and detectors, and so on.

An example from the plastics industry is polymer–clay composites, where exfoliated clay sheets (see Section 2.3.2) of about 1 nm thickness and 50 nm to some μm diameter are incorporated in organic polymers. The first commercialized polymer–clay nanocomposite was montmorillonite-reinforced nylon-6. Addition of clay nanosheets can make plastics less permeable to liquids and gases, more flame retardant and tougher. At the same time, the addition of small amounts of clay does not affect the transparency of plastics (because the clay layer is smaller than the wavelength of light). The reason for the reinforcing effect is the much higher interphase proportion compared to equal amounts of a conventional fillers (in the interphase, the polymer structure, perturbated by interaction with the filler surface, relaxes to that of the bulk polymer).

Preparation of nanocomposites comprises the synthesis of the nano-objects as well as that of the host phase. Synthesis methods have been outlined before. An important preparative option, however, is the sequence in which the host and guest phase are synthesized. The various possibilities are schematically shown in Figure 7.19 for nanoparticles.

The nanoparticles (same for nanowires, nanoplates, etc.) can be synthesized first and then incorporated in the host phase (Figure 7.19, left) either by synthesizing the host phase in the presence of the nanoparticles or, in the case of organic polymers, by mixing them with the host phase (dispersion of the nanoparticles in a polymer solution, mixing in an extruder, and so on). A challenge of this approach is to prevent agglomeration or aggregation of the nano-objects. An example is the polymer–clay composites mentioned before.

The more common approach is to create the nanoparticles within or on the preformed support (Figure 7.19, center). A wide-spread preparation method for metal–ceramic nanocomposites is impregnation of the support with the solution of a metal salt. Capillary forces favor a homogeneous impregnation and adsorption

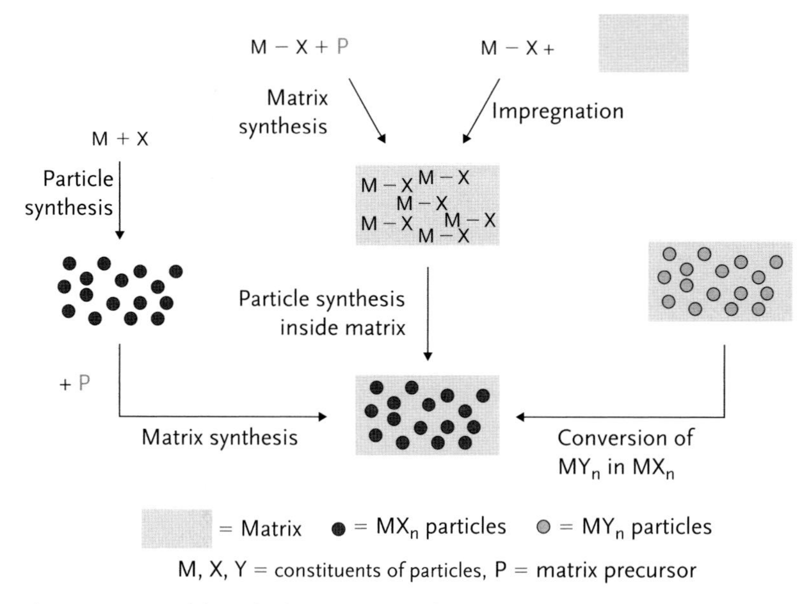

Figure 7.19 Possibilities for the preparation of nanocomposites.

on the inner surface. A variation is the so-called "incipient wetness" method, where the dried and powdered support material is "titrated" with the metal salt solution until it starts to lump together. Ion-exchange reactions are used to introduce catalytically active metals in zeolites (exchange of alkali-metal ions) or on the surface of oxides with a high concentration of surface −OH groups (exchange of H^+ ions). The metal-ion-doped materials are subsequently heated in a reducing atmosphere, and metal particles are formed. The metal particle-size distribution mainly depends on the surface properties and porosity of the support, the drying conditions, and the reduction temperature.

The matrix phase can alternatively be formed in the presence of the nanoparticle precursors. For example, when a metal compound is added to sol–gel precursor mixtures (see Section 4.5), the metal ions are incorporated into the formed gel network. The dispersion of the metal ions during sol–gel processing can be improved by coordination of the metal ions to alkoxysilanes with complexing groups. This is shown in Eq. (7.6) for an alkoxysilane substituted by an ethylene diamine group.

$$(RO)_3Si\text{–}(CH_2)_3\text{–}NH\text{–}(CH_2)_2\text{–}NH_2 + Pt^{2+} \longrightarrow [(RO)_3Si\text{–}(CH_2)_3\text{–}NH(CH_2)_2NH_2\text{–}Pt\text{–}NH_2(CH_2)_2NH\text{–}(CH_2)_3\text{–}Si(OR)_3]^{2+} \xrightarrow{+\,Si(OR)_4\,/\,+\,H_2O} [\equiv Si\text{–}(CH_2)_3\text{–}HN(CH_2)_2NH_2\text{–}Pt\text{–}NH_2(CH_2)_2NH\text{–}(CH_2)_3\text{–}Si\equiv]^{2+} \xrightarrow{+\,O_2/550\,°C} PtO/SiO_2 \xrightarrow{+\,H_2/100\,°C} Pt/SiO_2 \qquad (7.6)$$

A third, rather straightforward approach to create the nanoparticles of a certain chemical composition in or on a support is to convert one kind of nanoparticle to another by chemical reactions (Figure 7.19, right). Examples include the conversion of supported metal oxide nanoparticles in metal nanoparticles by reduction, as in Eq. (7.6), or of metal nanoparticles in metal sulfide nanoparticles by reaction with a sulfur source.

7.4
One-Dimensional Nanostructures

One-dimensional nanostructures are at the forefront of nanoscience and nanotechnology at the moment when this chapter was written. The most important representatives of these highly anisotropic nanostructures are nanorods, nanowires and nanotubes; nanobelts are rapidly gaining importance. In each of these structures, the dimensions of the cross section are in the nm range, while the length of these structures may range up to a few μm.

7.4.1
Nanowires and Nanorods

Most nanowires and nanorods are single crystalline. Their cross section is uniform and typically cylindrical, hexagonal, square, or triangular. The distinction between nanorods and nanowires is quite arbitrary; nanostructures with an aspect ratio (length/diameter) <20 are called nanorods, and those with an aspect ratio >20 nanowires. In the following discussion, no distinction is made between both.

As a consequence of the highly anisotropic shape, also the properties of nanorods are anisotropic. For example, the plasmon resonance band of gold nanorods (see Section 7.2.2) is split into two components, a transverse band (along the short axis) which is in the same wavelength region as that of gold nanoparticles ($\sim$530 nm), and a longitudinal band (along the long axis) at $\sim$600–1000 nm, depending on the aspect ratio. This anisotropy is also reflected in the chemical reactivity. Because regions with a higher positive curvature are higher in energy (see Section 4.1.2), dissolution or etching processes start at the ends of the rods. During the dissolution process, the aspect ratio becomes progressively smaller, and nanospheres (aspect ratio of 1) are formed as intermediate structures.

Two methods for the formation of nanorods were already discussed. The first is directed growth of (nano)crystals, as discussed in Section 4.2 (shape control). The second is growth of nanorods in templates with tubular holes ("nanocasting"), such as porous alumina (see Section 6.4.1) or phase-separated block copolymers (see Figure 7.15).

The most important method for the controlled growth of crystalline nanorods and nanowires is the so-called *vapor-liquid-solid (VLS) method*. This method proceeds in three stages (see Figure 7.20):

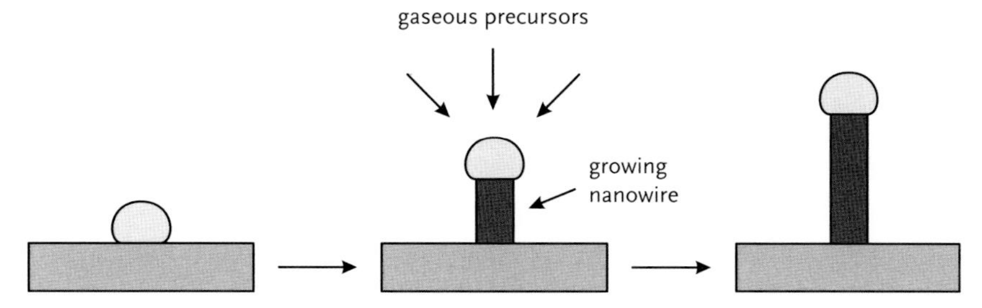

Figure 7.20 Simplified sketch of the growth of nanowires by the VLS method.

1. The process is initiated by melting of dispersed metallic nanoparticles or evaporating of metal targets on crystalline substrates.
2. In the second step, the gaseous reactants dissolve in the metallic nanodroplets. Crystalline nanowires start growing once nucleation occurs. The chemical reactions taking place in the nanodroplets are similar to that in CVD processes (see Section 3.2). The nanodroplets can be considered to be catalysts for the formation of the nanowires.
3. The single-crystalline nanowires grow away from the substrate. During this process, the nanodroplet is located at the end of wire. The diameter of the nanodroplet defines the diameter of the nanowire (typically 10–200 nm) and confines growth in one dimension. The length of the obtained wires can be in the μm and even mm range.

The VLS method allows a precise control over the chemical composition, size, structure, and growth direction of nanowires. These advantages of (mainly semiconductor) nanowires promise interesting applications such as field-effect transistors, p-n diodes, light-emitting diodes, nanoscale lasers, logical gates and others. Some of the advanced variations of the methods are as follows:

- Precise control of the crystallographic orientation of the nanowire is possible by epitaxial growth on correspondingly oriented substrate surfaces. For example, GaN nanowires (wurtzite structure) with a hexagonal diameter are obtained when the wires are grown in the [001] direction, and wires with a triangular diameter when grown in the [1–10] direction.
- Parallel arrangement of nanowires with tight control over size and diameter of the individual wire, and over the nanowire density is feasible (Figure 7.21). The production of nanowire arrays requires, as the first step, the (ordered) deposition of metallic nanoparticles on a substrate. Various methods have been tested, such as electron-beam lithography, self-organization, use of nanostructured substrates, and so on.
- Coaxial or longitudinal heterostructured nanowires can be grown. Coaxial structures are obtained by coating nanowires with a layer of a second material. Periodic longitudinal heterostructures (superlattice nanowires) can be generated by alternating delivery of different compounds in the nanowire growth reactor.

Figure 7.21 SEM image of a ZnO nanowire array on a sapphire wafer.

7.4.2 Nanotubes

Nanotubes of various inorganic compounds have been synthesized, such as boron nitride, binary and complex metal oxides, chalcogenides or halides, the synthesis procedures used being dependent on the chemical composition. The best investigated examples, however, are carbon nanotubes (Figures 7.22 and 7.23). Nanotubes can be either single walled (SWNT – single-walled nanotube), as the carbon nanotube in Figure 7.22, or multiwalled (MWNT – multiwalled nanotube), where several cylindrical layers surround the central tubule. They may be closed (as in Figure 7.22), or open at one or both ends. The intertube distance in multiwalled carbon nanotubes is 0.34 nm, which is also the distance between two graphene layers in graphite.

Carbon nanotubes possess an array of unprecedented structural, mechanical, and electronic properties. For example, they exhibit at the same time:

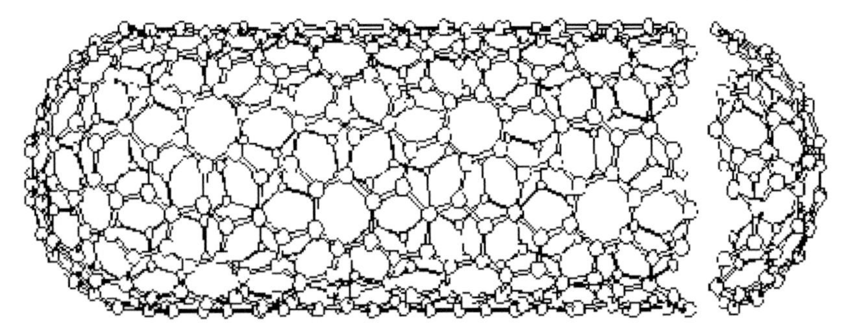

Figure 7.22 Schematic drawing of a closed single-walled carbon nanotube.

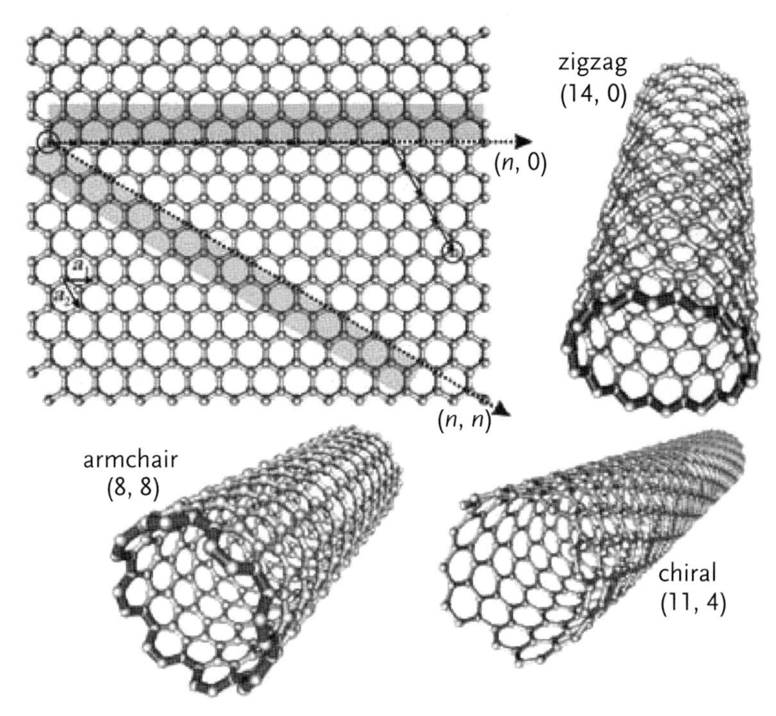

Figure 7.23 Roll-up of a graphene sheet leading to an armchair (8,8), chiral (11,4), or zigzag (14,0) tube, respectively. The dashed arrows represent the roll-up vectors.

- metallic or semiconducting behavior, in the case of metallic conduction: 100 times more conductive than copper;
- field emission properties;
- thermal conductivity matching that of diamond;
- chemical and thermal stability;
- extremely high tensile strength and elasticity (100 times stronger than steel; stiffer than diamond; highest Young's modulus and tensile strength);
- the ability to host molecules as nanocontainers; and
- the potential of chemical functionalization.

Many electrical, thermal, or mechanical applications based on these unique combinations of properties have been proposed, such as reinforcers in high-strength → composites, heat sinks, electronic field emission materials, transistors, or microelectronc interconnects. Carbon nanotubes are used as AFM and STM tips.

The electronic properties of carbon nanotubes depend primarily on how a graphene sheet, that is, a single layer of the graphite structure, is rolled into a cylinder (Figure 7.23). The roll-up vector (n,m) connects two crystallographically equivalent sites of the sheet, that is, the two points connected by the vector coincide in the nanotube. Limiting cases are tubes with $m = n$ ("armchair" tubes), and those with $m = 0$ ("zigzag" tubes). Armchair and zigzag tubes are achiral, while nanotubes with $n \neq m$ are chiral (Figure 7.23). The magnitude of n and m determines the diameter of the carbon nanotube, which is typically in the range of 1–3 nm for SWNT. The length can be up to a few cm. Armchair tubes are always metallic, while this is only the case with chiral or zigzag tubes if $(n\text{–}m)/3$ is an integer, otherwise they are semiconductors.

Carbon nanotubes are prepared by two methods:

1. *Generation of carbon atoms from graphite* by arc discharge between two graphite rods or by laser ablation of graphite in an inert-gas atmosphere. Nanotubes are formed in the presence of metal (Fe, Co, or Ni) nanoparticles as catalysts. These processes allow producing gram quantities per hour.
2. *Decomposition of gaseous feedstocks* (e.g., ethylene, methane or acetylene) in an inert gas atmosphere, typically at about 700–900°C in the presence of metal nanoparticles (Fe, Co, Ni, or alloys). The latter are produced by pyrolysis of suitable precursors [e.g., $Fe(CO)_5$, Cp_2M, MFe, Co]. Whether MWNT or SWNT grow is to some extent influenced by the particle size and morphology, as well as the reaction temperature. This process is related to the VLS process discussed in Section 7.4.1 and allows production of kg quantities per hour. The carbon compounds arriving at the catalyst surface initially assemble into a graphene cap, followed by lifting the cap and growth of the cylindrical part of the nanotube. Methods were developed to synthesize nanotube bundles by depositing the metal catalyst particles on a substrate. Two main growth models have been reported for deposited catalyst particles: (i) tip growth, where the nanotube lifts the catalyst particle from the substrate and growth below the catalyst, and (ii) base growth, where the nanotube growths above the catalyst particle.

As-obtained samples of carbon nanotubes are mostly very heterogeneous, because the SWNTs aggregate to bundles of different diameters, vary in length and diameter, possess a range of different roll-up vectors and may have defects. The raw materials prepared by either method usually also contain amorphous carbon, fullerenes, and catalyst particles. Purification is therefore required prior to further processing. This typical consists of several steps:

- Oxidation upon heating to ≈350°C in air to remove the amorphous carbon.
- Heating in non-oxidizing acids for removal of the metal particles.
- Ultrasonic treatment in a mixture of nitric and sulfuric acid leading to the opening of the tube caps (which are the most reactive regions due to the high curvature) as well as defects in the sidewalls.
- Thermal treatment at 1000°C in vacuum to anneal the defects created in the earlier steps.
- Debundling (optional). For many applications, it is advantageous to have individual nanotubes instead of bundles. To this end, ultrasonic treatment in an aqueous surfactant solution is very effective. This provides the individual nanotube with a surfactant shell adsorbed to the nanotube wall. Intensive sonification, however, may cause damage to the walls and/or shortening of the nanotubes.

- Semiconducting and metallic nanotubes can be separated by electrophoresis or by concentration of an octadecylamine dispersion. Bonding of the amine to semiconducting and metallic nanotubes is different because of the different charge-transfer properties.

7.4.2.1 Chemical Functionalization of Carbon Nanotubes

Non-destructive chemical functionalization of carbon nanotubes extends the scope of their application spectrum – without altering their intrinsic properties – by alterations of their electronic properties and tailoring their surface properties. The latter is important for rendering the nanotubes compatible with and dispersible in a surrounding medium and for implementing new functions. Four approaches are at hand:

1. *Defect functionalization* uses present or induced defects for covalent attachment of organic groups.
2. *Sidewall functionalization* is achieved by addition reactions.
3. *Non-covalent exohedral functionalization* involves adsorption of various organic and (rarely) inorganic compounds onto the sidewall of the nanotube through non-covalent interactions (π–π, van der Waals, or charge-transfer interactions, etc.)
4. *Endohedral functionalization* is concerned with the filling of the nanotubes with guest molecules. This is a special case that will not be further followed up in this section.

Defect functionalization Carbon nanotubes tolerate a limited number of defects before losing their special electronic or mechanical properties. Typical defects are shown in Figure 7.24. The functional groups created by oxidation damage, mainly COOH groups, can be used for functionalization. For example, carboxyl groups enable the covalent attachment of organic components through the formation of ester or amide linkages.

Sidewall functionalization Addition reactions to the partial carbon–carbon double bonds convert sp^2-hybridized in sp^3-hybridized (tetrahedral) carbon atoms. The positive curvature renders the sidewalls more reactive than a planar graphene sheet (the opposite is true for the inner surface having a negative curvature). Addition reactions enable the direct coupling of functional groups onto the carbon framework (Figure 7.25). C–F groups created by fluorination can be converted in other groups by nucleophilic substitution reactions with alcohols, amines, Grignard reagents, and so on. The presence of organic groups on the sidewalls (by either defect or sidewall functionalization) reduces non-covalent interactions between the nanotubes and thus strongly facilitates separation of nanotube bundles into individual tubes. Attachment of suitable groups renders the nanotubes soluble in water or organic solvents.

Exohedral functionalization This approach is completely non-destructive and preserves the intrinsic structure of the nanotube without changing its π-system and electronic properties. Wrapping of polymeric molecules (including natural

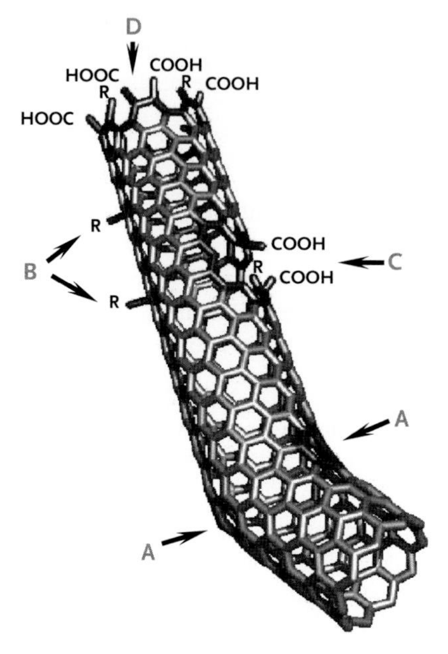

Figure 7.24 Typical defects in a SWNT. A: five- or seven-membered rings lead to a bend. B: sp^3-hybridized defects (R=H, COOH or OH); C: damage by oxidation; D: carboxyl groups terminating the open end of the tube (as a result from the oxidative work-up). Other terminal groups (NO_2, H, OH, O) are also possible.

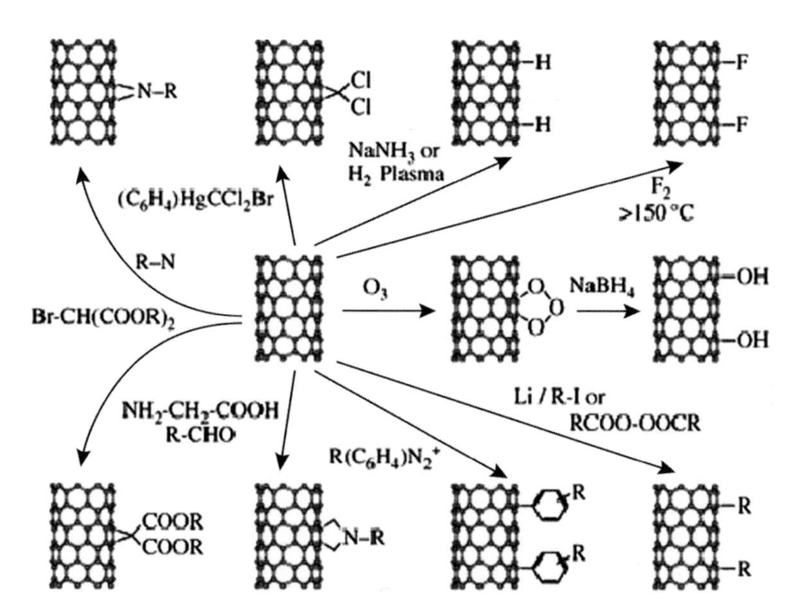

Figure 7.25 Selected addition reactions for sidewall functionalization.

polymers such as peptides or DNA) around carbon nanotubes or adsorbing surfactants to its sidewalls are versatile methods for their individual solubilization. The surfaces of carbon nanotubes can be readily derivatized through π–π interactions with polycyclic aromatic compounds, such as anthracene, perylene, phenanthrene, and so on. Especially pyrene derivatives have often been used as efficient dispersants due to the strong binding affinity between the pyrene group and the sidewall. Furthermore, pyrene groups can be used to anchor organic functionalities. An example is shown in Figure 7.26.

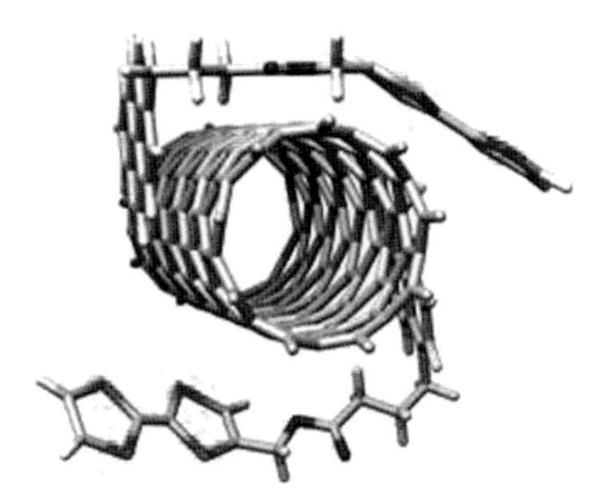

Figure 7.26 Example of exohedral functionalization of a SWNT by π–π interaction of a tetrathiofulvalene-substituted pyrene derivative.

7.5 Nanometer-Scale Layers

Layered or lamellar structures, in which the thickness of the layer is in the nanometer range, are two-dimensional nanostructures. There are essentially three types of two-dimensional nanomaterials:

1. **Self-supported two-dimensional structures**. Such structures have tremendous importance in biological systems, mainly as membranes. The most prominent artificial two-dimensional nanostructure is graphene, which consists of single sheets of the graphite structure and will be discussed in Section 7.5.1.
2. **Lamellar structures**, where nm-thick layers of different composition are stacked above each other. Lamellar structures with alternating surfactant and silica layers were already discussed in Section 6.3.2. The principle characteristic of multilayers of different inorganic materials is a composition modulation, that is, a periodic chemical variation. Multilayers can be deposited by PVD and CVD methods that allow the atomic-scale layering of materials (such as ALD) and have been discussed in Section 3.2. Multilayers composed of → topotactic single-crystal layers are called superlattices. Semiconductor superlattices

("heterostructures") were already mentioned in Section 7.2.2 with regard to the phenomenon of giant magnetoresistance (GMR).

3. **Nanometer-thick films on a support**. Inorganic layers are mainly prepared by PVD and CVD methods (Section 3.2) and are important for microelectronic applications (see examples in Figures 3.15 and 3.24). Thin layers have also been discussed in other chapters of this book. In passing, it should be mentioned that the layer thickness of normal commercial coatings (protective coatings, barrier coatings, etc.) is not in the lower nanometer range, but instead in the range of a few tenths to a few μm. In Section 7.5.2 the so-called self-assembled monolayers (SAM) will be discussed.

7.5.1 Graphene

Graphene is a single-atom-thick two-dimensional material of hexagonally arranged, covalently bonded carbon atoms, that is, a single layer of the graphite lattice (Figure 7.23). It is the thinnest and strongest known material. Unique properties originate additionally from its two-dimensional electron-gas behavior. Attractive applications are envisioned, such as for nanoelectronic devices, spintronics, ultracapacitors, transparent conducting electrodes, sensors, or composite materials. Additional interesting properties and applications may arise from the possibility of doping and of chemical modification, as discussed in Section 7.4.2 for carbon nanotubes. Graphene nanoribbons, that is, sections of the graphene structure, are also discussed as promising future materials.

Graphene can be synthesized by either bottom-up or top-down approaches. The bottom-up approach is the synthesis of graphene layers on a substrate from hydrocarbon sources (see also diamond synthesis in Section 3.2.3), such as high-temperature epitaxial growth of single layers on SiC or CVD on metal or Si surfaces.

Top-down approaches use graphite as a starting material. In order to exfoliate single graphene layers, the van der Waals interactions between the individual layers in graphite have to be overcome. The earliest method was to mechanically split graphene into individual layers by attaching an adhesive tape on highly oriented graphite and pulling the tape off. Although time consuming, this method provides graphene sheet of high structural and electronic quality, which can currently reach mm size. Instead of cleaving graphite manually, it is also possible to use ultrasonic cleavage. This leads to stable suspensions of submicrometer graphene crystallites. Sonification can be rendered more efficient by preceding intercalation (see Section 2.3), which widens the distance between the graphene layers in the graphite crystal.

A promising alternative route of weakening the interlayer forces in graphite is oxidation with strong oxidants, such as HNO_3, $KMnO_4$, or H_2O_2. The obtained oxidized material, with a typical C/O ratio of 2.5–3, contains carboxylate, peroxide, hydroxo, and epoxide groups. These groups create electrostatic repulsion and steric hindrance between the layers, thus facilitate exfoliation of single sheets of graphene oxide and prevent their reaggregation in water. The electronic structure of graphene oxide, however, is different from that of pristine graphene. Subsequent restoration of the graphene network by deoxygenation is therefore necessary ("reduced graphene oxide") to recover the original properties. This method has the greatest potential for large-scale production. The most advanced deoxygenation method is solvothermal (see Section 4.4) reduction by hydrazine. This process produces graphene in high yields and quality, but takes several days for completion. Furthermore, hydrazine is a hazardous chemical. For this reason, alternative deoxygenation methods are being developed. These include the replacement of hydrazine by other reductants, such as sodium borohydride, hydroquinone, some saccharides or especially ascorbic acid (vitamin C), and the development of methods that circumvent the use of reducing agents. The latter include electrochemical methods, heating of graphene oxide in strongly alkaline suspensions, or solvothermal treatment in water or organic solvents with or without microwave assistance.

7.5.2
Self-Assembled Monolayers

Self-assembled monolayers form by spontaneous chemisorption and dense packing ("self-organization") of long-chain, functionalized organic molecules on suitable substrates, when one end of the molecule, the "head group", has a special affinity for the substrate. Self-organization refers to the spontaneous assembly of molecules in a stable, well-defined structure by non-covalent interactions. The type of head group depends on the substrate which can be planar surfaces or curved surfaces, such as nanoparticles. Alkanethiols are the most commonly used molecules for SAMs because of the strong affinity of sulfur for metal surfaces (Figure 7.27). Among the best-investigated SAM systems are alkanethiols, X $(CH_2)_nSH$ (X=H or functional group), on gold surfaces. For $n > 11$, homogeneous

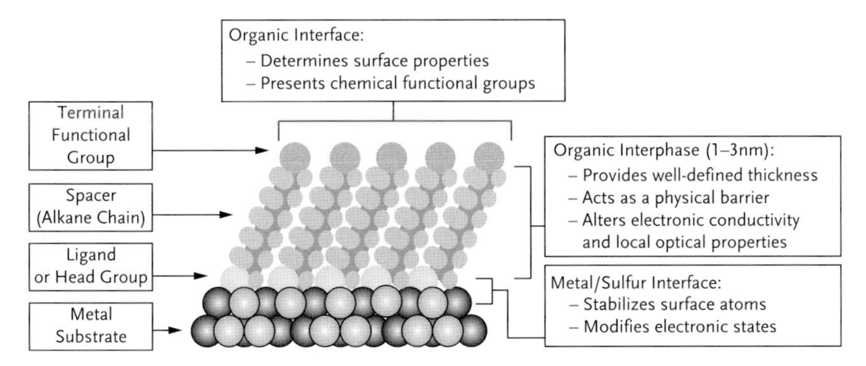

Figure 7.27 Schematic diagram of an alkanethiol SAM supported on a gold (111) surface.

2- to 3-nm thick layers are formed by dense packing of the parallel alkyl chains and the thiol groups interacting with the gold surface. Compounds with alkoxy- or chlorosilyl head groups are generally used on metal oxide surfaces. Contrary to SAM from alkanethiols, the chemisorbed silanes form a two-dimensional Si–O–Si network due to hydrolytic crosslinking of the alkoxy- or chlorosilyl groups. The terminal end of the self-organized molecules can be functionalized (group X) for the reasons discussed in Section 7.3.4. This opens up a wide field for the molecular engineering of surface properties.

Formation of self-assembled monolayers is a rather quick process when suitable substrates are immersed into a solution of the film-forming compound or treated with the vapor of such compounds.

SAMs offer an ideal platform for pattern formation on surfaces, with the potential of a high resolution, that is, on a molecular or nm level. The key requirements for a resistance material in nanolithography applications are high resolution, high contrast, and compatibility with the transfer of the pattern to the device. Nanostructuring is a particularly important issue for electronic devices. SAMs with functional end groups can be patterned by a number of lithographic techniques at least down to 20–100 nm. As the patterns become smaller, higher-energy beams must be used for lithography. Electron-beam, ion-beam, UV, or X-ray techniques were successfully applied for applications, although the high beam energy frequently causes unwanted ancillary damage. Such patterning techniques are outside the scope of this book.

Only two methods shall be mentioned here, where the patterning occurs during the formation of the SAM. The first is microcontact printing (μCP) which belongs to the group of "soft lithography" methods (Figure 7.28). In the first step, the pattern is stamped on the substrate surface with the self-assembling molecules as the "ink." SAMs are formed in the contact areas. The structured SAM can be used

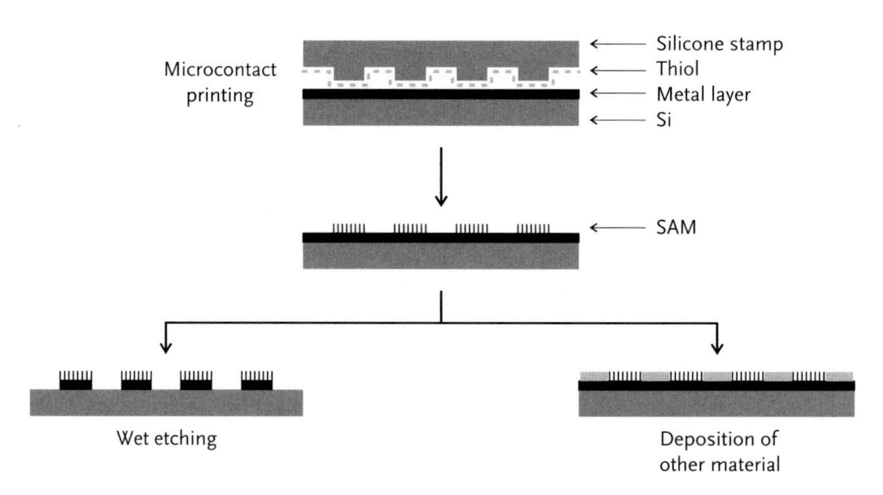

Figure 7.28 Microcontact printing in nanolithography.

as etch mask for wet etching processes (Figure 7.28, left) or to deposit selectively other materials on the substrate area not covered by the SAM (e.g., by selective CVD) (Figure 7.28, right).

The second method to be mentioned is dip-pen nanolithography (Figure 7.29). In this technique an atomic force microscope tip is used as a pen, that is, to transfer the organic molecules to the substrate. This is equivalent to writing a letter with a pen – but on a nm level. The molecular inks are delivered to the surface through a water meniscus. For this reason, alkane thiols are mostly used; chloro- or alkoxysilyl groups would not be compatible with the water meniscus. The choice of the most suitable ink is determined by the corresponding substrate.

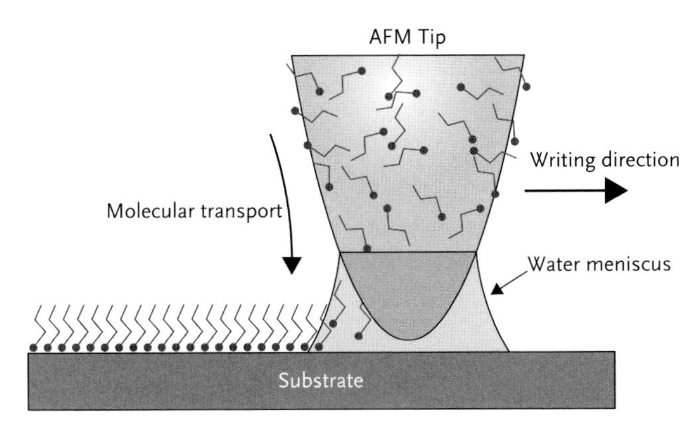

Figure 7.29 Dip-pen nanolithography.

Further Reading

1 Balasubramanian, K. and Burghard, M. (2005) Chemically functionalized carbon nanotubes. *Small*, **1**, 180–192.

2 Barth, S., Hernandez-Ramirez, F., Holmes, J.D., and Romano-Rodriguez, A. (2010) Synthesis and applications of one-dimensional semiconductors. *Prog. Mater. Sci.*, **55**, 563–627.

3 Cai, D. and Song, M. (2010) Recent advance in functionalized graphene/polymer nanocomposites. *J. Mater. Chem.*, **20**, 7906–7915.

4 Dresselhaus, M.S., Dresselhaus, G., and Avouris, P. (2001) *Carbon Nanotubes: Synthesis, Structure, Properties, and Applications*, Springer, New York.

5 Gaponik, N., Hickey, S.G., Dorfs, D., Rogach, A.L., and Eychmüller, A. (2010) Progress in the light emission of colloidal semiconductor nanocrystals. *Small*, **6**, 1354–1378.

6 Geim, A. (2009) Graphene: status and prospects. *Science*, **324**, 1530–1534.

7 Gleiter, H. (2000) Nanostructured materials: basic concepts and ideas. *Acta Mater.*, **48**, 1–29.

8 Green, M. (2010) The nature of quantum dot capping ligands. *J. Mater. Chem.*, **20**, 5797–5809.

9 Guldi, D.M. and Martin, N. (2009) *Carbon Nanotubes and Related Structure*, Wiley-VCH, Weinheim.

10 Hirsch, A. (2002) Functionalization of single-walled carbon nanotubes. *Angew. Chem. Int. Ed. Engl.*, **41**, 1853–1899.

11 Huber, D.L. (2005) Synthesis, properties, and applications of iron nanoparticles. *Small*, **1**, 482–501.

12 Inagaki, M., Kim, Y.A., and Endo, M. (2011) Graphene: preparation and structural perfection. *J. Mater. Chem.*, **21**, 3280–3294.

13 Köhler, M. and Fritzsche, W. (2007) *Nanotechnology – an Introduction to Nanostructuring Techniques*, Wiley-VCH, Weinheim.

14 Kumar, K.S., van Swygenhoven, H., and Suresh, S. (2003) Mechanical behavior of nanocrystalline metals and alloys. *Acta Mater.*, **51**, 5743–5774.

15 Love, J.C., Estroff, L.A., Kriebel, J.K., Nuzzo, R.G., and Whitesides, G.M. (2005) Self-assembled monolayers of thiolates on metals as a form of nanotechnology. *Chem. Rev.*, **105**, 1103–1169.

16 de Mello Donegá, C., Liljeroth, P., and Vanmaekelbergh, D. (2005) Physicochemical evaluation of the hot-injection method, a synthesis route for monodisperse nanocrystals. *Small*, **1**, 1152–1162.

17 Nalwa, H.S. (2004) *Encyclopedia of Nanoscience and Nanotechnology*, American Scientific Publishers, Stevenson Ranch, Ca., USA.

18 Neouze, M.-A. and Schubert, U. (2008) Surface modification and functionalization of metal and metal oxide nanoparticles by organic ligands. *Monatsh. Chem.*, **139**, 183–195.

19 Nessim, G.D. (2010) Properties, synthesis, and growth mechanisms of carbon nanotubes with special focus on thermal chemical vapor deposition. *Nanoscale*, **2**, 1306–1323.

20 Onclin, S., Ravoo, B.J., and Reinhoudt, D.N. (2005) Engineering silicon oxide surfaces using self-assembled monolayers. *Angew. Chem. Int. Ed.*, **44**, 6282–6304.

21 Pan, D., Wang, Q., and An, L. (2009) Controlled synthesis of monodisperse nanocrystals by a two-phase approach without the separation of nucleation and growth processes. *J. Mater. Chem.*, **19**, 1063–1073.

22 Rao, C.N.R., Satishkumar, B.C., Govindaraj, A., and Nath, M. (2001) Nanotubes. *Chem. Phys. Chem.*, **2**, 78–105.

23 Reich, S., Thomsen, C., and Maultzsch, J. (2004) *Carbon Nanotubes*, Wiley-VCH, Weinheim.

24 Schmid, G. (ed.) (2004) *Nanoparticles – from Theory to Applications*, Wiley-VCH, Weinheim.

25 Trindade, T., O'Brien, P., and Pickett, N.L. (2001) Nanocrystalline semiconductors: synthesis, properties, and perspectives. *Chem. Mater.*, **13**, 3843–3858.

26 Vollath, D. (2008) *Nanomaterials – an Introduction to Synthesis, Properties and Applications*, Wiley-VCH, Weinheim.

27 Walters, G. and Parkin, I.P. (2009) The incorporation of noble metal nanoparticles into host matrix thin films: synthesis, characterisation and applications. *J. Mater. Chem.*, **19**, 574–590.

28 Zhang, H., Chen, G., and Bahnemann, D.W. (2009) Photocatalytic materials for environmental applications. *J. Mater. Chem.*, **19**, 5089.

Glossary

Aerosol The suspension of very fine particles of a solid or droplets of a liquid in a gaseous medium.

Alloy A mixture of two or more metals.

Amphiphilicity An amphiphilic molecule (amphi = either) has two distinct regions: a hydrophilic ("water-loving") head group that is easily soluble in water and a hydrophobic ("water-hating") or oleophilic ("oil-loving") tail.

Bidentate ligand Bidentate → ligands can coordinate to only one metal (→ chelating ligands) or bridge two metals (bridging ligands). Correspondingly, monodentate ligands have only one coordinating site, and polydendate ligands several sites.

Calcination The heating of a solid to a high temperature, below its melting point, to create a condition of thermal decomposition or phase transition other than melting or fusing.

Ceramic yield The portion of a material (mostly an inorganic polymer) that is transformed to ceramics upon pyrolysis. Ceramic yield = [weight of ceramic residue] × 100/ [weight of pyrolyzed material].

Chalcogenide A compound made up of a chalcogen (the main group elements that are members of group VI: oxygen, sulfur, selenium, or tellurium) and a more electropositive element.

Chelating ligand When a → bidentate ligand is coordinated to one metal atom, a metallacycle (= metal-containing ring system) is formed. The ligand is then called a chelating ligand, the complex a chelate complex. Chelation (i.e., formation of a chelate system) is entropically favored.

Synthesis of Inorganic Materials, Third Edition. Ulrich Schubert and Nicola Hüsing.

Published 2012 by WILEY-VCH Verlag GmbH & Co. KGaA

Colloid — Colloids consist of a dispersed phase (mostly a solid) in a finely divided state that is uniformly distributed in a dispersion medium (or continuous phase; normally a liquid). The size of the dispersed phase is so small that gravitational forces are negligible and interactions are dominated by short-range forces, such as van der Waals attraction or surface charges.

Composite material — A materials system composed of a mixture or combination of two or more constituents that differ in form and (normally) chemical composition. There are distinct phase boundaries between the two constituents.

Crosslinking — Formation of bonds between chains of macromolecules (polymers).

Dielectric — An electrical insulator material.

Dielectric loss — The energy that is converted into heat in a dielectric material when the material is subjected to a changing electric field.

Dielectric strength — The maximum electric field that a dielectric can withstand without electric breakdown.

Ductility — Ductility is the ability of a material to change shape without fracture.

Electromigration — Transport of atoms induced by flowing electrons under high current densities.

Electrophoresis — A process in which electrically charged particles that are suspended in a solution move through the solution under the influence of an applied electric field.

Emulsion — A stable mixture of two or more immiscible liquids where one liquid, in the form of fine droplets, is dispersed in the other.

Enantioselectivity — Enantiomers are molecules of the same chemical composition that are mirror images of each other. A reaction that produces one of the two enantiomers in excess is called enantioselective.

Epitaxy — Two-dimensional structural similarity at the interface between two crystalline phases (a reactant and a product, the substrate and a crystalline film, etc.).

Eutectic mixture — An eutectic mixture is a solid solution consisting of two of more substances and having the lowest freezing point of any possible mixture of these components. The minimum freezing point for a set of components is called the eutectic point. At this point in the phase diagram, the solid phases and the liquid are in equilibrium.

Ferroelectricity Ferroelectric materials can be polarized by applying an electric field. They are distinguished from ordinary → dielectrics by their extremely large permittivities and the possibility of retaining some electric polarization after an applied voltage has been switched off. The crystal must belong to a non-centrosymmetric point group.

Ferromagnetism Ferromagnetic materials become highly magnetized when a magnetic field is applied. After the applied field is removed, the ferromagnetic material retains much of its magnetization. Elemental iron, cobalt or nickel are ferromagnetic materials. Ferromagnetic behavior is only exhibited below a certain temperature ("Curie temperature") and is caused by the spontaneous parallel alignment of the electron spins of neighboring atoms in small domains.

Filler A finely divided solid added to a liquid, semi-solid or solid composition (e.g., paint, plastics) to modify the composition's properties.

Goldschmidt process Johann Goldschmidt (1981–1923), a German chemist, discovered in 1898 the "thermite" reaction that produces molten iron from a mixture of iron oxide and aluminum that is ignited by a magnesium fuse.

Grain A single crystal in a polycrystalline aggregate.

Green body An unfired ceramic body. Green density is the density of a green body.

Hardness A measure of the resistance of a material to permanent deformation.

Heat capacity The heat energy required to raise the temperature of a specific substance by 1 K at constant pressure and volume.

Ligand Ions or neutral molecules bound to a central metal atom or metal ion in a coordination compound (metal complex) are called ligands.

	The nature of metal–ligand bonding can be covalent or ionic. Ligands are viewed as Lewis bases, which donate electron pairs to the Lewis-acidic metal center.
Liquid crystal	Certain organic materials exhibit states of intermediate order between the long-range order of solid crystalline compounds and the short-range order of ordinary liquids. These phases are called liquid crystals or mesophases since many of them can flow like a liquid but display birefringence and other properties reminiscent of the solid state.
Lyotropic liquid crystal	A liquid crystal phase that is dependent on the concentration of one component in another.
Mesogenic group	Compounds or substituents that form a mesophase (→ liquid crystalline phase) in a certain temperature range.
Mesophase	See: liquid crystals.
Metastable solid	Solid phases that are off the thermodynamic minimum but are kinetically stable to quite high temperatures.
Metathesis	A double displacement reaction of the type A-B + A′-B′ → A-B′ + A′-B in which partners are interchanged.
Micelle	Micelles are spherical aggregates of → amphiphilic molecules that assemble spontaneously. In water, the hydrophobic tails interact with each other and fill the interior of the micelle, while the hydrophilic head groups stick in the water phase. In non-polar solvents, reversed micelles are formed, in which the hydrophilic head groups are directed towards the interior of the micelle and the hydrophobic tails into the solvent.
Monodispersity	In monodisperse materials all polymer chains or particles have the same size.
Monomodal	A distribution (e.g., size distribution) having one maximum. Correspondingly, a bimodal distribution has two maxima, and a polymodal distribution several maxima.
Non-linear optics	The optical properties of materials are normally considered to be independent of the intensity of light. In non-linear optical materials, the optical

properties reversibly depend on the intensity of the incident light.

Nucleation Formation of thermodynamically stable solid particles (nuclei) from a continuous phase (gas, liquid) by aggregation of molecular or cluster species. Heterogeneous nucleation is nucleation on foreign nuclei, dust, or surfaces. Homogeneous nucleation is nucleation in the absence of any foreign nuclei.

Ostwald ripening A process in which larger particles grow at the expense of smaller ones. Larger, more stable particles are thus formed.

Paramagnetism A compound is paramagnetic if it contains unpaired electrons. Paramagnetic compounds or materials have a small positive magnetic susceptibility (= ratio of magnetization to applied magnetic field).

Permeability The capability of a porous substance to allow a fluid to pass through it. A permselective material (e.g., a membrane) allows only one type of compound to pass through it.

Perovskite A class of crystalline solids of the general formula $M^{2+}M^{4+}O_3$. The M^{2+} and oxide ions occupy the anion sites of the rock salt structure, while the M^{4+} ions are in one quarter of the octahedral sites (those only surrounded by oxide ions).

Photoresist A light-sensitive polymer used in photolithography.

Piezoelectricity Piezoelectric materials exhibit a linear relationship between electric and mechanical variables. They polarize under the action of an applied stress and develop electric charges on opposite crystal faces. By stretching or compressing a piezoelectric material, a voltage is generated. The reverse is also true: a voltage applied to the material causes it to become mechanically stressed.

Polycondensation Chemical reaction in which compounds of high molecular mass are formed from monomers by cleavage of a small molecule, typically water, hydrogen, and so on.

Polydispersity The polydispersity index is a measure of the breadth of the molecular mass distribution of polymers (i.e., the size distribution of the polymer chains). It has a minimum value of unity for → monodisperse systems.

Polymerization
Chemical reaction in which macromolecules (molecules of high molecular mass) are formed from monomers. In copolymerization reactions, two or more kinds of monomers are employed.

Polymodal
A distribution (e.g., size distribution) having several maxima. Correspondingly, a bimodal distribution has two maxima, and a monomodal distribution one maximum.

Polynuclear
A metal or semimetal complex is di- or polynuclear if it contains two or more central atoms. Metal clusters are typical polynuclear compounds.

Refractory material
Refractories are materials that resist the action of hot environments.

Resistivity
A measure of the difficulty of an electric current to pass through a unit volume of a material.

Rheology
The science of the flow and deformation of matter.

Semiconductor
A material whose electric conductivity is between that of a metal and an insulator. Conductivity increases with increasing temperature.

Solid solution
When one solid dissolves another solid without altering its crystal structure, these solids form a solid solution.

Spinel
A class of crystalline solids of the general formula MM'_2O_4. The M cations occupy tetrahedral, and the M′ ions octahedral interstices in a cubic close packed array of oxide ions.

Spinodal decomposition
A solution of two or more components can separate into distinct regions or phases. The spinodal decomposition mechanism differs from classical nucleation in that phase separation occurs uniformly throughout the material and not just at discrete nucleation sites. It starts with infinitesimal fluctuations of the composition.

Step coverage
The ratio of the film thickness on the step sidewall to the film thickness on the top surface.

Strain
Change in length of a sample divided by its original length.

Stress
Force divided by the area over which the force acts. The force can be applied by tension, compression or shear.

Superconductor
A material that conducts electricity with no resistance below a certain "critical" temperature, T_c.

Sustainable development
Sustainable development meets the needs of the present without compromising the ability of future generations to meet their own needs. Sustainability refers to the ability of a society, ecosystem, or any other ongoing system to continue functioning into the future without being forced into decline through exhaustion of key resources.

Tacticity
The regularity in which repeating units with a certain configuration are ordered in a polymer chain. In atactic polymers the side groups are randomly arranged on either side of the main chain, in isotactic polymers on only one side of the chain, and in syndiotactic polymers the groups alternate from one side of the chain to the other.

Template
A molecule or an assembly of molecules (e.g., a → micelle) serving as a pattern or mold for the synthesis of another compound.

Thermochromic compounds
A thermochromic compound changes its color when a certain temperature is passed.

Thixotropy
The viscosity of the system is reversibly decreased upon action of mechanical forces (shaking, stirring, etc.). When the systems comes to rest, the original (higher) viscosity is restored.

Topotaxy
Three-dimensional structural similarity between two crystalline phases.

Triboelectricity
Charging of solids by friction.

Tribology
The study of mechanical phenomena in frictional processes, lubrication, and wear on surfaces.

Vesicle
Vesicles consists of a detergent (phospholipid) *bilayer* that encloses a small volume of a solvent (water, buffer solution, etc.). The hydrophilic heads of the → amphiphilic molecules point towards the solvent both inside and outside the vesicle. The hydrophobic tails hold the vesicle together.

Vulcanization — A chemical reaction that causes → crosslinking of polymer chains.

Wave guide — An optical wave guide is a thin fiber along which light can propagate by total internal reflection and refraction.

Yield strength — The → stress at which a specific amount of → strain occurs in a tensile test.

Index

Synthesis of Inorganic Materials, Third Edition. Ulrich Schubert and Nicola Hüsing.

Published 2012 by WILEY-VCH Verlag GmbH & Co. KGaA